S0-CBI-115

Persistent, Bioaccumulative, and Toxic Chemicals II

About the Book

The book covers for Volumes I and II of *Persistent, Bioaccumulative, and Toxic Chemicals* were prepared in contrasting blue and green backgrounds, respectively. The design motif consists of some examples of PBT molecules from chapters in the two volumes that are "floating" above a globe of the earth representing the global nature of the environmental problem posed by the persistence and long-range atmospheric transport of such chemicals. The initial draft of the globe and PBT molecules was prepared by Karen Boswell, a graphic artist within EPA's Office of Pollution Prevention and Toxics, using 3-dimensional chemical structures generated by Bo Jansson, one of the editors of Volume II. The final book cover design was prepared at the direction of Margaret Brown, Senior Production Specialist, ACS Books Department, by the graphic artist Howard Petlack of A Good Thing Inc., with the guidance of STM Publication Services.

ACS SYMPOSIUM SERIES **773**

Persistent, Bioaccumulative, and Toxic Chemicals II

Assessment and New Chemicals

Robert L. Lipnick, EDITOR
U.S. Environmental Protection Agency

Bo Jansson, EDITOR
Stockholm University

Donald Mackay, EDITOR
Trent University

Myrto Petreas, EDITOR
California Environmental Protection Agency

American Chemical Society, Washington, DC

Library of Congress Cataloging-in-Publication Data

Persistent, Bioaccumulative, and toxic chemicals II : assessment and new chemicals / Robert L. Lipnick [et al.], editor.

p. cm.—(ACS symposium series ; 773)

"Developed from a symposium...at the 217th National meeting of the American Chemical Society, Anaheim, California, March 21–25, 1999"

Includes bibliographical references and index.

ISBN 0–8412–3675–5

1. Persistent pollutants—Environmental aspects—Congresses. 2. Persistent pollutants—Bioaccumulation—Congresses. 3. Persistent pollutants—Toxicology—Congresses.

I. Lipnick, Robert L. (Robert Louis), 1941– . II. Series.

TD196.C45 P46523 2000
628.5′2—dc21 00–58629

The paper used in this publication meets the minimum requirements of American National Standard for Information Sciences—Permanence of Paper for Printed Library Materials, ANSI Z39.48–1984.

Distributed by Oxford University Press

PRINTED IN THE UNITED STATES OF AMERICA

Foreword

THE ACS SYMPOSIUM SERIES was first published in 1974 to provide a mechanism for publishing symposia quickly in book form. The purpose of the series is to publish timely, comprehensive books developed from ACS sponsored symposia based on current scientific research. Occasionally, books are developed from symposia sponsored by other organizations when the topic is of keen interest to the chemistry audience.

Before agreeing to publish a book, the proposed table of contents is reviewed for appropriate and comprehensive coverage and for interest to the audience. Some papers may be excluded in order to better focus the book; others may be added to provide comprehensiveness. When appropriate, overview or introductory chapters are added. Drafts of chapters are peer-reviewed prior to final acceptance or rejection, and manuscripts are prepared in camera-ready format.

As a rule, only original research papers and original review papers are included in the volumes. Verbatim reproductions of previously published papers are not accepted.

ACS BOOKS DEPARTMENT

Contents

Preface xi

1. **Identification of Persistent, Bioaccumulative, and Toxic Substances** 1
   Bo Jansson, Robert L. Lipnick, Donald Mackay, and Myrto Petreas

Persistence and Modeling

2. **Defining the Bioaccumulation, Persistence, and Transport Attributes of Priority Chemicals** 14
   Donald Mackay, Eva Webster, Andreas Beyer, Michael Matthies, and Frank Wania

3. **CART Screening Level Analysis of Persistence: A Case Study** 29
   Deborah H. Bennett, Thomas E. McKone, and W. E. Kastenberg

4. **Environmental Categorization and Screening of the Canadian Domestic Substances List** 42
   Roger Breton and Robert Chénier

5. **Relationship between Persistence and Spatial Range of Environmental Chemicals** 52
   Martin Scheringer, Deborah H. Bennett, Thomas E. McKone, and Konrad Hungerbühler

6. **Application of Negligible Depletion Solid-Phase Extraction (nd-SPE) for Estimating Bioavailability and Bioaccumulation of Individual Chemicals and Mixtures** 64
   Joop L. M. Hermens, Andreas P. Freidig, Enaut Urrestarazu Ramos, Wouter H. J. Vaes, Willem M. G. M. van Loon, Eric M. J. Verbruggen, and Henk J. M. Verhaar

7. **Modeling Historical Emissions and Environmental Fate of PCBs in the United Kingdom** 75
   Andrew J. Sweetman and Kevin C. Jones

8. **Estimating Accumulation and Baseline Toxicity of Complex Mixtures of Organic Chemicals to Aquatic Organisms: The Use of Hydrophobicity Dependent Analytical Methods** **89**
Eric M. J. Verbruggen and Joop L. M. Hermens

9. **Selection and Evaluation of Persistent, Bioaccumulative, and Toxic Criteria for Hazard Assessment** **100**
Han Blok, Froukje Balk, Peter Okkerman, and Dick Sijm

10. **An Environmental Protection Agency Multimedia Strategy for Priority Persistent, Bioaccumulative, and Toxic Pollutants** **114**
Sam Sasnett, Thomas Murray, Sheila Canavan, John Alter, Kathy Davey, and Paul Matthai

11. **Dioxin Pollution Prevention Inventory for the San Francisco Bay** **124**
Greg Karras

12. **U.S. Environmental Protection Agency New Chemicals Program PBT Chemical Category: Screening and Risk Management of New PBT Chemical Substances** **138**
Kenneth T. Moss, Robert S. Boethling, J. Vincent Nabholz, and Charles M. Auer

13. **The U.S. Environmental Protection Agency Minimization Prioritization Tool: Computerized System for Prioritizing Chemicals Based on PBT Characteristics** **151**
Mark D. Ralston, Daniel L. Fort, Jay H. Jon, and James K. Kwiat

New Chemicals

14. **Short Chain Chlorinated Paraffins: Are They Persistent and Bioaccumulative?** **184**
Derek Muir, Don Bennie, Camilla Teixeira, Aaron Fisk, Gregg Tomy, Gary Stern, and Mike Whittle

15. **Polycyclic Musk Fragrances in the Aquatic Environment** **203**
Hermann Fromme, Thomas Otto, and Konstanze Pilz

16. **Polychlorinated Naphthalenes in the Atmosphere** **223**
Tom Harner, Terry F. Bidleman, Robert G. M. Lee, and Kevin C. Jones

17. Polybrominated Diphenyl Ethers (PBDEs) in Human Milk from Sweden 235
S. Atuma, M. Aune, P. O. Darnerud, S. Cnattingius, M. L. Wernroth, and A. Wicklund-Glynn

18. Characterization of Q1, an Unknown Major Organochlorine Contaminant in the Blubber of Marine Mammals from Africa, the Antarctic, and Other Regions 243
W. Vetter

Author Index 260

Subject Index 261

Preface

This monograph, *Persistent, Bioaccumulative, and Toxic Chemicals II: Assessment and New Chemicals,* represents one of two books derived from a symposium sponsored by the American Chemical Society (ACS) Division of Environmental Chemistry, Inc., March 21–25, 1999 in Anaheim, California. The symposium was initially inspired by the U.S. Environmental Protection Agency's 1998 national program on persistent, bioaccumulative, and toxic chemicals (PBTs). EPA's national PBT program based its priority chemical list on those already agreed to by the United States and Canada in 1997 for "virtual elimination" of PBTs in the Great Lakes. This action was preceded by a 1995 agreement signed by Canada, Mexico, and the United States to develop and implement North American Regional Action Plans (NARAPs) under the North American Free Trade Agreement (NAFTA) to phase out or reduce PBT emissions.

Other regional agreements that are responsible for the surge in interest in PBTs include the protocol for Long Range Transport of Air Pollutants (LRTAP) under the United Nations Economic Commission for Europe (UN-ECE). Globally, such reduction and phase out of PBTs is being led by the United Nations Environmental Program (UNEP) through global Persistent Organic Pollutants (POPs) negotiations.

The Anaheim symposium provided a much-needed means for research scientists and policy-makers to exchange information regarding the state of scientific knowledge and theory on this subject, and how these findings can be employed in the decision-making process. With the encouragement of the ACS Books Department, the Editors also recruited additional papers for the monographs to fill gaps. These monographs provide a means of disseminating this information to a larger audience.

This volume includes papers from Canada, Germany, The Netherlands, Sweden, the United Kingdom, and the United States, covering persistence, modeling, and assessment of PBTs in the environment, as well as the identification of new members of this class. An introductory chapter provides an overview of the properties and identification of PBTs, and current national and international efforts directed toward their reduction and elimination.

Following this introduction are papers from academia, industry, and government on characterizing the PBT properties for selecting candidate chemicals, concluding their long-range environmental transport. These are followed by coverage of estimation of bioavailability and toxicity of individual PBTs and mixtures from physicochemical properties; case studies of emissions and fate of polychlorinated biphenyls; PBT partitioning between air and vegetation; PBT hazard and risk assessment; and new categories of PBTs including chlorinated paraffins, musk fragrances, polychlorinated naphthalenes, polybrominated diphenyl ethers, and a recently isolated chlorinated contaminant in Antarctic seals.

The companion volume to this monograph, entitled *Persistent, Bioaccumulative, and Toxic Chemicals I: Fate and Exposure*, includes chapters covering theory of movement of hydrophobic chemicals through sediments, uptake by plants, bioconcentration (theoretical framework and bioaccumulation of toxaphene, DDT, and dieldrin), soil degradation, and remediation. Additional chapters cover emissions at both local and regional levels related to polychlorinated biphenyls (PCBs), butyltin, toxaphene, and tetrachlorodioxin. These are followed by chapters pertaining to human exposure to PCBs (generic model, soil concentrations, and concentrations in breast milk), dioxin (food from animal sources and body burdens in California populations), and mercury (hair analysis and atmospheric deposition studies in the United States).

The Editors gratefully acknowledge the continuing support and encouragement provided by Anne Wilson, Senior Product Manager, Kelly Dennis, Editorial Assistant, and Margaret Brown, Senior Production Specialist, ACS Books Department, in completing this ACS Symposium Monograph.

ROBERT L. LIPNICK
Office of Pollutant Prevention and Toxics (7403)
U.S. Environmental Protection Agency
401 M Street, S.W.
Washington, DC 20460

BO JANSSON
Institute of Applied Environmental Research
Stockholm University
SE–10691 Stockholm
Sweden

DONALD MACKAY
Canadian Environmental Modelling Centre
Trent University
Peterborough, Ontario K9J 7B8
Canada

MYRTOS PETREAS
Hazardous Materials Laboratory
Department of Toxic Substances Control
California Environmental Protection Agency
2151 Berkeley Way
Berkeley, CA 94704

Chapter 1

Identification of Persistent, Bioaccumulative, and Toxic Substances

Bo Jansson[1], Robert L. Lipnick[2], Donald Mackay[3], and Myrto Petreas[4]

[1]Institute of Applied Environmental Research, Stockholm University, SE–10691 Stockholm, Sweden (bo.jansson@itm.su.se)
[2]Office of Pollution Prevention and Toxics, U.S. Environmental Protection Agency, 401 M Street, SW, Washington, DC 20460 (lipnick.robert@epa.gov)
[3]Canadian Environmental Modelling Centre, Trent University, Peterborough, Ontario K9J 7B8, Canada (dmackay@trentu.ca)
[4]Hazardous Materials Laboratory, Department of Toxic Substances Control, California Environmental Protection Agency, 2151 Berkeley Way, Berkeley, CA 94704 (mpetreas@dtsc.ca.gov)

Increased sensitivity in the analyses of environmental media, including air, water, soils, sediments and biota, has revealed the presence of a number of toxic chemicals at significant concentrations and at considerable distances from known sources. Identification of these persistent, bioaccumulative and toxic (PBT) substances has often been fortuitous, suggesting that a more thorough examination of the universe of chemicals of commerce will reveal more PBTs. There is a need to predict the behavior of candidate PBTs using a combination of structure-property and structure-activity relationships, experimental data and mass balance modelling. Clearly, there is an incentive for international cooperation in this task, given the large number of chemicals. Recent activities in a variety of national and international fora are reviewed, providing a report on the status of global efforts to identify and, ultimately, regulate PBT substances.

INTRODUCTION

Compounds that are slowly degraded in the environment may be subsequently distributed over large areas of our planet by a variety of transport processes, such as winds, water currents and migrating biota. As a result, critical concentrations can appear far from the source(s) and it may be difficult to connect resulting toxicological effects with the actual emission(s). The risk that critical levels of such persistent, bioaccumulative and toxic chemicals (PBTs) may be reached in biota increases with the substance's tendency to bioaccumulate. Compounds that are both persistent and bioaccumulative may therefore achieve high concentrations at locations distant from the regions where the compounds are used, especially at higher trophic levels because of further biomagnification.

Developments in analytical chemistry have enhanced the identification of "new" PBT compounds in the environment as the sensitivity and specificity of the methods have increased. The development of ionization detectors for gas chromatography (*1*) made possible the analysis of 2,2-bis(p-chlorophenyl)-1,1,1-trichloroethane (DDT) and other chlorinated pesticides in biological samples. A further advance was gained by the combination of gas chromatography and mass spectrometry, GC-MS, (*2*) in which a separation was coupled directly to a spectroscopic analysis, which permitted the separation of polychlorinated biphenyls (PCBs) that had previously interfered with the analysis of pesticides. Following this pioneering work, continuing improvements have taken place in separation and quantification methods leading to the identification of new PBT contaminants in the environment.

In addition, it is noteworthy that many of the "classical" PBT compounds in the environment have been detected serendipitously, as in the case of PCBs mentioned above that interfered with the analysis of other compounds. Other such examples so detected are pentabromotoluene, which previously co-eluted with DDE during gas chromatographic analysis (*3*); the methyl sulphone metabolites of PCBs, found when the chart recorder of a gas chromatograph was accidentally left on overnight (*4*); and octachlorostyrene, which is not produced industrially, but is a manufacturing by-product (*5*).

Concerns about PBT compounds have prompted a reexamination and a rediscovery of whole classes of chemicals. These "chemicals of emerging concern" include a number of unintentional industrial byproducts, as well as industrial products with widespread use and applications as dielectrics, lubricants, flame retardants, fragrances, etc. Chapters 14-18 in this volume discuss measurements of such chemicals in a variety of media.

It is desirable that the PBT-like behavior of chemicals be foreseen or predicted so that their release and distribution into the environment can be avoided, since degradation may be very slow. Bioaccumulation can be measured in the laboratory either directly with fish and other aquatic organisms or estimated from the lipophilicity of the substance. Lipophilicity or hydrophobicity are generally characterized by measurement of the distribution of the substance between n-octanol and water (K_{ow}) which can be used to estimate the biomagnification potential.

It is more difficult to measure the persistence of compounds because they can be degraded in many different ways in the environment (biological, chemical, and photolytic), and the conditions under which degradation takes place vary widely with respect to geographic location and time. Several standardised tests exist for assessing biodegradation potential. While these may be reliable for identification of non-persistent compounds, they are less useful for assessing those substances that prove to be persistent in the environment. Although the half-life values of compounds of potential PBT concern should be available for water, air, soil, and sediment, this database of half-lives is very limited.

Nevertheless, certain physicochemical and chemical properties, such as partitioning and reactivity behavior, have been correlated with chemical structure through regression analysis based upon linear free energy relationships and other types of quantitative structure property relationships (QSPRs), permitting the estimation of partitioning and reactivity properties of chemicals from a knowledge of molecular structure alone; predicting rates of biodegradation is still limited. Such predicted and measured data are currently being used to increase our understanding of environmental processes. We can now formulate mass balance models describing the fate of chemicals as they enter the environment and are transported and transformed in the air, water, soils, sediments and biota of our ecosystem. Although such ecosystem models can be quite sophisticated for well-defined systems such as lakes or agricultural soils, they become less reliable when applied to larger and more heterogeneous geographic regions. Attempts are underway to test models for regions, countries, continents and even the global environment, and there is a compelling incentive to improve these models. The ability to identify potential PBT compounds in advance of their commercial production, by using our acquired knowledge of chemical properties and environmental fate of other substances, may enhance pollution prevention by avoiding the introduction of such chemicals into the environment. Many of the chapters in this book are devoted to the development and application of such methodologies by government, academia, and industry.

The toxicity aspect of PBTs corresponds to a broad set of adverse effects, both acute and chronic, for a variety of organisms and ecosystem populations, including those effects as yet undefined. Given the large number of organisms present in many environments, it is not feasible to test all possible toxic effects for all organisms. Nor is it possible to investigate all possible combinations of chemical mixtures with respect to organism and toxicological endpoint. In practice, toxicologists have developed sets of basic test data to perform environmental risk assessments. While the basic test set differs somewhat among various national programs, efforts towards test protocol harmonization are being made through international bodies such as the Organisation for Economic Cooperation and Development (OECD).

Given the limitations regarding measured data and modelling of PBT properties and toxic effects, expert judgment continues to play an important role in the assessment of candidate substances, although international agreements have encouraged more systematic approaches, as discussed in the sections below.

UNEP

International negotiations on a Persistent Organic Pollutant (POP) protocol are being led by the United Nations Environmental Program (UNEP). As an initial step, UNEP participants have developed a list of 12 well-established POP compounds (6). A group is developing criteria and key cut-off values for addition of other PBT candidates to this list. These criteria have not been developed yet, but the following values have been proposed for persistence: Half-life in water greater than 2 or 6 months; or half-life in soil or sediment greater than 6 months; or evidence that the substance is otherwise sufficiently persistent to be of concern within the scope of the convention. For bioaccumulation the following criteria have been proposed: Bioconcentration factor (BCF) in aquatic species greater than 5,000 or, in the absence of BCF, log K_{ow} greater than 4 or 5. Additionally, monitoring data in biota indicating high bioaccumulation potential may qualify the substance to be of concern within the convention. Monitoring data or model predictions indicating potential for long-range transport are also needed as well as evidence based upon toxicity or ecotoxicity data for the potential for adverse effects to human health or to the environment.

Table 1. POP compounds negotiated within UNEP

Pesticides	*Industrial chemicals*	*Unintended by-products*
Aldrin	Hexachlorobenzene	Dioxins
Chlordane	Polychlorinated biphenyls	Furans
DDT		
Dieldrin		
Endrin		
Heptachlor		
Mirex		
Toxaphene		

UN-ECE LRTAP

The UNEP list above is derived from a protocol for Long Range Transport of Air Pollutants (LRTAP) within the United Nations Economic Commission for Europe (UN-ECE) framework (*7*). This protocol was signed in 1998 and includes an initial list of 17 established POP compounds. The protocol also includes measures to be taken to reduce emissions of these substances, ranging from a total ban to specific emission control. The regions covered by this agreement include Europe and North America, although as of mid 1999 only one country has ratified the protocol. Negotiations towards ratification by other countries are in progress.

Table 2. UN-ECE list of compounds included in the POP protocol

Scheduled for elimination	*Scheduled for restrictions*	*Scheduled for measures*
Aldrin	DDT	PAHs
Chlordane	Hexachlorocyclohex	Dioxins/furans
Chlordecone	Polychlorinated	Hexachlorobenzen
DDT		
Dieldrin		
Endrin		
Heptachlor		
Hexabromobiphenyl		
Hexachlorobenzene		
Mirex		
Polychlorinated		
Toxaphene		

North American Great Lakes

A region of particular concern has been the North American or Laurentian Great Lakes of Superior, Michigan, Huron, Erie and Ontario which straddle the border region between the US and Canada. The binational International Joint Commission (IJC) was organized to coordinate boundary activities by the US and Canada, including environmental issues. Within the scope of its Great Lakes Water Quality Agreement, IJC seeks to maintain and improve the quality of shared water resources. Persistent and bioaccumulative chemicals were early recognized as affecting fish and especially fish-eating birds and mammals with a corresponding threat to human health from fish dietary exposure. As part of the 1997 U.S.-Canada

Binational Agreement, voluntary efforts were developed to promote the "virtual elimination" of specific PBTs and other substances of concern (Level I) and to list additional chemicals (Level II) for future consideration.

Table 3. US-Canada Binational Agreement: Compounds for "virtual elimination" (Level I)

Aldrin/dieldrin	Mirex
Benzo(a)pyrene	Octachlorostyrene
Chlordane	PCBs
DDT (+DDD+DDE)	PCDD (Dioxins)
Hexachlorobenzene	PCDF (Furans)
Alkyl-lead	Toxaphene
Mercury and mercury compounds	

Table 4. US-Canada Binational Agreement: Compounds for future consideration (Level II)

Cadmium and cadmium compounds	Pentachlorobenzene
1,4-Dichlorobenzene	Pentachlorophenol
3,3'-Dichlorobenzidine	Tetrachlorobenzene (1,2,3,4- & 1,2,4,5-)
Dinitropyrene	Tributyl tin
Endrin	PAHs as a group, including but not limited to Anthracene, Benzo(a)anthracene, Benzo(g,h,i)perylene, Perylene, and Phenanthrene
Heptachlor (+Heptachlor epoxide)	
Hexachlorobutadiene (+Hexachloro-1,3-butadiene)	
Hexachlorocyclohexane	
4,4'-Methylenebis(2-chloroaniline)	

This effort is directed towards a goal of "zero discharge" of selected PBTs, particularly certain organochlorines. Water quality and levels in fish in the Great Lakes have declined due to reduced emissions, but in recent years it has become apparent that levels in water and fish are increasingly controlled by atmospheric sources. This exposure route demonstrates the need for comprehensive actions to control all sources to the air, water and soil environments on an international scale. Local remediation effects, while partially effective, can not entirely eliminate these substances because of their ability to persist and migrate through the ecosystem regardless of political boundaries.

European Union

While no activity exists at present within the European Union (EU) specifically directed towards the PBT compounds, the proposed Water Framework Directive would serve as a tool to characterize the quality of the aquatic environment. A method entitled Combined Monitoring-based and Modelling-based Priority Setting Scheme (COMMPS) would be used to prioritize chemical parameters, leading to a ranking of exposure based on both monitoring and model predicted data. Toxicity data are also ranked and the product of these rankings is used in the final priority setting. This scheme yields a list (Table 5) containing a number of well-established PBTs as indicator compounds.

Table 5. Compounds proposed to be used as chemical indicators to describe the quality of the aquatic environment in the EU

Alachlor	Di(2-ethylhexyl)phthalate	Nickel and its compounds
Anthracene	Dichloromethane	Nonylphenols
Atrazine	Diuron	Octylphenols
Benzene	Endosulfan	Pentachlorobenzene
Brominated diphenyl ethers	Hexachlorocyclohexane	Pentachlorophenol
Cadmium and its compounds	Hexachlorobenzene	Polyaromatic hydrocarbons
Chlorfenvinphos	Hexachlorobutadiene	Simazine
Chloroalkanes, C10-13	Isoproturon	Tributyltin compounds
Chlorpyrifos	Lead and its compounds	Trichlorobenzenes
1,2-Dichloroethane	Mercury and its compounds	Trichloromethane
	Naphthalene	Trifluralin

OSPAR

In 1995, countries bordering the North Sea declared that within one generation (the Esbjerg Declaration (*8*)) no emissions of hazardous substances should occur into the marine environment, with the goal that concentrations of such compounds in the North Sea should be close to zero. One generation was set as 25 years, with persistent, bioaccumulating and toxic (PBT) chemicals being those defined as hazardous.

In 1998, a strategy with regard to hazardous substances was adopted under the OSPAR Convention for the Protection of the Marine Environment of the North-East Atlantic, which sets out goals and objectives similar to those agreed in the Esbjerg Declaration. One essential requirement in this strategy is the identification of hazardous substances, belonging either to the PBT group or representing a similar level of concern. The OSPAR Commission established a working group on the development of a dynamic selection and prioritization mechanism for hazardous substances (DYNAMEC) for this task. This group has now been working for 2 years and decided on a procedure depicted in Figure 1.

An objective of the DYNAMEC process is to tackle as much as possible of the chemical universe. In the initial selection process, measured and calculated data for some 180,000 substances or groups of substances were used. Data on biodegradation in water from the OECD test methods were used to indicate the persistence of the compounds. In some cases half-lives or BOD/COD were available. For the bioaccumulation potential, bioconcentration factors (BCFs) were used when available, otherwise log K_{ow} was used. Data on acute or chronic toxicity were used, as well as R-phrases corresponding to the proposed cut-off values.

A Danish QSAR Database containing estimates for over 46,000 compounds that may be on the European market was used for the model identification of PBTs. This is built on estimates from the Syracuse programs BIODEG for persistence and BCFWIN and/or LOGKOW for bioaccumulation. The aquatic toxicity, assumed to be base-line toxicity or non-polar narcosis, is calculated from BCF. Mammalian toxicity is predicted with TOPKAT.

With the presently proposed PTB cut-off values this results in a preliminary list of chemicals of possible concern containing some 400 compounds. Further ranking, taking use and/or exposure indications into account, of these have identified five substances of very high concern and another seven with less severe PTB profile (Table 6). The OSPAR Commission will make the final decision on new priority substances in June 2000.

North American Regional Action Plans (NARAPs)

Within the framework agreement of the 1995 Council Resolution 95-5, the Sound Management of Chemicals (SMOC) was developed to provide a regional means of reducing persistent, bioaccumulative, and toxic chemicals in North America. SMOC was established under the Commission for Environmental Cooperation (CEC), which was created as an environmental side agreement to the North American Free Trade Agreement (NAFTA). Under the SMOC framework, trilateral North American Regional Action Plans (NARAPs) can be developed when a determination is made that mutual benefit and interest exists for Canada, Mexico, and the United States. NARAPs have been developed and approved by the CEC Council for PCBs, DDT, and chlordane. Phase 1 of a more comprehensive plan for mercury has also been approved. Additional NARAPs are currently being developed for dioxins and furans, and for hexachlorobenzene. Lindane and lead are being evaluated as possible NARAP candidates for regional action. A NARAP on environmental monitoring and assessment is also under development (*9*).

The United States

The United States Environmental Protection Agency (USEPA) has provided additional guidance with respect to PBTs for its Premanufacture New Chemicals

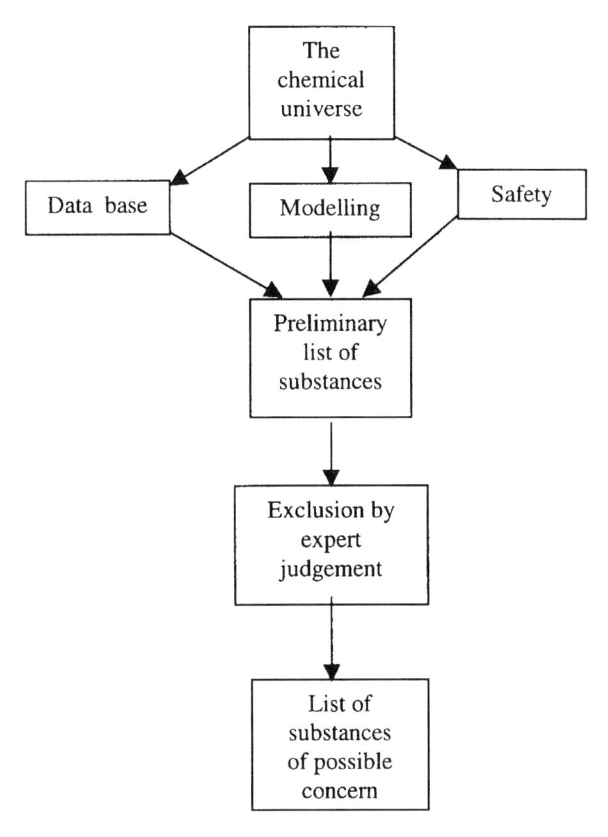

Figure 1. General scheme of the DYNAMEC process.

Table 6. Proposed new priority PBT substances in the OSPAR framework

Substances of very high concern	*Compounds with less severe PTB profile*
Decylphenol	Hexamethyldisiloxane
Dicofol	Tetrabromobisphenol A
Endosulfan	1,2,4-Trichlorobenzene
Methoxychlor	1,2,3-Trichlorobenzene
Octylphenol	1,3,5-Trichlorobenzene
	4-*tert*-Butylphenol
	Hexachlorocyclopentadiene

Program, under Section 5 of the Toxic Substances Control Act, which is discussed in chapter 12 of this volume. The Office of Pollution Prevention and Toxics, which is responsible for this effort, has led the Agency in the development of a Multimedia Strategy for PBT pollutants, discussed in chapter 10. Additionally, USEPA's Office of Solid Waste has developed a computerized database and decision algorithm to prioritize candidate PBT chemicals based on various characteristics, as described in chapter 13. Several States and local authorities have implemented programs to collect and replace thermometers and manometers containing mercury, in an effort to reduce exposures.

Canada

The Canadian Federal Government was among the first to formulate criteria for identifying PBTs as part of its Toxic Substances Management Policy. These criteria were influential in the development of the UNEP criteria described above, and are similar in approach. In Canada, chemicals are controlled by the Canadian Environmental Protection Act. About 60 substances or groups of substances have been listed as high in priority and may have been assessed for sources, fate and effects on the environment and human health. Another 22,500 substances already in use have been scheduled for assessment in the next seven years. Those substances deemed to pose a threat may be subject to regulation. Generally, the list of regulated substances is similar in content to the UN-ECE list. Canada as a circumpolar nation is particularly concerned about contamination of its arctic ecosystems as a result of long range transport from the south. Residents of the Canadian north, including the Inuit tend to rely heavily on locally caught animals for food. Often these animals are rich in lipids and have remarkably high contaminant levels causing unacceptable exposure. The high levels appear to be caused in part by the global fractionation or distillation effect in which "semivolatile" chemicals tend to partition preferentially into cold ecosystems where they are well preserved. Details of current activities can be found at the Environment Canada website (*10*).

Sweden

The Swedish Government decided in 1999 that "The environment shall be free from man-made substances and metals that represent a threat to health or biological diversity. This means that the levels of substances that occur naturally in the environment must be close to background levels, while the levels of man-made substances must be close to zero." To achieve this objective it is believed that goods produced 10 to 15 years from now have to be essentially free from synthetic organic compounds which are persistent and bioaccumulative. Application of this precautionary principle results in the inclusion of compounds with unknown toxicological activity.

The Netherlands

The National Institute of Public Health and the Environment has developed criteria for the evaluation of PBTs for hazard assessment, described in chapter 9.

The Dutch Ministry of Housing, Spatial Planning and Environment undertook a study to identify chemicals with hazardous inherent properties, viz. persistence, toxicity and potential for bioaccumulation. This study focussed on compounds with experimental toxicity data and those were further investigated for the other two properties. ISIS/Riskline contains some 180,000 substances and is based on the European Inventory of Existing Chemical Substances (EINECS), supplemented by the Registry of Toxic Effects of Chemical Substances (RTECS), the International Registry of Potentially Toxic Chemicals (IRPTC), and the European List of Notified Chemical Substances (ELINCS). Toxicity data are also imported from AQUIRE and a Dutch database with information on newly notified substances (KNS).

Persistence was estimated from degradability data and for compounds that are expected to be distributed into the air (>10%) atmospheric degradation was also taken into account. When no experimental data were available, the Syracuse BIOWIN program was used to estimate the biodegradation. The cut-off values for toxicity and bioaccumulation are given in Table 7.

The overlap between databases made it possible to select from about 65,000 compounds. The final outcome was a list of 165 substances.

Table 7. Toxicity and bioaccumulation criteria used to identify BT compounds in the Netherlands

Parameter	*Criteria**
Mammalian acute toxicity	
- Oral (LD50 rat mg/kg)	R22 or LD50 < 2000
- Inhalation (LC50 rat, mg/l/4h)	R20 or LC50 < 20
- Dermal (LD50 rat/rabbit, mg/kg)	R21 or LD50 < 2000
Carcinogenicity	R45 or R49 or IARC 1 or IARC 2A
Ecotoxicity	
- Acute toxicity (mg/L)	L(E)C 50 < 1
- Chronic toxicity (mg/L)	NOEC < 0.1
Bioaccumulation	log K_{ow} >5

*R-phrases are cut-off values for labeling hazardous chemicals (*11*).

Scientific Organizations

In addition to these government efforts to identify PBT substances, professional societies such as the Society of Environmental Toxicology and Chemistry (SETAC) have sponsored studies on behalf of the scientific community. An example is the report on "Chemical Ranking and Scoring: Guidelines for

Relative Assessments of Chemicals" edited by Swanson and Socha (*12*). In 1998, SETAC sponsored a workshop devoted to establishing sound scientific criteria for persistence and the potential for long range transport. The proceedings of this workshop will be published in 2000 and will contain a review of the state of the art in assessing degradation rates in air, water, soils and sediments, as well as the role of partitioning and mass balance models. Recommendations will be made for the tiered assessment of persistence. Details of the availability of these and other related reports can be found at the SETAC website (*13*). Environmental groups such as the World Wildlife Fund (*14*) have also been active in promoting the broader and more complete assessment of pesticides and chemicals of commerce.

CONCLUSIONS

Presently there are several fora in which attempts are being made to identify PBT substances before they are detected in the environment and cause adverse effects. Several lists are already available and several more will be produced in the near future. Discharges of these compounds must be controlled so that effect levels are not reached in the environment. Because of the persistent character of these substances the emissions over a very long time period must be considered, as well as life-long exposure. Since we often have incomplete data on toxic effects we must be careful with all persistent bioaccumulating substances to avoid future surprises.

The challenge is to create an international system in which new and existing chemicals are manufactured and used for the benefit of society, but with an assurance that they will not cause adverse effects on humans, wildlife or in general on the ecosystem of which we are part, and on which we depend.

REFERENCES

1. Lovelock, J.E.; Lipsky, S.R. J. Am. Chem. Soc. 1960, 82, 431.
2. Jensen, S. New Scientist 1966, 32, 612.
3. Mattsson, P.E.; Norström, Å; Rappe, C. J. Chromatog. 1975, 111, 209-213.
4. Jensen, S.; Jansson, B. Ambio 1976, 5, 257-260.
5. Kuehl, D.W.; Kopperman, H.L.; Veith, G.D.; Glass, G.E. Bull. Environ. Contam. Toxicol. 1976, 16, 127-132.
6. http://irptc.unep.ch/pops/newlayout/aitstdefaultbody.html
7. http://www.unece.org/env/lrtap/pops_hl.htm
8. http://odin.dep.no/nsc/esbjerg
9. http://www.cec.org
10. http://www.ec.gc.ca/cceb1
11. http://www.hse.gov.uk/hthdit/chip/chip8.htm
12. Swanson M and Socha. A Chemical Ranking and Scoring: Guidelines for relative assessments of chemicals. SETAC publication sp97-1 Pensacola FL 1997
13. http://www.setac.org
14. http://www.worldwildlife.org/toxics/progareas/pop/ipen2.htm

Persistence and Modeling

Chapter 2

Defining the Bioaccumulation, Persistence, and Transport Attributes of Priority Chemicals

Donald Mackay[1], Eva Webster[1], Andreas Beyer[2], Michael Matthies[2], and Frank Wania[3]

[1]Environmental Modelling Centre, Trent University, Peterborough, Ontario K9J 7B8, Canada
[2]Institute of Environmental Systems Research, University of Osnabrück, Osnabrück, Germany
[3]Department of Chemistry and Division of Physical Sciences, University of Toronto at Scarborough, Toronto, Ontario M1C 1A4, Canada

Persistence, tendency to bioaccumulate and potential for long range transport are three attributes of an organic chemical which, in combination with toxicity, can result in classification as a priority pollutant. These three attributes are largely consequences of the fundamental physical-chemical, degradation and transport properties of a substance. Methods by which these fundamental properties can be translated into expressions of persistence, bioaccumulation and long range transport as part of the classification system are discussed. It is argued that mass balance models must play a central role in any such classification systems. Strategies by which models can be used in this context are reviewed, suggested and discussed.

Introduction

Regulatory agencies, industry and the scientific community frequently face the task of selecting from among the universe of commercial chemicals and by-products. Those chemicals which require priority for assessment and potential management. The key undesirable properties are Persistence, Bioaccumulation tendency, Toxicity (which we do not address here), and potential for long range transport to remote locations. These chemicals are often referred to as "PBTs". The scientific and regulatory issues surrounding their control have been reviewed by Vallack et al. (1998) with a more recent update on United National Environmental Program (UNEP) activities by Schmidt (1999). The issue, which we address in this Chapter, is the identification of chemicals, which have these undesirable attributes, and the assignment of priorities within such a group.

To assist in this process we first need a mental picture of the environment and the fate of chemicals within that environment. For most purposes the environment can be viewed as consisting of approximately 10 compartments or phases such as air, aerosols, water, suspended solids, bottom sediments, terrestrial soils, vegetation, and groups of biota which occupy these compartments. If equilibrium partition coefficients between these media, as well as reaction half lives or rate constants in each medium, and all intermedia transport rate parameters were known for each chemical, then, in principle at least, the environmental behavior of the chemical could be fully described and quantified. If absolute, as distinct from relative, concentrations are desired then emission rates must also be supplied. Temperature variation in the environment introduces a complexity that could be accommodated by introducing temperature coefficients such as enthalpies of phase transfer or activation energies for each process. Since the total number of such chemical properties that control environmental fate is probably of the order of a hundred, this raises at least two problems.

First, it is impractical to gather data for all these properties for every chemical. Seeking relationships between these properties and more fundamental physical-chemical descriptors can alleviate this problem. For example, octanol-water partition coefficient can be used to estimate lipid-water and organic carbon-water partition coefficients. Most problematic are that such methods do not currently exist to estimate reaction rate constants, especially those involving biodegradation. A major pursuit of environmental chemists is the search for these quantitative structure-activity and structure-property relationships.

Second, even if all the data were available it is beyond the capacity of the human mind to assimilate all the data and arrive at an estimate of persistence, bioaccumulation, or long-range transport. This is due in part to the fact that these properties vary over such a wide range of magnitude. For example air-water partition coefficients vary over a range of at least 1010. The human mind has difficulty grasping these numbers.

Given these difficulties, it is clear that two research thrusts are required relating to the environmental chemistry of chemical substances. First is a search for improved methods by which molecular structure or readily measurable properties of chemicals can be used to estimate partitioning, transport and transformation in the environment. Second is the development of procedures, usually some form of decision framework or mass balance model, which will accept these data and use them to calculate the integrated properties of concern such as persistence or transport distance. Given the potential risk to humans and ecosystems and the considerable investment required to

research the fate of chemicals in the environment it is unthinkable to consider adopting a "wait and see" strategy to determine if chemicals entering the environment can cause effects as was the case with DDT and PCBs. Nor is it acceptable to allow the process to be driven only by political pressures. A sound, transparent, science-based process is needed. It is thus essential that we build up a predictive capability for application to the many thousands of existing chemicals and any proposed new chemicals. Furthermore, we should seek to validate the capability by predicting and confirming that the fate of existing chemicals is in accord with our present understanding of their likely behavior.
As a contribution to this goal we describe here applications of mass balance models for evaluating bioaccumulation, persistence and long-range transport.

Bioaccumulation

Fish are the most convenient organism for measuring bioaccumulation for reasons of both expense and experimental convenience. The results are applicable in principle to other organisms with appropriate modification of physiological and exposure conditions. International conventions do not explicitly state that fish should be the subject organism, but it is usually assumed that fish will be the test species.

It is now widely accepted that a measured bioaccumulation factor (BAF) or ratio of fish and water concentrations exceeding 5000 is cause for concern. This corresponds approximately to a lipid-adjusted bioconcentration factor of 10^5 and a lipid content of 5%. The usual approach is to assume that lipid-water and octanol-water partition coefficients (K_{OW}) are equivalent, thus if simple equilibrium partitioning is assumed to apply a K_{OW} exceeding 10^5 is indicative of excessive bioaccumulation. In reality 10^5 should not be regarded as a "bright line" separating bioaccumulative and non-bioaccumulative chemicals. The attribute of bioaccumulation is continuous in nature.

There are several additional factors which complicate this simple approach and may require consideration in borderline cases. The first two factors tend to reduce the BAF thus the use of a K_{OW} criterion is conservative.

If the rate of metabolism within the fish is comparable to, or larger than, the rate of uptake, the fish will achieve a sub-equilibrium condition in which the actual BAF is less than the equilibrium value deduced from K_{OW}. Notable in this respect are the polycyclic aromatic hydrocarbons and phthalate esters. Unfortunately metabolic half-lives are species dependent. It is only with the aid of a bioaccumulation model that the metabolic half-life can be deduced from experimental data.

If K_{OW} is large, the fraction of the chemical in the water column which is truly dissolved and is bioavailable to partition can become considerably less than 1.0. This effect is specific to the sorbing capacity of the water in question, thus it is an attribute both of the chemical and the local water chemistry.

When food is the primary route of uptake there may be a biomagnification factor (BMF) in which beyond-equilibrium conditions are achieved. The extent of this effect depends on the nature of the food source or food web and K_{OW}. From experiments exploring the mechanism of biomagnification it appears that a prey to predator BMF may be up to a factor of 3 to 4 with each trophic level change (Gobas 1993). Long food webs are thus susceptible to large BMFs. This effect becomes most important for substances with the K_{OW} exceeding 10^5 which are already designated as bioaccumulative.

Other chemicals may display attributes which confound the simple partitioning paradigm. Certain substances of high molar mass and volume may be inefficiently absorbed in the gill or gut thus retarding uptake. If the substance ionizes, or it has a tendency to form strong links to enzymes, or it is highly surface active it may not conform to simple partitioning.

Despite these complications the attribute of bioaccumulation is well described for non-polar substances by K_{OW}, where necessary being modified by metabolic half-life as determined by uptake/clearance experiments. Reliable mechanistic models exist which can describe the uptake by single organisms and entire food webs (Gobas 1993, Clark et al. 1990, Thomann 1989, Thomann et al. 1992) Campfens and Mackay (1997). The obvious strategy is thus to use K_{OW} as a first estimator of BAF and where justified modify the BAF to obtain a more accurate value using a mass balance model of a single fish of standardized species and size, or even a standardized food web.

Persistence

Persistence (τ_R h) is an important attribute because in principle it is the quantity which links steady state discharge rate into the environment (E kg/h) to amount in the environment (M kg).

$$M\ (kg) = \tau_R\ (h) \times E\ (kg/h) \tag{1}$$

Alternatively, persistence links the total mass in a multi-media system to the total rate of reactive loss R kg/h. It thus describes the rate at which chemical is removed from the environment, and indirectly, the recovery time of a contaminated system.

$$\tau_R\ (h) = M\ (kg)\ /\ R\ (kg/h) \tag{2}$$

The usual descriptor of persistence is half-life; a concept borrowed from radioactive decay, but in the case of chemical half-life this time is a function both of the chemical and its reactive environment, for example the concentration of OH radicals in air. It can be shown that residence time under steady state conditions as defined above τ_R and half-life $\tau_{1/2}$, which is usually measured under dynamic conditions, are related by

$$\tau_{1/2} = 0.693\tau_R \quad (3)$$

As has been discussed by Webster et al. (1998) there are formidable difficulties in estimating half-lives in a single medium such as air or water. Half-lives vary diurnally and seasonally at any location and with climate and latitude. Half-lives exceeding a month are difficult to measure in the laboratory because constant conditions should be maintained for several half-lives. Measurement in the environment is even more difficult because of competing loss processes and variation in temperature. The issue of temperature is problematic. Those who live in cold climates may favor adoption of lower temperatures, and longer half lives. It can be argued that is the half-life in the region of chemical use that is most relevant. For experimental reasons it is likely that a standard temperature of 25°C may be adopted.

The approach taken by Environment Canada (1995) is to set half-life criteria of 2 days in air, 182 days (6 months) in water and soil and 365 days in bottom sediments. If any one criterion is achieved or exceeded the substance is deemed persistent. Presumably these half-lives apply under standard conditions. The major difficulty with this approach is that it ignores environmental partitioning. For a non-volatile chemical the half-life in air is irrelevant, as is the bottom sediment half-life for a highly volatile substance. Since chemicals can migrate throughout the environment and they tend to have most adverse effects in the media into which they partition most strongly. Webster et al. (1998) have argued that the key persistence attribute is the overall persistence, i.e. the average residence time that a molecule exists in the environment as a whole. It can be shown that under steady-state conditions the overall rate constant for loss by reaction k_R (which is related reciprocally to the overall residence time) is the weighted mean of the single media rate constants, each medium being weighted according to the mass in that medium, e.g. for three media

$$k_R = k_1 m_1 + k_2 m_2 + k_3 m_3 \quad (4)$$

where m_i is the mass fraction in a medium, i.e. the mass in the environmental medium M_i kg divided by the total mass in the environment M. It follows that the overall half-life is given by

$$1/\tau_{1/2} = m_1/\tau_{1/2,1} + m_2/\tau_{1/2,2} + m_3/\tau_{1/2,3} \quad (5)$$

This equation explicitly expresses the importance of both partitioning (which largely controls the individual values of m_i) and individual half-lives as contributions to the overall half- life.

What then is a reasonable overall half-life criterion to separate persistent and non-persistent substances? One approach is to suggest that if less than about 10% of chemical survives over one year, this provides assurance that there is unlikely to be continued accumulation of chemical as a result of prolonged usage. If "standard" reaction conditions are assumed to apply for 8 months of the year, with 4 months of winter with low reaction rates, then a half-life of 2 months will result in loss of 15/16^{ths} or 92% of the chemical. There will thus be little carry-over from year to year, except possibly in media which contain relatively little of the chemical burden. A 4 month half-life, which could leave a carry-over of 25%, seems excessively long.

If regulatory penalties are to be applied to substances which exceed such criteria it seems inevitable that there will be legal controversy about the criteria and the assignment of these half- lives. This is unfortunate and largely pointless because chemicals experience a distribution of half-lives in the environment, and any realistic expression of half-life must include this distribution. Indeed, the distribution will be widened by uncertainties in chemical and environmental properties.

Overall persistences are not directly measurable in the environment except under carefully controlled conditions such as mesocosm experiments. Consequently they must usually be determined by some form of mass balance model. Two options present themselves.

The simpler is to use a Level II model in which the chemical is assumed to exist at equilibrium between defined environmental media and a steady state system is set up such that a constant emission rate is balanced by an equal reaction rate. For details of the assumptions and structure of Level I, II, III and IV models the reader is referred to Mackay (1991) or Mackay and Paterson (1982). This approach has been discussed by Müller-Herold (1996) and Müller-Herold et al. (1997). No advective losses are permitted since these losses merely move the chemical to another region. Such models are widely described (Mackay 1991) and available at the website of the Environmental Modeling Centre (www.trentu.ca/envmodel) . Reasonable volumes of relevant media are assumed with typical compositions such as organic carbon contents of soils. Here we calculate the output from such a model, specifically a modified Level II version of the EQC model (Mackay et al. 1996) in which there is no advective loss. The properties of hypothetical chemicals which illustrate these concepts are given in Table 1.

As shown in Figure 1, for Chemical A, the total mass in the system is 168 kg and since the input (and output) rates are 1.0 kg/h, the persistence or residence time is 168 h or 7 days. The half-life is 4.8 days. Clearly the important loss processes are reaction in air and soil thus it is the half-lives in these media which should be known most accurately. Even if infinite half-lives are assumed in water and sediment the persistence rises negligibly by only one hour to 169 h. The relative importance of reaction in each medium is clearly controlled by both mass fraction in each medium and the prevailing, local half-life.

If the partitioning properties are changed to those of chemical B in Table 1 but the same half-lives are retained, the overall persistence rises by a factor of 11 to 1867 h or 77.8 days because more of the chemical partitions into media with longer half-lives. Clearly, partitioning profoundly affects persistence. Chemical C has identical partitioning properties to B but it has shorter half-lives and would be declared "non-persistent" by the Environment Canada criteria whereas both A and B are "persistent". This "non-persistent" chemical has an overall persistence of 58.3 days, thus it is more persistent by a factor of 8 than A which is "persistent" by the single media criteria

Table 1. Properties of hypothetical chemicals at 25°C.

	A	*B*	*C*
Molar mass g/mol	200	200	200
Solubility in water g/m^3	2	0.5	0.5
Vapor pressure Pa	1	0.1	0.1
Log K_{OW}	4.5	5.5	5.5
Half-life in air (days)	2	2	1.5
Half-life in water (days)	100	100	80
Half-life in soil and sediment (days)	200	200	150
K_{AW}	0.040	0.016	0.016
Overall Persistence (days) (Level II)	7	77.8	58.3

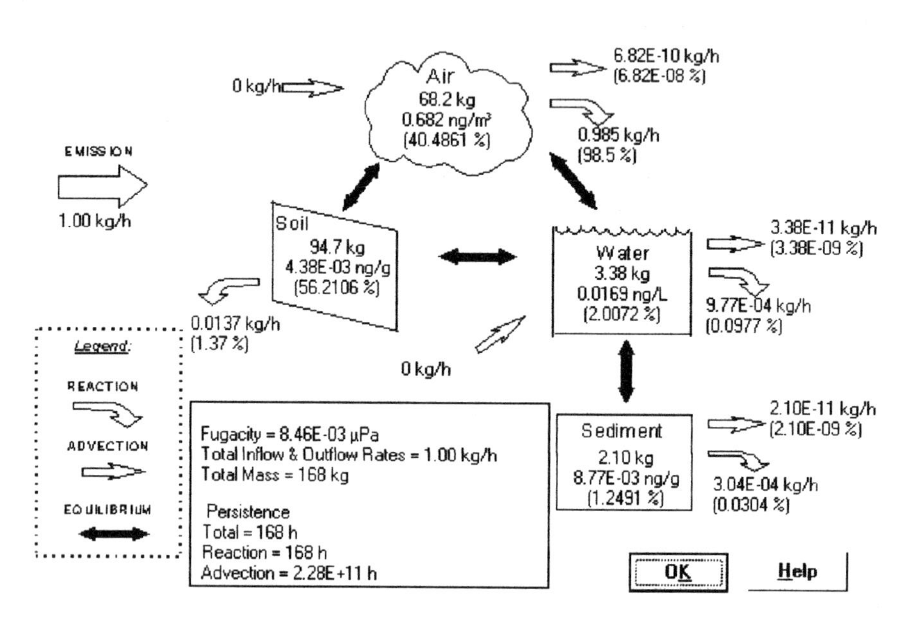

Figure 1. Output of a Level II calculation for chemical A in which equilibrium is assumed to apply between all media.

An even simpler and more transparent calculation is to consider the environment as being equivalent in partitioning capacity to defined volumes of air, water and octanol. For the EQC model (Mackay et al. 1996) these are air 10^{14} m^3, water 2 x 10^{11} m^3 and octanol (representing organic carbon soils and sediments) 181 x 10^6 m^3. If these are designated V_A, V_W and V_O, respectively then the mass fractions of chemical A in each medium are respectively.

$$m_A = V_A K_{AW}/(V_W + V_A K_{AW} + V_O K_{OW}) = 0.405 \quad (6)$$

$$m_W = V_W/(V_W + V_A K_{AW} + V_O K_{OW}) = 0.020 \quad (7)$$

$$m_O = V_O K_{OW}/(V_W + V_A K_{AW} + V_O K_{OW}) = 0.575 \quad (8)$$

The overall half-life can then be calculated using Equation 4 earlier and is 4.8 days as before.

This simple calculation could be extended to differentiate between the organic matter in soil and sediment. If a half-life is not known, the effect of assuming an infinite value can be explored. If the substance passes a criterion under these conditions, it can only be less persistent with additional information on the missing actual half-life is provided, thus the additional information may not be needed. It is suggested that this simple approach is suitable for screening large numbers of chemicals for which very limited data are available. The major weakness of this approach is that it does not take into account how the chemical enters the environment, or the effects of inter-media transport rates or resistances.

A variation on this theme is to estimate the "three solubilities" of the chemical in air, water and octanol as discussed by Cole and Mackay (1999). The partition coefficients are ratios of these solubilities, for example K_{AW} is S_A/S_W where S_A is solubility in air (actually vapor pressure/RT) and S_W is solubility in water. In many cases these solubilities are hypothetical and unmeasurable. Fundamentally each is $1/v\gamma$, where v is molar volume of the solution and γ is the activity coefficient. The corresponding equation for m_A takes the form

$$m_A = V_A S_A/(V_W S_W + V_A S_A + V_O S_O) \quad (9)$$

with similar expressions for m_W and m_O.

The more complex and realistic approach is to use a Level III model in which the media are not at equilibrium and intermedia transport resistances are included. It is now essential to define the "mode of entry", i.e. into which media the chemical is introduced. Fortunately fairly robust expressions exist for intermedia transport processes which do not usually require additional data for the chemical, thus a Level III calculation makes no demand for more chemical data than the approach given above, except for mode of entry.

This approach has been illustrated and advocated by Webster et al. (1998) who used a model similar to EQC. EQC is available for downloading free of charge from our website at http://www.trentu.ca/envmodel. It accepts as input, the following information on the properties of the chemical: name, molecular mass, solubility in water, vapor pressure, octanol-water partition coefficient and half-lives in air, water, soil and sediment. Alternatively it may be more convenient to input partition coefficients directly for chemicals which have extreme properties such as total miscibility with water, or those which do not follow common structure-property relationships such as the K_{OW} - K_{OC} - K_D correlation's for sorption to soils or sediments.

Level III calculations for chemical A emitted to air, water and soil give persistences of 5.5, 39.6, and 375 days respectively. The mode of entry clearly affects persistence, especially when the chemical is emitted into a medium in which the half-life is long and the rate of transport to other media is slow. The simpler Level II calculation avoids this problem by assuming instantaneous equilibrium throughout the system, but this simplification results in loss of realism and some risk of misclassifying persistent chemicals as non-persistent.

There are several implications of these illustrations.

- The overall persistence is a single number more truly indicative of environmental behavior than the four individual half-lives.
- The overall persistence depends on the chemical's partitioning properties.
- The overall persistence depends on the mode of entry or emission pattern (i.e. to air or water or both), but not on the emission rate, except under the infrequent conditions that non-first-order processes occur rendering the model and behavior non-linear.
- Since the individual half-lives can be very difficult to measure or estimate, a sensitivity or Monte Carlo analysis should be used to determine which are most important and thus devote most effort to determining these half-lives.
- Some half-lives can be unimportant because of the tendency of the chemical not to partition into the corresponding medium. A mass balance model reveals such situations in which it may be acceptable to assume an infinite half-life.
- It must be recognized that the environmental reality is that chemicals display a distribution of individual and overall half-lives as a function of time, place, and other factors. It is thus likely that a substance can exceed a criterion for some percentage of the time. A pass-fail system based on a "bright line" is thus fundamentally flawed. A more valid approach is to generate a distribution and compare chemicals on a distribution versus distribution basis, or express persistence on a probabilistic basis, e.g. 20% of the time it is expected that persistence will exceed 100 days with a likely 5% exceedence of 200 days (Webster et al. 1998). The persistence is, of course, a function of the nature of the environment or geographic location.

In summary, it is suggested that for screening purposes a Level II model or calculation be used which exploits partitioning information which is more accurate and readily available than degradation rate data. If the persistence proves to be of potential concern a Level III calculation can be done incorporating mode of entry information.

Long Range Transport (LRT)

This attribute is important as an indicator of the potential for a substance to contaminate regions far from sources. There is an ethical dimension to this attribute in that it seems inherently unjust for one group to enjoy the benefits of chemical use, for example as a pesticide to increase agricultural yields, while also inflicting adverse effects on distant groups who derive no such benefits. A notable example of this disproportionate sharing of benefits and disbenefits arises from the phenomenon of global fractionation as discussed by Wania & Mackay (1993, 1996, 1999). Substances tend to partition from the atmosphere to solid and liquid components of the ecosystem at low temperatures when K_{OA} increases or K_{AW} decreases, or in general when the solubility in air S_A decreases. This has contributed to unusually high levels of contamination in the arctic where degrading reactions are slower. Residents who rely on diets rich in locally gathered fats from fish or sea mammals are especially vulnerable and can be highly exposed (Dewailley et al.1989, 1993)

Quantifying this attribute is the topic of considerable recent and current research with notable contributions having been made by Scheringer 1994, 1997, van Pul et al. 1998, and Bennett et al 1998. The primary medium of LRT is the atmosphere, but for some compounds that have low K_{AW}, ocean transport can be important and even be the dominant transport route (Wania and Mackay (1999). It has also been suggested that biotic transport in migrating birds, fish and marine mammals such as whales can be significant. Several approaches can be taken to estimate the potential for LRT and it is not clear at present which will prove most appropriate.

Multi-Box Models

Scheringer (1997) has used multiple connected boxes or unit worlds representing meridional zones to evaluate a chemical's potential for LRT. The boxes consist of air, soil and ocean water and intermedia transport follows a Level III approach. Transport from box to box can be quantified and boxes can be at different temperatures. Brandes et al. 1996 have applied a related approach with nested boxes which simulates the wider dispersal of a chemical from a point source. An alternative approach is to use a single box where the fraction leaving the system by advection used indicates of potential for LRT.

Continuous Models

van Pul (1998) and Bennett (1998) have suggested calculating the decline in concentration in air as a function of distance over a soil surface and using this information to indicate potential for LRT. Beyer (1999) has extended this to treat transport in air over water and mixed surfaces, and in water under air and over sediment. Mathematically the equations are similar to the traditional river rearation expressions or river "die-away" models. Beyer (1999) has also explored the possibility of using the results of box models to parameterize continuous decay models.

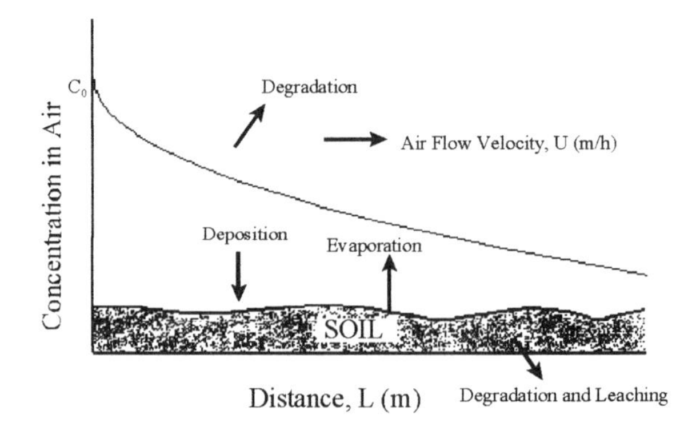

Figure 2. Schematic diagram of the model system in which the decline in air concentration is calculated as a result of degradation in air and reversible deposition to soil.

The general principle underlying such models is depicted in Figure 2 for transport in air over soil. At distance zero the concentration in air is C_O. The air flows undirectionally with a constant vertical mixing height with no lateral dispersion and over a soil of constant composition at steady-state with respect to the air locally above it. If the air velocity is U m/h and the rate constant of reactive decay is k_R, then in the absence of loss to the soil the concentration C at any distance L will be C_o exp(-Lk_R/U) or C_O exp(- k_Rt) where t is the transit time L/U. A characteristic distance of travel can then be defined as U/k_R at which the term in the exponent is -1 and C has fallen to 37% of C_O. A "half distance" can also be defined as 0.693 U/k_R or $U\tau_{1/2}$ where $\tau_{1/2}$ is the half-life in air. These distances are essentially expressions of the half-lives in air.

If the substance also reversibly deposits to soil and degrades there, the expression can be shown (Beyer 1999) to become:

$$C = C_O \exp(-L(k_R + k_D F)/U) \quad (10)$$

where k_D is the rate constant for gross deposition and F is the fraction of the chemical which is deposited to the soil which remains there, degrades or leaches and therefore, does not return to the air. k_DF is the rate constant of net deposition. F is a "stickiness" which can vary from zero for a very volatile, persistent substance to 1.0 for an involatile or a non-persistent substance which does not evaporate. It is noteworthy that k_D depends on the assumed height of the atmosphere, whereas k_R is independent of height.

The rate constant k_D can be deduced from rates of absorption and wet and dry deposition. F is a function of the competitive rates of evaporation, degradation and leaching in the soil and obviously requires a calculation of fate in the soil.

For transport over water a similar expression applies but F is defined differently and requires that the fate of the chemical in the water be evaluated.

These rate constants and F can be deduced from the results of existing multimedia box models, but it may be advantageous to design a model specifically for this purpose, as has been described by Beyer (1999). Here we present selected results from that study to illustrate the nature of the findings.

Table 2. Half-life in air (τ1/2)h attributable to photochemical degradation, overall environmental persistence (τ_{OV}) days and characteristic travel distance in air (L_A) km. Substances are ranked according to L_A and grouped into three classes. The assumed air velocity is 4 m/s or 14.4 km/h.

Substance	$\tau_{1/2}$ *(h)*	τ_{OV} *(days)*	L_A *(km)*
HCB	7350	2300	130000
α-HCH	1420	220	13000
Tetrachlorobiphenyl	1700	4100	7600
Chlorobenzene	170	15	5100
γ-HCH	1040	1300	4700
pp'-DDE	170	1500	3500
Hexachlorobiphenyl	5500	4700	3200
Toxaphene	170	2100	2900
Biphenyl	55	5.1	1600
Heptachlorobiphenyl	5500	4700	1500
Dieldrin	55	240	1500
2378 TCDD	170	1400	750
Decachlorobiphenyl	55000	4800	540
Benzene	17	1.5	510
OCDD	550	4600	470

Table 2 gives calculated overall persistences for 15 chemicals as well as characteristic transport distances in air over soil and half-life in air. These data are also displayed in Figure 3. It is encouraging that the substances which are calculated as having LRT characteristics are detected in remote regions, thus the model appears to be reproducing LRT phenomena. Figure 3 is useful as a graphical representation of the phenomena. A substance which does not partition to soil, or for which F is zero will lie along the maximum line which corresponds to Equation 9. If F is one, i.e. the substance is totally "sticky" the lower limit applies and the rate of degradation in air is relatively unimportant because in most cases deposition is the dominant removal process. The value of F determines the location of the point corresponding to the chemical between these limits.

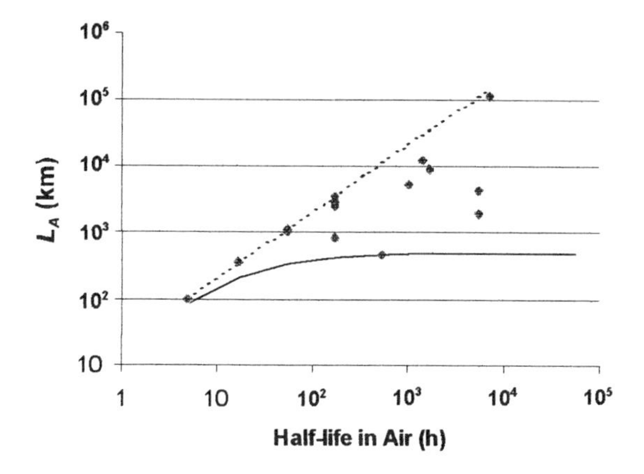

Figure 3. Plot of log characteristic travel distance (L_A) versus log half-life in air. "Sticky" substances with F close to 1.0 lie close to the lower limit.

Hexachlorobenzene is notable as being "non-sticky" and is thus susceptible to LRT. Octachlorodibenzo *p* dioxin is very "sticky" and can travel only a relatively short distance. It thus appears that defining a LRT attribute in air requires information on k_R , k_D and F. The selection of U the air velocity is arbitrary and affects the absolute values, but not the ranking. The selected atmospheric height affects the absolute values and does influence the ranking. Increasing temperature can have conflicting effects since k_R is generally increased, but k_D may be decreased, and F is generally, but not always, decreased.

Again it is more indicative of actual environmental behavior to present the LRT distances as distributions rather than single numbers. A similar approach can be used for transport in water. The data in Table 2 suggest that a CTD of over 2000 km (as defined in this model) is a strong indication of likely LRT. A CTD of less than 600 km suggests little potential for LRT. An intermediate CTD indicates that more detailed evaluation is needed.

It must be stressed that these absolute distances are products of one model reflecting one set of conditions and parameter values. The distances are dependent on these assumptions and should not be interpreted as having much real world significance. They can be used, however, for relative ranking of chemicals.

Mode of Entry

This analysis of LRT applies to a chemical which is initially present in air, possibly as a combustion product or emission to air. If discharge to the environment is to a municipal waste water treatment system, surface water or soil (e.g. as a pesticide) then only a fraction of the chemical will enter the air by evaporation. Assessment of LRT potential in such cases thus requires an additional evaluation of the fraction of the chemical which will reach the atmosphere. The overall statement of susceptibility to LRT may then take the form that when the chemical is used under prescribed conditions (e.g. as a pesticide or in a "down-the-drain" consumer product) 10% may evaporate to the atmosphere in which it has a CTD of 1000 km.

Conclusions and Implications

When assessing the inherent properties of PBTs as they influence persistence and potential for long range transport, mere inspection of raw property data is not adequate, and comparison of a set of property values *versus* corresponding criteria values can be very misleading. Evaluative mass balance models can play an invaluable role in accomplishing this task. The models do not purport to describe exactly how the chemical will behave, but they can provide an indication of how the many contributing physical-chemical and reactivity properties and environmental factors combine to determine bioaccumulation, persistence and transport potential by applying the laws of nature as we currently understand how they control the environmental fate of chemicals.

Using the bioaccumulation, persistence and LRT models enables these attributes to be assigned numerical values or ranges. By building up ranked lists of existing and new chemicals of concern it should be possible to identify those substances which merit international regulation and in extreme cases, outright bans. Other factors such as toxicity, value to society, quantities used and the availability of more benign substitutes will play a role in the final evaluation. Other effects such as potential to cause ozone depletion or contribute in other ways to affect ecosystem or human health adversely must be considered. Perhaps the use of "bright line" criteria such as a BAF of 5000 will fall into disuse except as a means of grouping substances with similar attributes. Rather than set "bright lines" now it is better to assign priorities to chemicals using all available approaches, select the best evaluation methods and identify the chemicals most worthy of more detailed assessment. Given the limited resources available for chemical evaluation it seems more appropriate to rank chemicals than rigidly classify them, as for example persistent vs non-persistent. Undoubtedly there will be difficult decisions ahead when substances display one attribute but not another. Lindane, for example is clearly persistent and subject to LRT but it is not very bioaccumulative. We can only hope that such difficult assessments will be fully informed by realistic evaluations of chemical fate using state of the art knowledge of environmental chemistry.

References

1. Bennett, D.H.; McKone, T.E.; Matthies, M.; Kastenberg, W.E. *Environ. Sci. Technol.* **1998,** 32, 4023-4030.
2. Bennett, D.H.; Kastenberg, W.E.; McKone, T.E. *Environ. Sci. Technol.* **1999**, 33, 503-509.
3. Beyer, A. thesis, Inst. of Environ. Systems Research, University of Osnabrück, Germany, 1999.
4. Brandes, L.J.; van Hollander, H.A.; van de Meent, D. 1996, RIVM Report 719101-029, Bilthoven, Netherlands.

5. Campfens, J., Mackay, D. *Environ. Sci. Technol.* **1997**, 31, 557-583.
6. Clark, K.E.; Gobas, F.A.P.C., Mackay, D. *Environ. Sci. Technol.* **1990**, 24, 1203-1213.
7. Cole, J.G.; Mackay, D. *Environ. Toxicol. Chem.* **1999** (in press).
8. Dewailly, É.; Nantel, A.; Weber, J.P.; Meyer, F. *Bull. Environ. Contom. Toxicol.* **1989**, 43, 641-646.
9. Dewailly, É.; Ayotte, P.; Bruneau, S.; Laliberté, C.; Muir, D.C.G.; Norstrom, R.J. *Environmental Health Perspectives* **1993**, 101, 620.
10. Environment Canada Toxic Substances Management policy: Persistence and bioaccumulation criteria. **1995**, En 40 499/2-1995E. Final Report, Ottawa, ON, Canada.
11. Gobas, F.A.P.C. *Ecological Modelling* **1993**, 69, 1-17.
12. Gobas, F.A.P.C.; Zhang, X.; and Wells, R. *Environ. Sci. Technol.* **1993**, 27, 2855-2863.
13. Mackay, D.; Paterson, S. *Environ. Sci. Technol.* **1982**, 654A-660A.
14. Mackay, D. *Environ. Sci. Technol.* **1982**, 16, 274-278.
15. Mackay, D. *Multimedia Environmental Models - The Fugacity Approach*; Lewis Publishers Inc. **1991.**
16. Mackay, D.; Di Guardo, A.; Paterson, S.; Cowan, C.E. *Environ. Toxicol. Chem.* **1996**, 15, 1627-1637.
17. Müller-Herold, U. *Environ. Sci. Technol.* **1996**, 30, 586-591.
18. Müller-Herold, U.; Caderas, D.; and Funck, P. *Environ. Sci. Technol.* **1997**, 31, 3511-3515.
19. Scheringer, M.; Berg, M. *Fresenius Enviro. Bull.* **1994,** 3, 493-498.
20. Scheringer, M. *Environ. Sci. Technol.* **1997,** 31, 2891-2897.
21. Schmidt, C.W. *Environmental Health Perspectives* **1999,** 107, A18-A25.
22. Thomann, R.V. *Environ. Sci. Technol.* **1989,** 23, 699-707.
23. Thomann, R.V.; Connolly, J.P.; Parkerton, T. *Chemical Dynamics in Aquatic Ecosystems;* F.A.P.C. Gobas, J.A. McCorquodale Eds. Lewis Publishers: Chelsea, MI, **1992**, pp 153-186.
24. Vallack, H.W.; Bakker, D.J.; Brandt, I.; Broström-Lundén, E.; Brouwer, A.; Bull, K.R.; Gough, C.; Guardans, R.; Holoubek, I.; Jansson, B.; Koch, R.; Kuylenstierna, J.; Lecloux, A.; Mackay, D.; McCutcheon, P.; Mocarelli, P.; Taalman, R.D.F. *Environ. Toxicol. Pharmacol.* **1998**, 6, 143-175.
25. van Pul, W.A.J.; de Leeuw, F.A.A.M.; van Jaarsveld, J.A.; van der Gaag, M.A.; Sliggers, C.J. *Chemosphere* **1998**, 37, 113-141
26. Wania, F.; Mackay, D. *Ambio* **1993**, 22, 10-18.
27. Wania, F.; Mackay, D. *Environ. Sci. Technol.* **1996**, 30, 390A-396A.
28. Wania, F.; Mackay, D. *Environ. Toxicol. Chem.* **1999**, 18, 1400-1407.
29. Webster, E.; Mackay, D., Wania, F. *Environ. Toxicol. Chem.* **1998**, 17, 2148-2158.

Chapter 3

CART Screening Level Analysis of Persistence: A Case Study

Deborah H. Bennett[1], Thomas E. McKone[1,2], and W. E. Kastenberg[3]

[1]Environmental Energy Technologies Division, Lawrence Berkeley National Laboratory, Berkeley, CA 94720
[2]School of Public Health and [3]Department of Nuclear Engineering, University of California at Berkeley, Berkeley, CA 94720

For the thousands of chemicals continuously released into the environment, it is desirable to make prospective assessments of those likely to be persistent. Persistent chemicals will remain in the environment a long time. Based on specific criteria for persistence, a binary logic tree can be developed to classify a chemical as "persistent" or "non-persistent" based on the chemical's properties. In this approach, the classification is based on the results of a standardized multimedia model. Thus, the classifications are more comprehensive for multimedia pollutants than classification using single media half-lives. A case study using twenty-six chemicals for three modes of entry into a unit world environment compares the characteristic time calculated from the multimedia model to the classification resulting from the tree.

Introduction

Some chemical pollutants released into the environment are rapidly degraded while others persist for years or even decades. Those that persist can have higher environmental concentrations per unit release and are removed more slowly from the environment after use has ceased. Many persistent chemicals partition into multiple environmental media. Because persistent chemicals pose a greater potential concern per unit release than non-persistent pollutants, there is a need for simple but reliable methods for identifying potentially persistent pollutants.

It is desirable to create screening level qualifications to determine which chemicals are likely to be non-persistent and which may persist (1, 2). The ability to assess large numbers of chemicals provides a distinct regulatory advantage by eliminating the need to assess persistence by running a complex model for each chemical under consideration. If a chemical is classified as "non-persistent" using a screening method, we eliminate the need for further analysis. If a chemical is classified as persistent, it would likely undergo a more thorough evaluation before deciding if the use of the chemical should be restricted. Ideally, the screening method should minimize both false negatives (preventing the use of persistent chemicals) and false positives (minimizing the need for a higher level evaluations).

To provide a more robust screening method for classifying chemicals as persistent or non-persistent, we use multi-media assessments, without requiring extensive modeling and data for each chemical. To do this, we link the results of a simple multimedia model to a classification system, in this case: Classification and Regression Tree (CART) (3). A multi-media model is more effective than a single medium model if the chemical partitions among multiple media. A classification approach is linked to the model because modeling the persistence of numerous new and existing chemicals would be a major undertaking. The results of simple models can be used with Monte Carlo simulations as tools to develop effective strategies to classify chemicals without requiring an explicit simulation for each chemical.

We develop a classification tree for each of three modes of entry into the environment. These trees are used to classify twenty-six chemicals as persistent or non-persistent. We compare the classifications to the characteristic time calculated by a multimedia model. The results of this analysis give us an idea about the effectiveness of CART and the limits on the precision of the method.

Methods

Multimedia Model

We use a closed unit world system to approximate the actual distribution of environmental media found across the earth as presented in Figure 1 (4). The model uses fugacity principles, a common approach for describing partitioning in multimedia systems (5, 6). The fugacity capacities (the chemical concentration per unit chemical fugacity) and dimensions for each compartment of the evaluation unit are listed in Table 1 (4, 6-9). The chemical exchange rates between compartments are based on both chemical properties and landscape properties and are defined as inventory-based mass transfer coefficients, "T" values, which can be found in Bennett et al., along with landscape property values used in the calculation (10).

We have chosen to use a steady state mass distribution because it accounts for the shift from equilibrium resulting from the source and advective processes, while retaining sufficient simplicity to complete calculations in a tractable form, such as a spreadsheet (7).

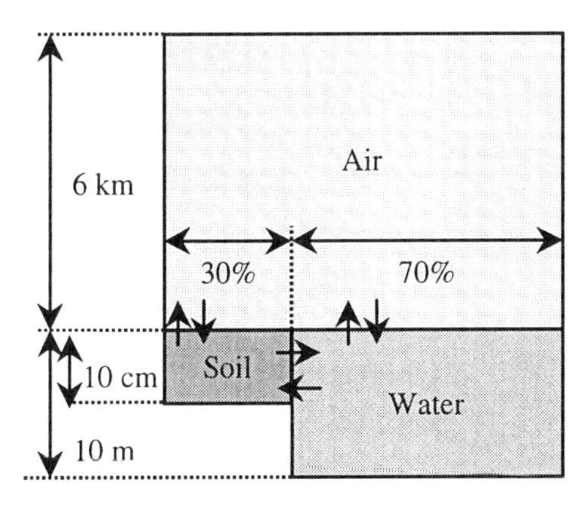

Figure 1: Configuration of the unit world environmental model.

Table 1: Fugacity Capacities and Model Dimensions

Fugacity Capacity, Air; $Z_a = Z_{air} + Z_{ap}$; $Z_{air} = \dfrac{1}{RT}$;

If Tm>T, $Z_{ap} = \dfrac{0.17 \times SA}{RT \times VP \times e^{\left(-6.81\left(1-\frac{Tm}{T}\right)\right)}}$ If Tm<T, $Z_{ap} = \dfrac{0.17 \times SA}{RT \times VP}$

Fugacity Capacity, Water; $Z_{water} = 1/H$

Fugacity Capacity, Soil;

$Z_s = Z_{air}\alpha_s + Z_{water}\beta_s + (1 - \alpha_s - \beta_s)Z_{sp}$ $\quad Z_{sp} = \dfrac{K_d \times \rho_s}{H \times 1000}$

universal gas constant (Pa-m^3/mol-K), R=8.31
ambient environmental temperature (K), T=288
surface area of particulars (m^2/m^3), SA=1.5×10^{-4}
soil particle density (kg/m^3), ρ_s=2.60×10^3
water content of soil (vol. fraction), β_s=0.28
air content of soil (vol. fraction), α_s=0.17
partition coefficient in soil layer, $K_d = 0.41 \times K_{ow} \times f_{oc}$
organic carbon fraction in soil zone, f_{oc}=0.02

Depth of Air, $d_a = 6000$ *m*
Depth of Water, $d_w = 10$ *m*
Depth of Soil, $d_s = 0.1$ *m*
Fraction of Area in Surface Water, fw = 0.70

Characteristic Time

Persistence in the environment can be calculated from the total mass in a steady state system divided by the total loss from the system. This is also considered a residence or characteristic time in the environment. For a steady state multimedia system, this can be calculated from mass distribution between environmental media and the loss rate in each media. The characteristic time, τ, can be calculated as (7):

$$\tau = \frac{M_1 + M_2 + M_3 + ... + M_i}{M_1 k_1 + M_2 k_2 + M_3 k_3 + ... + M_i k_i} \qquad (1)$$

where M_j is the chemical mass and k_j is the transformation rate in compartment j, $j = 1, 2, 3, ..., i$.

The characteristic time is used to classify a chemical as persistent or non-persistent by comparison to a specified cutoff value of τ. The cutoff value defines the boundary between characteristic times that are considered non-persistent and those considered persistent. If six months were selected, all chemicals with a characteristic time of less than six months would be classified as non-persistent while all chemicals with a characteristic time of more than six months would be classified as persistent. Policy or appropriate regulations should define the actual cutoff value. However, in the examples that follow, a cutoff value of six months is used for illustrative purposes.

The CART Approach

The CART algorithm is a non-analytic, computationally intensive, statistical procedure that classifies data by producing a tree structure (3). At each node in the tree, CART uses a parametric decision based on an inequality for a specific input parameter. The CART algorithm compares all possible splits at each node, considering all parameters and all values for that parameter as possible inequality expressions, and then determines which maximizes the reduction in variance between the parent node and the resultant two sub-nodes. The tree is grown until either there is no variance or the sample size is small at each node. When the tree is complete, CART assigns each terminal node a class outcome (3).

The results of the CART approach are concise and easy to understand, which should assist decision making. CART has been applied in many fields, including multimedia modeling (11, 12). Eisenberg and McKone applied CART to the output from a multimedia exposure model to determine which chemical properties result in high levels of human exposure for a unit input soil concentration (12). They used multiple realizations of chemical property sets, covering a broad range of property values (12). We use a similar approach, applying CART to a different environmental model and evaluating a different environmental outcome (in this case persistence).

CART is used with a multimedia model to identify the range of input values that give rise to classifications of "persistent" or "non-persistent". This study focuses on chemical realizations (i.e. hypothetical sets of chemical properties), rather than specific chemicals. The range of each chemical property is selected to fall within the plausible range for synthetic organic chemicals. The chemical parameter value ranges used in the simulations are presented in Table 2. Also listed are examples of chemicals with property values near the minimum and maximum of each range. Most of the distributions are log-uniform (a uniform distribution in log space), yielding the same number of simulations in each decade of the range.

Table 2: Ranges of Chemical Properties (Reproduced from reference (10). Copyright 2000 Society for Environmental Toxicology and Chemistry)

Property		Distribu-tion type	Lower End	Upper End	Example of Chemical with Property at Lower End of Range	Example with Property at Upper End of Range
Henry's law const. ($Pa\text{-}m^3/mol$)	H	log uniform	1×10^{-3}	1×10^{5}	Phenol	Nitrogen gas
octanol-water partition coeff.	K_{ow}	log uniform	1	1×10^{9}	Butanol, Methylchloride	Di-n-octyl-phthalate
decay rate in air (1/day)	k_a	log uniform	4×10^{-4}	1×10^{2}	Toxiphene, Bromodichloro-methane	Benzo(a) Pyrene
decay rate in water (1/day)	k_w	log uniform	1×10^{-5}	1×10^{2}	hexachloroethane	Pyrene
decay rate in soil (1/day)	k_s	log uniform	1×10^{-5}	1×10^{2}	PCB	Anthracene
vapor pressure (Pa)	VP	log uniform	1×10^{-6}	1×10^{5}	Chrysene, TCDD	Atmospheric Pressure
melting point (K)	Tm	uniform	100	600	Vinyl Chloride	Chrysene, beta – HCH, TCDD
diffusion coeff. in pure air (m^2/s)	D_{air}	uniform	.2	1.7	Hexachloroethane	2,4 – Dinitrotoluene
diffusion coeff. in pure water (m^2/s)	D_{water}	uniform	3×10^{-5}	1×10^{-4}	Endrin	Vinyl Chloride

The method of linking the CART analysis to a multimedia model have been presented in Bennett et al. (10) and are only summarized here. Using the Monte Carlo package Crystal Ball (13), 10,000 chemical realizations are developed from the chemical parameter distributions listed in Table 2. For each realization, the characteristic time is calculated by inputting the parameter values into the unit world model. Each realization is classified as persistent or non-persistent by comparing the calculated characteristic time (τ) to the cutoff value (6 months). The classification and the input parameters for each chemical along with the resulting classification are entered into the CART program. A CART analysis is then completed to determine which chemical parameter sets lead to pollutants defined as "non-persistent". A commercial implementation of the CART program was used (14). The resulting tree can then be used to classify chemicals as persistent or non-persistent.

CART assigns the classification of non-persistent only when a specified percent of the chemicals in that grouping are non-persistent. This percentage can be varied so as to minimize the false negatives and hence the number of chemicals improperly classified as non-persistent. However, those chemicals given the persistent classification are not necessarily persistent and some persistent chemicals will slide through the screening undetected. Because of our preference for false positives over false negatives, each "persistent" node includes both persistent and non-persistent chemicals. A chemical may be classified as persistent using the classification tree even though the same multimedia model used to generate the CART tree will determine it is non-persistent. Typically this occurs because the misclassified chemicals have similar properties to those identifies as persistent.

When a screening level approach is used, plans for a more rigorous evaluation need to be in place for chemicals that do not pass the screening level evaluation. This approach has been outlined by a SETAC committee on evaluating persistent pollutants (15). The CART framework should be considered a first tier, with a second tier of testing using a full model and complete data set for "persistent" chemicals.

Case Study

Classification trees are developed for each of three modes of entry (emissions to air, soil, and water) into a three-compartment unit world system. The cutoff point between persistent and non-persistent chemicals is a characteristic time of more than six months. We evaluate twenty-six real chemicals using both a multimedia model and the CART trees presented below for each of three modes of entry.

Classification Trees

We apply CART to the results from emission to air in a unit world environment, with the resulting tree shown in Figure 2. We use this example to explain how the tree should be read. The CART algorithm and splitting rules first found that splitting the data based on whether the half-life in air was more or less than 26 days maximizes the variance reduction between the whole data set and the two resultant sub-sets.

To read the tree, the user first asks, "is the half-life in air less than 26 days?" If the answer is yes, the user follows the left branch of the tree and asks, "is the Henry's law constant less than 0.407 Pa m^3/mol?" If the answer is no, the user follows along the right branch and learns that 98% of the chemicals within the specified value ranges for these two properties are non-persistent. In this case, the classification was based on only two inputs. The half-life in air is important because the release is to the air compartment and thus a significant portion of the chemical can be found in this phase. With a high Henry's law constant, the chemical is more likely to remain in the air and thus is influenced by the short half-life in air (rather than partition to the water).

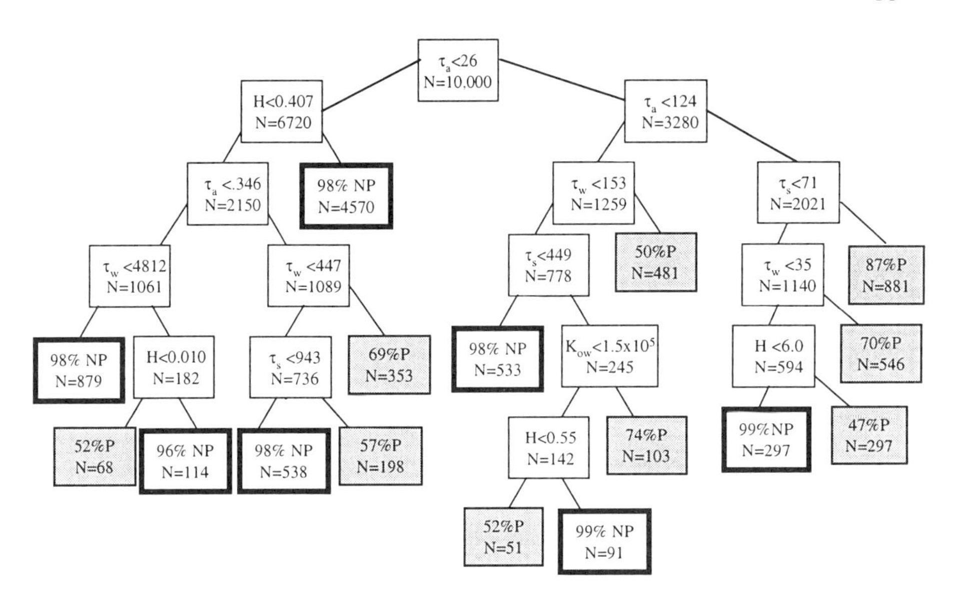

Figure 2: CART tree for emissions to air, persistence defined as six months, all chemical and landscape properties used as predictor variables. The tree is read by asking a question regarding the inequality listed. If the answer is yes, the user follows the left branch while if the answer is no, the user follows the right branch. The user follows this procedure until a terminal node is reached, which indicates if the chemical can be classified as non-persistent or if the chemical needs to be evaluated in a second tier evaluation. N defines the number of chemical realizations in that node. The terminal nodes indicate the percent of the realizations that are either persistent, P, or non-persistent, NP. (Reproduced from reference (10). Copyright 2000 Society for Environmental Toxicology and Chemistry)

Classification trees were generated for emissions to water and soil using 6 months as the cut-off time and are presented in Figures 3 and 4. The qualifications defining a non-persistent chemical differ from those for emissions to air and demonstrate the importance of classifying the mode of entry of the chemical into the environmental system. This technique is not applicable for speciating or dissociating chemicals. Additionally, CART can be used in cases where there is incomplete data, an example of this can be found in Bennett et al. (10).

Evaluation Using Real Chemicals

To evaluate the effectiveness of the classification trees, the classification resulting from the trees for twenty-six chemicals are compared to the predicted

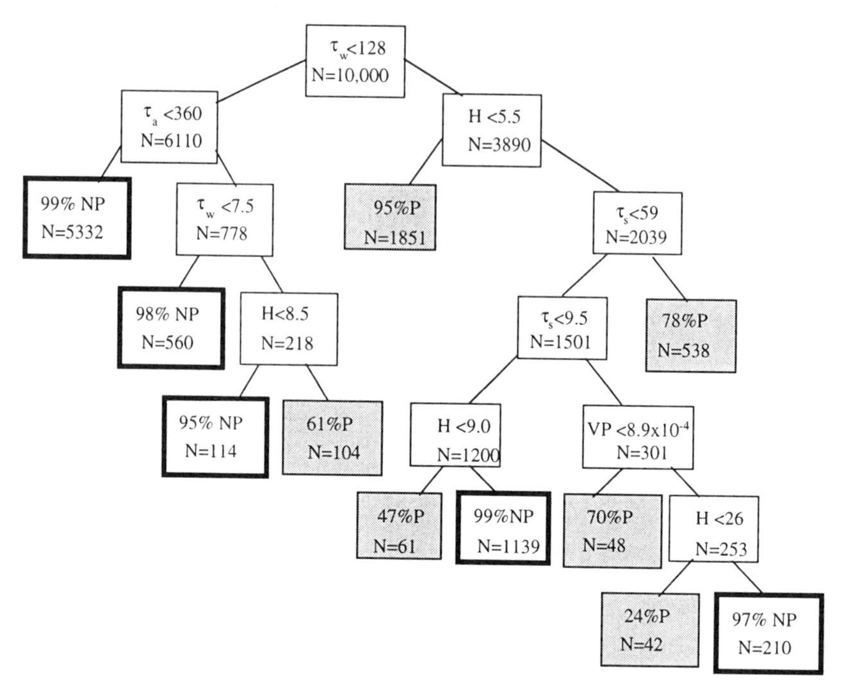

Figure 3: CART tree for emissions to water, persistence defined as 6 months. The tree should be read as in Figure 2.

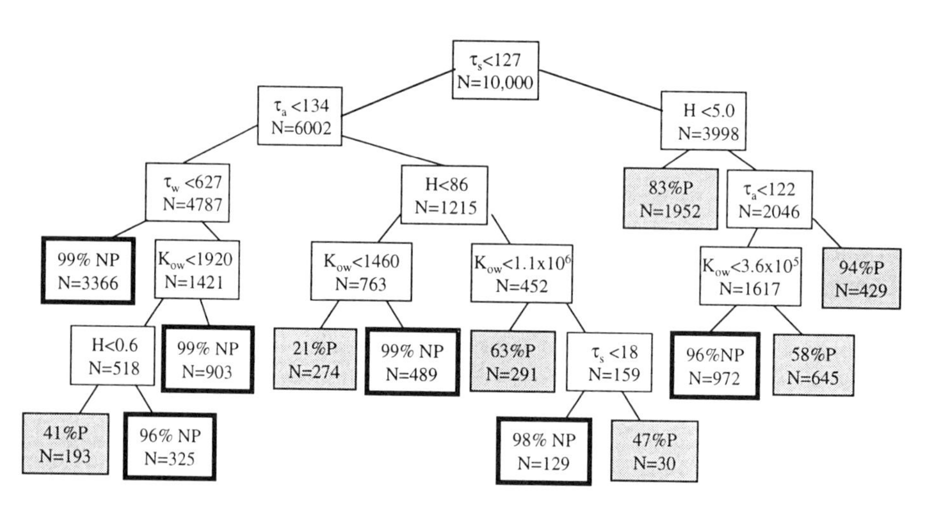

Figure 4: CART tree for emissions to soil, persistence defined as 6 months. The tree should be read as in Figure 2.

characteristic time from a multimedia model. A representative set of chemicals was selected that exhibit various half-lives and partition coefficients.

For each chemical, the characteristic time was determined for each of the three modes of entry, soil, water, and air, in the unit-world multimedia model. Additionally, using the three trees in Figures 2-4, each chemical was classified as persistent or non-persistent in the environment for each mode of entry. The chemicals and their corresponding properties are listed in Table 3. The results for each mode of entry are listed in Table 4, with the first column indicating the calculated persistence using the multimedia model and the second column indicating the classification resulting from the classification tree for each chemical.

The asterisks indicate the two cases where the CART analysis and the calculated characteristic time result in different classifications. Heptachlor emissions to soil are classified as non-persistent while the model results in a characteristic time of 642 days, which should be a persistent classification. To evaluate the reasons for the difference, we need to examine the classification tree in Figure 4. The first inequality on the tree is whether the half –life in soil is more or less than 127 days. The half-life in soil is 1100 days, much greater than this value. The next inequality is whether the Henry's law constant is more or less than 5 Pa-m^3/mol. The Henry's law constant for heptachlor is 22 Pa-m^3/mol, which is slightly greater. The next inequality is whether the half-life in air is more or less that 122 days. Heptachlor rapidly degrades in air and thus the half-life in air is significantly less then this value. Finally, the K_{ow} for heptachlor is 1.5 x 10^5, slightly less then the value we compare to, 3.5 x 10^5, leading to a non-persistent classification. The chemical is classified as non-persistent because both values for the partitioning coefficients indicated that a high enough percentage of the chemical would be in the air compartment where it would degrade rapidly, decreasing the characteristic time significantly. However, for this particular chemical, the percentage in air was not high enough to compensate for the extremely high value for the half-life in soil, leading to a misclassification.

Chlordane released into the water compartment results in a misclassification for a similar reason. The Henry's law constant was slightly higher than the value in the inequality, indicating enough chlordane would be in the air to reduce the overall characteristic time. Thus resulting in a non-persistent evaluation. Again, the combination of an extremely long half-life in water and a partitioning coefficient that did not favor air strongly enough leads to a long characteristic time in the environment. If CART were being used for an overall evaluation, chlordane would still be flagged because the soil mode of entry results in a persistent classification.

The calculated characteristic times for aldrin and lindane were both close to the cutoff value of 182.5 days. These chemicals also had a property that was close to a bounding property value in an inequality that would have lead to a persistent classification. These examples indicate that there is some uncertainty in values obtained by CART, and that the present values should be considered as indicators of the correct order of magnitude.

Table 3: Chemicals used in case study evaluations and their corresponding property values (16)

Compound name	half-life, soil	half-life, water	half-life, air	K_{ow}	Henry's Law Const.	vapor pressure	melting point	diffusion coeff. in air	diffusion coeff. in water
units	(d)	(d)	(d)		(Pa-m^3/mol)	(Pa)	(K)	(m^2/d)	(m^2/d)
symbol	τ_s	τ_w	τ_a	K_{ow}	H	VP	Tm	D_{air}	$D_{water} \times 10^5$
γ-HCH (lindane)	332	127	2.12	5.29E3	3.50E-1	6.15E-3	386	0.152	5.51
Toluene	28.4	13	2.38	4.82E2	6.63E+2	3.77E+3	178	0.752	8.51
Fluoranthene	852	1.75	0.463	1.27E5	1.02	1.19E-3	384	0.261	5.90
Chrysene	385	0.363	0.184	5.64E5	1.59E-1	1.50E-6	529	0.214	5.41
Carbon tetrachloride	197	4032	3667	5.27E2	2.70E+3	1.50E+4	250	0.674	8.73
Benzo(a)pyrene	229	2.34	0.063	2.20E6	9.20E-2	7.13E-7	451	0.436	5.26
Benzene	190	11.15	5.91	1.51E2	5.74E+2	1.27E+4	279	0.756	9.63
Hexachloro-benzene	253	1529	860	3.51E5	1.07E+2	1.73E-3	501	0.468	5.83
2,3,7,8-TCDD	6657	429	30	4.62E6	2.47	1.60E-6	578	0.421	5.11
Styrene	45.0	21.6	0.171	8.92E2	3.26E+2	8.32E+2	243	0.613	7.92
Acenaphthene	57.1	6.31	0.201	9.26E3	4.03E+1	9.47E-1	368	0.364	6.76
Atrazine	83	1.01	3.57	5.13E2	2.90E-4	4.00E-5	446	0.640	6.60
Pentachlorophenol	210	23.1	23.1	1.12E5	5.16E-2	4.15E-3	447	0.640	6.60
Benzo(a) anthracene	878	0.083	0.083	4.97E5	8.13E-1	2.45E-5	433	0.441	5.44
Pyrene	819	0.057	0.057	1.04E5	1.40	6.12E-4	429	0.235	5.95
Butyl benzyl phthalate	4	4	1.38	4.29E4	1.69E-1	1.65E-3	238	0.149	4.45
Di-n-butyl phthalate	233	7.5	1.70	3.64E4	1.63E-1	6.40E-3	236	0.378	4.45
Di-n-octyl phthalate	17.5	17.5	1.03	3.27E9	1.17E-1	8.96E-4	246	0.130	3.42
Aldrin	867	306	0.208	7.20E6	1.09E+1	3.61E-3	379	0.114	4.74
Chlordane	5699	812	1.19	1.21E6	1.22E+1	1.64E-2	298	0.102	4.55
Dieldrin	991	627.5	0.927	5.13E5	6.25E-1	3.10E-4	448	0.108	4.64
Methoxychlor	131	0.158	0.257	4.53E4	5.17E-1	1.66E-4	366	0.135	4.37
Heptachlor	1095	3.18	0.225	1.67E5	2.17E+1	4.37E-2	369	0.097	5.09
Diethyl phthalate	29.5	29.5	4.85	2.22E2	5.46E-2	2.20E-1	233	0.221	5.19
Dimethyl phthalate	4	4	25.625	4.70E1	1.74E-1	3.7	276	0.491	5.83
α-HCH	42.7	74.4	2.12	6.31E3	4.91E-1	5.69E-3	432	0.152	5.51
1,3 butadiene	17.5	11.2	0.058	9.77E1	7.46E+3	2.81E+5	164	0.929	8.21

Table 4: Calculated characteristic time using the unit world model and CART classification based on a cutoff time of 6 mo. (182 days) for three modes of entry

Mode of Entry	Soil		Air		Water	
Calculation Method	τ_{ss} (days)	CART	τ_{ss} (days)	CART	τ_{ss} (days)	CART
γ-HCH (lindane)	348	P	6.89	NP	178	NP
Toluene	3.49	NP	3.43	NP	10.56	NP
Fluoranthene	1,009	P	2.59	NP	2.53	NP
Chrysene	547	P	0.87	NP	0.52	NP
Carbon tetrachloride	5,272	P	5,273	P	5,275	P
Benzo(a)pyrene	327	P	0.54	NP	3.37	NP
Benzene	8.61	NP	8.54	NP	12.13	NP
Hexachlorobenzene	684	P	769	P	787	P
2,3,7,8-TCDD	7,354	P	1,192	P	992	P
Styrene	0.34	NP	0.25	NP	11.75	NP
Acenaphthene	8.26	NP	0.30	NP	7.52	NP
Atrazine	101	NP	29.84	NP	1.46	NP
Pentachlorophenol	299	P	84.53	NP	33.51	NP
Benzo(a)anthracene	1,158	P	0.81	NP	0.12	NP
Pyrene	909	P	0.27	NP	0.08	NP
Butyl benzyl phthalate	5.77	NP	2.15	NP	5.77	NP
Di-n-butyl phthalate	320	P	6.65	NP	10.81	NP
Di-n-octyl phthalate	25.25	NP	1.98	NP	25.22	NP
Aldrin	1,205	P	0.77	NP	174	NP
Chlordane	4,639	P	10.35	NP	238	*NP*
Dieldrin	1,391	P	5.96	NP	762	P
Methoxychlor	184	P	0.51	NP	0.23	NP
Heptachlor	642	*NP*	0.53	NP	4.48	NP
Diethyl phthalate	38.90	NP	10.98	NP	42.52	NP
Dimethyl phthalate	12.68	NP	27.53	NP	5.81	NP
α-HCH	59.21	NP	4.15	NP	105	NP
1,3 butadiene	0.17	NP	0.08	NP	8.15	NP

To evaluate the precision of the CART analysis, a second set of 10,000 chemical realizations were generated for emissions to air and a second CART tree was generated. The format of the tree was very similar to the previous tree, indicating that 10,000 simulations is probably a sufficient number for creating an evaluation tree. However many of the chemical property values in the inequalities were slightly shifted, confirming that there is some uncertainty in the values of the split point. Additionally, one should consider the effect of using a slightly different cutoff value (e.g. 5 or 7 months) to examine the robustness of the criteria relative to a changing cutoff value. If CART were to be used in policy decisions, one would do additional simulations to reduce some of the uncertainty in the values, but ultimately, the exact values would be chosen as a policy decision.

Discussion

We have previously proposed the utilization of a screening level classification system (CART) for determining if a chemical pollutant is persistent or non-persistent (10). Classification of persistence is determined by comparing the overall residence time of a compound in a multimedia evaluation environment to a reference time value. The multimedia model is used with 10,000 simulations to capture a broad range of chemical-property data sets, allowing us to identify ranges of property values that result in a "non-persistent" residence time in the multimedia model results.

The screening method was applied to twenty-six chemicals with three emission scenarios. Among all emission scenarios, only one chemical was misclassified as non-persistent that had been determined to be persistent by the multimedia model. We found that in cases where there was a misclassification, the chemicals had one property value close to the value of a boundary between persistence and non-persistence, indicating that the boundaries should be taken as approximate, rather than strict values. Additionally, there are model uncertainties in the multimedia model used and we acknowledge that there is a defined reliability limit or uncertainty level within which the model can identify the overall residence time associated with a chemical. However, these reliability limits are sufficient for preliminary analysis, especially if they offer an improvement over other available methods.

One advantage of CART for developing a classification method is that although the process of generating the classification trees requires an understanding of multimedia interactions, these results are reduced to a classification diagram that does not require an understanding or use of multimedia models for interpretation. This is beneficial for a policy maker because they can use the classification trees to evaluate a chemical and can see the causative factor that precluded the chemical from meeting the policy requirements. By using a multimedia model to develop the classification tree, the partition coefficients and half lives in the various environmental media are implicitly included in the screening criteria, providing for a much more robust classification than one using single media data.

Acknowledgments

This work was supported in part by the US Environmental Protection Agency National Exposure Research Laboratory through Contract # DW-988-38190-01-0. The EPA STAR Fellowship program also provided funding for D.H. Bennett.

References

1. U.N.E.P., *Executive Body Decision 1998/2; internet version {http://irptc.unep.ch/pops/CEG-1/WG52AIIE.html}*, 1998.
2. US EPA., *Persistent Bioaccumulative Toxic Chemicals; Proposed Rule*, in *US Federal Resister*. 1999. p. 688.
3. Breiman, L., *et al.*, *Classification and Regression Trees*; Wadsworth, Inc.: Monterey, CA, 1984.
4. Klein, A. In *Handbook of Environmental Chemistry*, Hutzinger, O., Ed. Springer: Berlin. 1985, pp 1-28.
5. Mackay, D.; Paterson, S. *Environ Sci Technol* **1991**, *25*, 427-436.
6. Mackay, D., *Multimedia Environmental Models, The Fugacity Approach*; Lewis Publishers: Chelsea, MI, 1991.
7. Bennett, D.H.; Kastenberg, W.E.; McKone, T.E. *Environ Sci Technol* **1999**, *33*, 503-509.
8. Mullerherold, U.; Caderas, D.; Funck, P. *Environ Sci Technol* **1997**, *31*, 3511-3515.
9. Scheringer, M. *Environ. Sci. Technol.* **1997**, *31*, 2891-2897.
10. Bennett, D.H.; McKone, T.E.; Kastenberg, W.E. *Environ Tox Chem* **1999**, *19*, Issue 3, in press.
11. Eisenberg, J.N.S.; Bennett, D.H.; McKone, T.E. *Environ Sci Technol* **1998**, *32*, 115-123.
12. Eisenberg, J.N.S.; McKone, T.E. *Environ Sci Technol* **1998**, *32*, 3396-3404.
13. Decisioneering, *Crystal Ball*, 4.0. 1996: Boulder, CO.
14. Steinberg, D.; Colla, P., *CART--Classification and Regression Trees: Supplementary Manual for Windows*. 1997, Salford Systems: San Diego, CA.
15. Van De Meent, D., *et al.* In *Evaluation of Persistence and Long-Range Transport of Organic chemicals in the Environment: Report of a SETAC Pellston Workshop*, Klecka, G.M.;Mackay, D., Eds. SETAC Press: Pensacola, FL. in press.
16. McKone, T.E., *et al.*, *Intermedia Transfer Factors for Contaminants found at Hazardous Waste Sites -- Exectutive Report*, . 1995, Risk Sciences Program, University of California, Davis: Sacramento, CA.

Chapter 4

Environmental Categorization and Screening of the Canadian Domestic Substances List

Roger Breton and Robert Chénier

Commercial Chemicals Evaluation Branch, Chemicals Evaluation Division, Environment Canada, Hull, Québec K1A 0H3, Canada

Under the revised *Canadian Environmental Protection Act*, Environment Canada and Health Canada must "categorize" and "screen" 23 000 substances that are considered to be in Canadian commerce and therefore listed on the Domestic Substances List (DSL) to determine whether they are "toxic" as defined in the Act. This paper focuses on the environmental aspect of this project only. The environmental categorization identifies persistent and/or bioaccumulative and inherently toxic (PBT) substances on the DSL. The criteria used to categorize these substances are those set out in the Toxic Substances Management Policy. An Advisory Group and several working groups have been set up to provide recommendations on how to resolve scientific and technical issues that emerge from implementation of this project. A pilot project has been initiated which will identify 100 substances representative of several chemical classes of concern.

Introduction

Concerns relating to persistent bioaccumulative substances have been expressed for the past few decades. Concerted actions to deal with these have been seen in particular over the past ten years. Canada has participated in activities at several different levels. The federal government adopted the Toxic Substances Management Policy in 1995 (*1*). The policy promotes a preventative precautionary approach to substance management in all federal initiatives, and calls for the virtual elimination of persistent bioaccumulative substances that satisfy specific criteria (Track 1 substances). Recently, 12 substances were identified as satisfying criteria for Track 1

(*2*). The policy recognizes the importance of on-going domestic and international efforts to deal with these substances. These include actions taken with the provinces, such as the Canada-Ontario Accord, the Great Lakes Action Plan and the Saint-Laurent Vision 2000, which all target persistent bioaccumulative substances. More recently, the Canadian Council of Ministers of the Environment adopted a Policy for the Management of Toxic Substances that calls for cooperative work in the prioritization, assessment and management of toxic substances, with particular focus on actions for persistent bioaccumulative substances. Canada has also been working with the United States as reflected in the Canada-United States Binational Strategy on Toxic Substances. Continental efforts include the Sound Management of Chemicals initiatives pursuant to the North American Free-Trade Agreement that identifies substances of mutual concern that warrant continental management actions. Regional initiatives include the United Nations Economic Council for Europe protocol for Persistent Organic Pollutants (POPs) under the Long Range Transboundary Air Pollutants Convention. Most recently, Canada is actively participating in the development of a globally binding convention on POPs through the UN Environmental Protection Program.

Federal powers to support Canadian actions in these initiatives are derived from various federal legislation. Key among these is the *Canadian Environmental Protection Act* (CEPA), that provides for the assessment and management of substances that can be released into the Canadian environment.

The CEPA provides for the protection of the environment and of the health of Canadians from toxic substances and other pollutants. The legislation was enacted in 1988. It provided for actions on toxics that focused on the assessment, management and clean up of existing pollution. It has had a positive impact on how Canada deals with environmental challenges and has made possible many environmental achievements.

Provisions in CEPA call for a review of the Act after 5 years. The review that has been on-going for the past few years has focused on new expectations of Canadians and developments in environmental law and scientific knowledge since 1988. The revised CEPA has been modernized and strengthened by making pollution prevention the cornerstone of national efforts to reduce toxic substances in the environment. It shifts the focus from cleaning up environmental problems to preventing them in the first place. The Act received Royal Assent on September 14, 1999, and should be promulgated in early 2000.

The revised legislation requires the government to assess more substances more quickly, and set firm deadlines for action to control toxic substances. Several new mandates have been introduced in the Act which proposes a more efficient process of identifying, screening, assessing and managing toxic substances.

Mandate under CEPA

Although the revised CEPA is still not promulgated, Environment Canada has initiated the implementation of some of the new initiatives. One of these initiatives involves the identification of persistent, bioaccumulative and inherently toxic substances (PBTs) that are in Canadian commerce. It requires the Minister of the Environment and the Minister of Health to "categorize" and then "screen" substances listed on the Domestic Substances List (DSL) to determine whether they are "toxic" as defined in the Act. Under the Act, a substance is "toxic" if it is entering or may enter the environment in a quantity or concentration or under conditions

> (a) having or that may have an immediate or long-term harmful effect on the environment;
> (b) constituting or that may constitute a danger to the environment on which human life depends; or
> (c) constituting or that may constitute a danger in Canada to human life or health.

The DSL includes substances that were, between January 1, 1984, and December 31, 1986, in Canadian commerce, used for manufacturing purposes, or manufactured in or imported into Canada in a quantity of 100 kg or more in any calendar year. The List has been amended from time to time and currently contains approximately 23 000 substances. Types of substances on the DSL include simple organic chemicals, pigments, organometallic compounds, surfactants, polymers, metal elements, metal salts and other inorganic substances, as well as substances that are of "Unknown or Variable Composition, complex reaction products, or Biological materials" (referred to as UVCBs) (Figure 1). Substances identified in Figure 1 under "Other Programs" include new substances that have been added to the DSL since its development.

Figure 2 outlines the various uses for which these substances were notified in 1984-1986. Substances could be identified under any of 50 different functional use groups, including an "other" category (6%). The largest represented use group is "other industrial uses", which is a combination of over 30 use patterns such as absorbents, abrasives, analytical reagents, catalysts, flame retardants, fuel additives and many more. The three single largest uses were fragrances, perfumes, deodorizers and flavouring agents (14%), formulation component (14%) and polymers (11%).

Many different industrial sectors notified these substances (Figure 3). The single largest sectors were the organic chemicals sector (12%), the pigment, dye and printing ink sector (10%) and the paint and coating sector (9%). The "other" sector represents over 30 combined industry sectors, including adhesive and sealant production, construction materials, mining, metal and non-metal and organometallic chemicals.

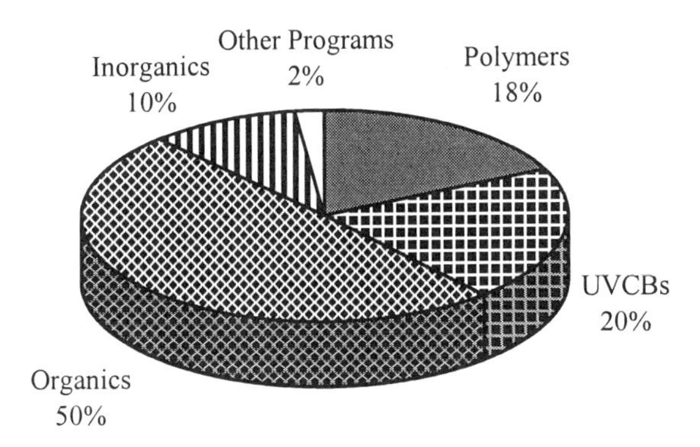

Figure 1. Types of substances on the Domestic Substances List.

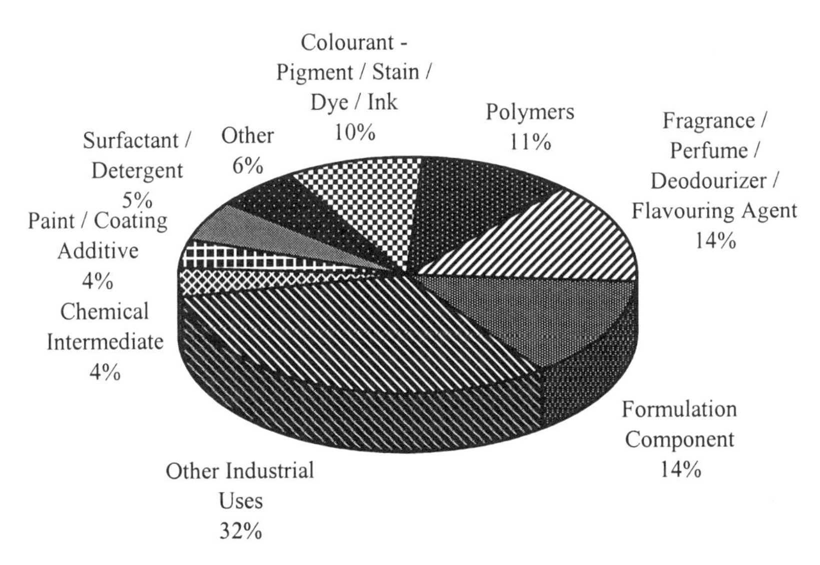

Figure 2. Reported use patterns for substances on the Domestic Substances List in 1984-1986.

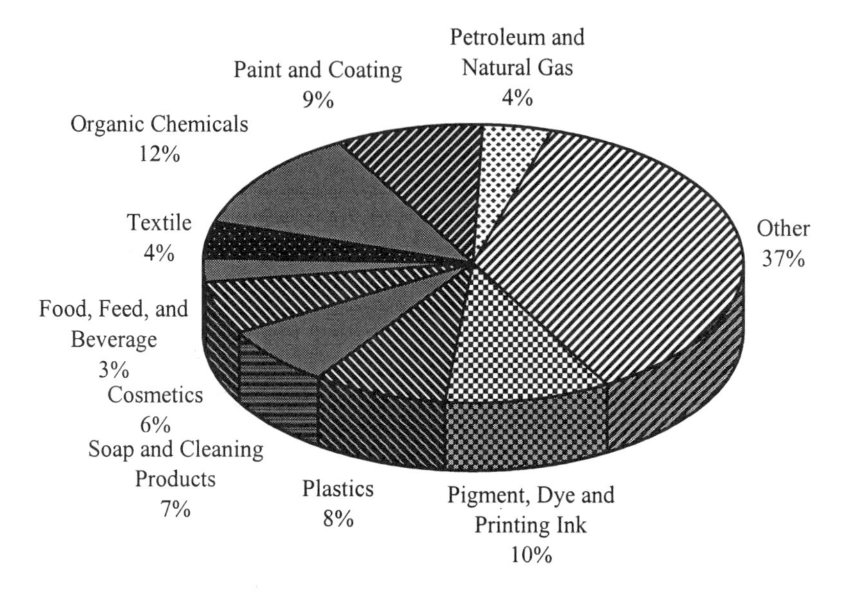

Figure 3: Reported industrial sectors for substances on the DSL in 1986.

Since most of the substances on the DSL have not undergone any environmental or human health assessment, the revised CEPA provides for the systematic assessment of substances on the DSL that are to be carried out in two phases. The initial phase, referred to as the categorization of substances, identifies substances that will proceed to the second phase, a screening level risk assessment (Figure 4). The environmental categorization of substances on the DSL requires the Minister of the Environment to assess the substances on the basis of their persistence, bioaccumulation and inherent toxicity. The criteria for persistence (P) and bioaccumulation (B) are the same as those in the Toxic Substances Management Policy (TSMP) and, as stated in CEPA, will be included in regulations that are currently being developed. The definition and interpretation of inherent toxicity for this exercise is currently under development as are the approaches for categorizing substances based on inherent toxicity.

Criteria for persistence and bioaccumulation set out in the TSMP were selected by an expert panel of scientists, based on knowledge of the properties that are most characteristic of persistent organic pollutants (*3*). The criteria were chosen to identify substances that are likely to be of greatest concern with regards to persistence and bioaccumulation. Persistence is based on a consideration of all environmental media. A substance is considered persistent if its transformation half-life satisfies the criterion in any one medium as identified in Table I. A substance is considered as bioaccumulative if any of the bioaccumulation criteria in Table I are satisfied.

As identified in the revised CEPA, all substances on the DSL must be categorized within 7 years after the Bill receives Royal Assent (September 14, 1999).

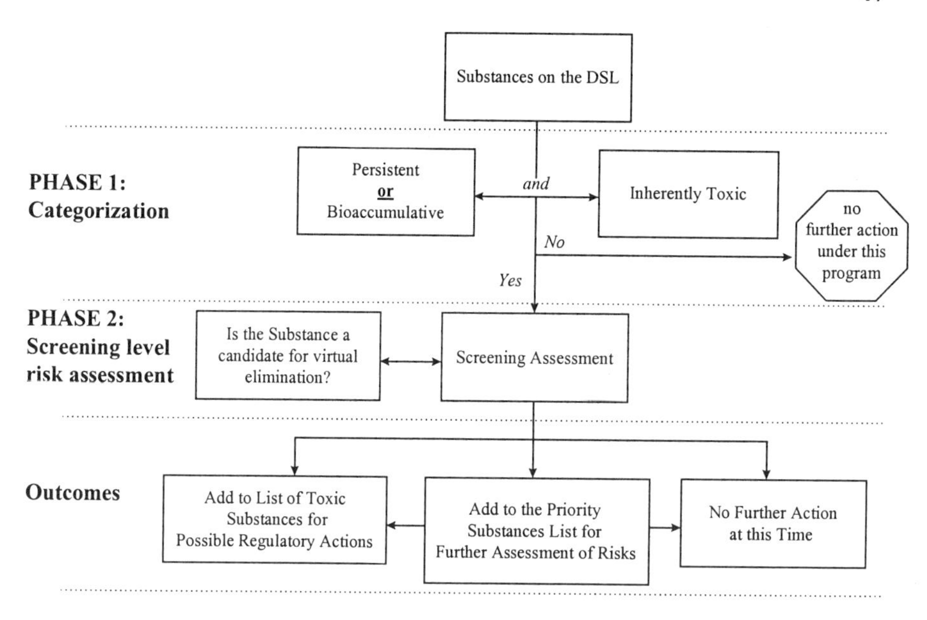

Figure 4. Environmental categorization and screening of substances on the Domestic Substances List.

Table I. Criteria for Persistence and Bioaccumulation

Persistence [a]		Bioaccumulation [c]
Medium	Half-life	BAF [d] ≥5,000
Air	≥ 2 days [b]	or
Water	≥ 6 months	BCF [e] ≥ 5,000
Sediment	≥ 1 year	or
Soil	≥ 6 months	Log K_{OW} ≥ 5

[a] A substance is considered persistent when the criterion is met in any one medium.
[b] A substance may be considered as persistent in air if it is shown to be subject to atmospheric transport to remote regions such as the Arctic.
[c] Bioaccumulation Factors (BAF) are preferred over Bioconcentration Factors (BCF); in the absence of BAF or BCF data, the octanol-water partition coefficient (log K_{OW}) may be used.
[d] Bioaccumulation factor means the ratio of the concentration of a substance in an organism to the concentration in water, based on uptake directly from the surrounding environment and food.
[e] Bioconcentration factor means the ratio of the concentration of a substance in an organism to the concentration in water, based only on uptake directly from the surrounding environment.

If a substance does not meet the criteria, then no further action is required for this substance under this categorization and screening exercise. When a substance is categorized as satisfying the criteria for persistence and/or bioaccumulation and inherent toxicity, then a screening level risk assessment is required (Figure 4). The screening level risk assessment involves a more in-depth analysis of a substance to determine whether the substance is "toxic" or capable of becoming "toxic" as defined in CEPA. A screening level risk assessment results in one of the following outcomes:

- no further action is taken at this time, if the screening level risk assessment indicates that the substance does not pose a risk to the environment or human health;

- the substance is added to the CEPA Priority Substances List in order to assess more comprehensively the possible risks associated with the release of the substance, if the substance is not already on the Priority Substances List but the screening level risk assessment indicates possible concerns; or

- it is recommended that the substance be added to the List of Toxic Substances in Schedule 1 of CEPA, if the screening level risk assessment indicates clear concerns, whether these are associated or not with the persistence or bioaccumulation properties of the substance; substances on Schedule 1 can be considered for regulatory or other controls.

As mandated under the revised CEPA, where, based on a screening level risk assessment, a substance is determined to be "toxic" or capable of becoming "toxic", and a substance may have a long-term harmful effect on the environment because it is persistent, bioaccumulative, inherently toxic and predominantly anthropogenic, then the substance will be added to the List of Toxic Substances. If, in addition, the substance is not a naturally occurring inorganic substance or a naturally occurring radionuclide, then the substance will be proposed for virtual elimination under CEPA (Figure 4).

In addition to certain substances that have undergone screening level risk assessments, substances that have not been identified as being of concern through the categorization and screening exercise can nonetheless be assessed. The Priority Substances Assessment Program is mandated under CEPA. Under this mandate, the Minister of the Environment and the Minister of Health develop a list of substances, known as the Priority Substances List, that should be given priority for assessment to determine whether they are "toxic" as defined under CEPA. Substances that are assessed as "toxic" may be placed on the List of Toxic Substances (Schedule 1 of the Act), and be considered for possible risk management measures such as regulations, guidelines, or codes of practice to control any aspect of their life cycle, from the research and development stage through manufacture, use, storage, transport and ultimate disposal.

Issues, Challenges and Approach

Experimental data on persistence, bioaccumulation and inherent toxicity needed to conduct the categorization are only available for a small number of the substances on the DSL. Models such as Quantitative Structure Activity Relationships (QSARs) used to predict these data are not available for many of the classes or types of substances on the DSL (*e.g.*, pigments, polymers, ionic surfactants). Some of the challenges regarding modeling include the estimation of media-specific half-lives, the estimation of the bioaccumulation potential for some classes of substances (*e.g.*, pigments), the estimation of toxicity of many classes of substances (*e.g.*, organometallic substances, surfactants), and the use and development of fate and exposure models for the screening level risk assessments. In addition, the nature of some of these substances, such as polymers, UVCBs, and metal elements and metal salts poses a challenge when applying the criteria for the categorization. One of the challenges, considering the many hundreds of inorganic substances on the DSL, will involve determining the inherent toxicity of such substances while taking into account concepts such as bioavailability and transformation.

The development of approaches and methods for conducting a screening level risk assessment is currently underway. A risk based approach will be used and will consist of four major parts: characterization of entry, exposure, effects and risk. The objective of the entry characterization will be to determine the uses, types of releases and amounts of the substance entering the Canadian environment. These data will then be used as input to the characterization of exposure which is most likely to involve modeling. The objective of the exposure characterization will be to determine estimated exposure concentrations to biota. Characterization of effects will first consider experimental data. However, in most cases, due to the lack of experimental data, QSARs will be used to estimate toxicity. The final part, risk characterization, involves determining the likelihood of adverse effects to biota. This will be carried out by combining the results of the characterization of entry, exposure and effects. As stated in the revised CEPA, when conducting and interpreting the results of a screening level risk assessment, a weight of evidence approach and the precautionary principle will be applied.

An Advisory Group that consist of experts from government, industry, environmental organizations and consultant groups, have been set up to provide an expert resource to Environment Canada for identifying and assisting in resolving issues of a scientific, technical and process nature. The group will assist in identifying approaches to conduct the categorization and screening level risk assessments and in the preparation of technical guidance documents. The Advisory Group will also assist in establishing technical Working Groups as required to carry out specific tasks.

Two Working Groups and a workshop are currently being set up. The first Working Group will be responsible for developing an approach to fragment or divide the DSL into manageable groups of similar substances. The second Working Group will be responsible for recommending a practical approach for categorizing inorganic substances against the criteria for persistence, bioaccumulation and inherent toxicity. This Group will also recommend an approach for conducting screening level risk assessments for inorganic substances. A workshop on QSARs is being organized by Environment Canada. The workshop will bring together about 30 international experts on estimating persistence, bioaccumulation and toxicity. The objectives of the workshop include recommending which QSAR to use for the categorization of different types of substances, identifying alternatives to QSARs for substances for which QSARs are not available and recommending research for data gaps identified. The workshop is scheduled for November 11-12, 1999, in Philadelphia, Pennsylvania.

Environment Canada has initiated a pilot project which will identify 100 substances representative of several chemical classes of concern with regards to persistence, bioaccumulation and inherent toxicity. This will involve the evaluation of a list of over 7 000 organic substances. The categorization of these substances will be completed by the winter of 1999-2000. Environment Canada will be carrying out screening level risks assessments as substances are identified that meet the categorization criteria.

The methods, approaches and results of the categorization and screening level risk assessments will be collated in technical guidance documents and reports. In addition, a database containing the technical data used to conduct the categorization and screening assessments is currently being developed. Decisions and documents will be publicly available. A web site for this project is currently being developed and will be linked to the Commercial Chemicals Evaluation Branch web site which can be found at http://www.ec.gc.ca/cceb1/.

Conclusion

One of the new initiatives under the proposed revised *Canadian Environmental Protection Act* states that Environment Canada must "categorize" and "screen" substances that are listed on the Domestic Substances List (DSL) to determine whether they are "toxic" as defined in the Act. The environmental categorization identifies persistent and/or bioaccumulative and inherently toxic (PBT) substances on the DSL. The criteria used to categorize these substances are those set out in the Toxic Substances Management Policy. Because of the difficulty in determining the persistence, bioaccumulation and inherent toxicity of the various classes of substances on the DSL and in conducting screening level risk assessments on these substances, Environment Canada has set up several mechanisms including an Advisory Group and several working groups to provide guidance on how to resolve scientific and technical issues. A pilot project has been initiated which will identify 100 substances representative of several chemical classes of concern.

Because many of the experimental data on persistence, bioaccumulation and inherent toxicity are only available for a small number of the substances on the DSL, models such as Quantitative Structure Activity Relationships (QSARs) will be used to predict these data. However, current models are not sophisticated enough to predict reliable estimates for all classes of substances on the DSL. Some of the challenges regarding modeling include the estimation of media-specific half-lives, the estimation of the bioaccumulation potential for some classes of substances (*e.g.*, pigments), the estimation of toxicity of many classes of substances (*e.g.*, organometallic substances, surfactants). In addition, the nature of some of these substances, such as polymers, UVCBs, and metal elements and metal salts poses a challenge when applying the criteria for the categorization. One of the topics of discussion for the workshop on QSARs organized for November 1999 will include recommendations for research and development of models for data gaps that are identified.

References

1. Government of Canada. *Toxic Substances Management Policy*; Environment Canada: Ottawa, ON, Canada, June 1995.
2. Department of the Environment. *Canada Gazette, Part I.* July 4, 1998, pp 1568-1569.
3. Government of Canada. *Toxic Substances Management Policy - Persistence and Bioaccumulation Criteria*; Environment Canada: Ottawa, ON, Canada, June 1995.

Chapter 5

Relationship between Persistence and Spatial Range of Environmental Chemicals

Martin Scheringer[1], Deborah H. Bennett[2], Thomas E. McKone[2], and Konrad Hungerbühler[1]

[1]Laboratory of Chemical Engineering, ETH Zentrum, CH–8092 Zürich, Switzerland
[2]Lawrence Berkeley National Laboratory, Berkeley, CA 94720

Several approaches to calculate the spatial range or travel distance of environmental chemicals have been proposed in the literature. Here we evaluate the relationship between different definitions of spatial range and travel distance and between these quantities and the chemical's atmospheric residence time. We show that the results from a simple global multimedia fate and transport model can account for the analytical relationship. In contrast, the relationship between a chemical's overall persistence and spatial range cannot be described by an analytical expression. A plot of the spatial range versus the overall persistence does not show a well defined relationship between these two measures. The deviations from the analytical relation between atmospheric residence time and spatial range are caused by differences in the phase partitioning of the chemicals. In addition, deviations are strongly influenced by the release media. These effects are demonstrated by correlating the deviations from the analytical relation with the octanol-air partitioning coefficient for each chemical.

Introduction

The spatial range or characteristic travel distance of an environmental chemical has been introduced as a measure of the chemical's mobility and potential for long-range transport (LRT) *(1–7)*. In analogy to persistence, which quantifies the duration of the environmental exposure, the spatial range is used as a measure of the spatial extent or "effective length" of the spatial distribution of a chemical in the environment. Indicating the extent of an exposure pattern, the spatial range helps to distinguish among local, regional, and global pollution problems. Local, regional, and global scale pollutants require different distribution models, exposure models, measurement programs, and regulatory approaches.

Spatial range and persistence are attributes that can be used to compare and classify chemicals. These two quantities provide exposure screening indicators which can be seen in analogy to hazard indicators such as toxicity or mutagenicity on the effect level. Spatial range and persistence are of particular importance for the characterization and identification of Persistent Organic Pollutants (POPs) *(8)* but they can also be used in a general framework for the assessment of chemicals *(9, 10)*.

Since a chemical cannot be transported in the environment if it is not persistent for some time, it seems to be obvious that persistence and spatial range are correlated. However, this is not an ideal correlation for several reasons. First, there are different approaches to calculate a chemical's spatial range and persistence, leading to a variety of persistence-spatial range relations. Second, persistence and spatial range are influenced in a complex way by the chemical's partitioning between the environmental media soil, water and air. These media differ in their mobility and in the chemical's degradation rate constants, which can lead to different spatial ranges for a given persistence. Additionally, if the chemical is released to a different environmental medium, the distribution of the chemical between the environmental media will change, affecting the overall persistence.

In this paper, we first investigate the relationship between three definitions of spatial range (denoted by R, ρ, and L) and between spatial range and the chemical's residence time in the mobile medium (usually air), denoted by τ_a. Based on advective and macro-diffusive transport processes, we develop analytical relationships between R, ρ, L, and τ_a. We next compare the analytical relationships between atmospheric residence time and spatial range obtained for macro-diffusive flow with numerical results from a simple model of the global circulation of chemicals. Then the relation between spatial range and overall persistence, which is not accessible in an analytical form, is investigated with the model. Finally, the significant influence of the medium the chemical is released into is demonstrated by comparing release to soil and release to air.

Approaches to Calculating Spatial Range

Several definitions of the spatial range or travel distance have been proposed and are currently used:

- Müller-Herold and Nickel *(5)* calculate the Shannon entropy S of the spatial concentration or exposure distribution and subsequently the entropy rank $\rho = \exp\{-S\}$. ρ is a measure of the spread of a distribution.
- Bennett et al. *(6)* and van Pul et al. *(7)* define the travel distance L of a chemical by the point where a monotonously decreasing function $c(x)$ has dropped to $c_0/e \approx 0.368 \cdot c_0$.
- Scheringer *(3)* introduced the spatial range R as the 95% interquantile distance of the spatial concentration distribution. This quantity can be determined for any kind of spatial distribution, e.g. for nearly uniform distributions that are obtained for CFCs in the troposphere and for distributions with more than one maximum, such as the accumulation of POPs in polar regions.

All of these definitions were introduced in the context of multimedia models that are used to calculate the environmental concentrations as functions of location. However, spatial range and persistence are not part of a specific model but can be determined from a variety of models or measured concentration data.

Calculating Spatial Range for Advective and Diffusive Air Flow

The results for spatial range or travel distance presented in the literature differ with respect to the underlying multimedia models and the assumed transport mechanisms. There are several regional models that have advective flow in the atmosphere as the basic transport mechanism *(6, 7)*. There are also global models that include macro-diffusive transport processes in water and air *(4, 11, 12, 13)*. The use of both types of models in calculating spatial range and travel distance is described in the following subsections.

Transport through Advective Flow

The regional models consist of a multimedia environment with advective inflow and outflow. Air (or water) moves through the system with a constant wind speed of 4 m/s *(6)* or 5 m/s *(7)*. The transport process is driven by this advective flow, and a certain parcel of air travels without mixing with the surrounding air (plug flow). The direction of the transport is defined by the direction of the air flow.

The assumption of a steady-state mass distribution and first-order removal processes (atmospheric degradation and transfer to, and subsequent degradation in, soil and water) with a constant advective flow results in the following spatial concentration profile *(6)*

$$c(x) = c(x_0)\cdot\exp\left\{-\frac{k}{u}\cdot x\right\}. \quad (1)$$

k (in s^{-1}) is the effective atmospheric removal rate constant which is given by the net flow through the air compartment (in kg/s) divided by the mass in air (in kg) *(6)*. $\tau_a = 1/k$ is then the atmospheric residence time. u is the wind speed in m/s, and x is the distance from the point of inflow into the system, x_0. The quantity $L = u/k$ denotes the point where the concentration has dropped to c_0/e, which is defined as the characteristic travel distance *(6)*. Accordingly, a linear relation

$$L = u\cdot\tau_a \quad (2)$$

between travel distance and residence time is obtained for this model.

The spatial range R is given by the point $q_{0.95}$ with

$$\int_0^{q_{0.95}} c(x)\,dx \Big/ \int_0^{\infty} c(x)\,dx = 0.95, \quad (3)$$

see ref *(3)*. Inserting the exponential function from eq 1 for $c(x)$ in eq 3, we obtain

$$R = q_{0.95} \approx 3.00 \cdot u/k = 3.00 \cdot u \cdot \tau_a = 3.00 \cdot L. \tag{4}$$

For atmospheric residence times of 100 days or greater, the assumption of a single advective flow leads to travel distances larger than the circumference of the earth. Accordingly, models with advective flow are appropriate for local and regional transport processes, but are not predictive on the global scale, only indicating if a chemical is likely to be dispersed globally.

Transport through Macroscopic Diffusion

Macroscopic diffusion in water and air is the relevant transport mechanism in global models *(4, 11, 12, 13)*. The transport is driven by the concentration gradient between adjacent air or water compartments and finally leads to uniform concentration distributions. The direction of the transport is determined by the concentration gradient, which means that the chemical can travel in two or more directions at the same time (depending on the model geometry). Combining the second law of diffusion with a first-order removal term yields the following relationship:

$$\dot{c}(x,t) = D_{\text{eddy}} \frac{\partial^2 c(x,t)}{\partial x^2} - k \cdot c(x,t). \tag{5}$$

D_{eddy} is the macroscopic diffusion coefficient (in m^2/s) and k is the atmospheric removal rate constant in s^{-1} (the same k as used in the advective model). Typical values for D_{eddy} in air are in the range of $5 \cdot 10^5 \text{m}^2/\text{s}$ to $4 \cdot 10^6 \text{m}^2/\text{s}$ *(3, 11, 14)*.

The steady state ($\dot{c}(x,t) = 0$) relationship obtained from eq 5 is:

$$D_{\text{eddy}} \frac{d^2 c(x)}{dx^2} = k \cdot c(x)$$

and the steady state solution follows as

$$c(x) = c(x_0) \cdot \exp\left\{ -\sqrt{\frac{k}{D_{\text{eddy}}}} \cdot x \right\} \tag{6}$$

(with the boundary conditions $c(0) = c_0$ and $c(x \to \infty) = 0$).

In this case, the travel distance L is given by

$$L = \sqrt{D_{\text{eddy}}/k} = \sqrt{D_{\text{eddy}}} \cdot \sqrt{\tau_a}. \tag{7}$$

Corresponding to the case of advective transport, the spatial range follows as

$$R \approx 3.00 \cdot L = 3.00 \sqrt{D_{\text{eddy}}} \cdot \sqrt{\tau_a}. \tag{8}$$

As shown in ref *(5)*, the result for the entropy rank is

$$\rho = e \cdot \sqrt{D_{\text{eddy}}/k} \approx 2.72 \sqrt{D_{\text{eddy}}} \cdot \sqrt{\tau_a}. \tag{9}$$

For two-sided distributions, a factor of 2 applies to all of these quantities.

Spatial Range–Persistence Diagrams

Analytical Expressions and Results from a Model for Diffusive Flow

As described in the previous section, the analytical relationships between spatial range R or travel distance L and atmospheric residence time τ_a are $L = u \cdot \tau_a$ for advective transport in one direction and $R = 6.00 \sqrt{D_{eddy}} \cdot \sqrt{\tau_a}$ for diffusive transport in two directions. These relationships are plotted in Figure 1 with $u = 5\,m/s$ and $D_{eddy} = 2 \cdot 10^6 m^2/s$. This plot illustrates how the travel distance L and the spatial range R can be calculated from the atmospheric residence time and how they compare for two different transport mechanisms.

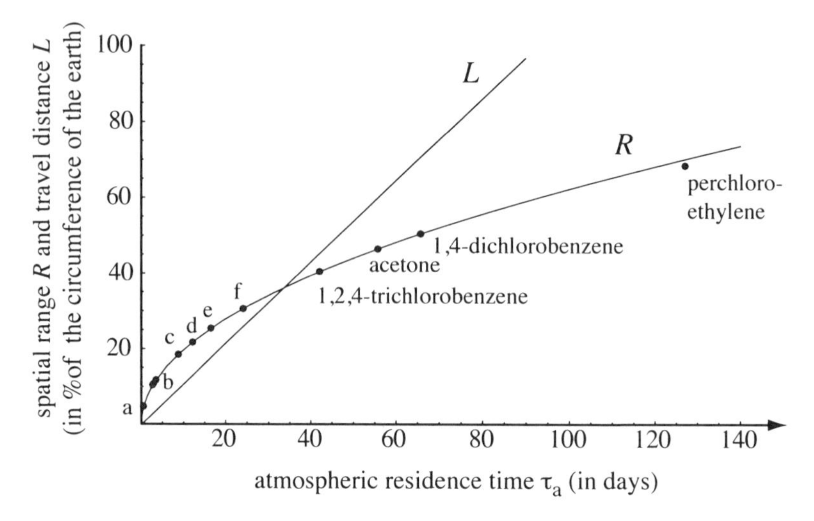

Figure 1. Relationships between atmospheric residence time τ_a and atmospheric spatial range R and travel distance L. The lines show the analytical expressions for advective and macro-diffusive transport. The dots indicate the model results obtained for selected chemicals with a global model for diffusive transport described in ref (3); the chemicals are released to air. The letters indicate: a) p-cresol; b) dioxane, cyclohexane, toluene, octane (similar results); c) methyl t-butyl ether; d) chlorotoluene; e) benzene; f) chlorobenzene.

Additionally, the results for the atmospheric residence time and the spatial range calculated with a global fate model are shown for selected chemicals (dots). This model, which is described in ref *(3)*, is based on the analytical diffusive transport model, and is a steady-state multimedia model including diffusive transport in air and water. In Figure 1, the comparison of the model results with the analytical expression $R = 6.00 \sqrt{D_{eddy}} \cdot \sqrt{\tau_a}$ is based on release to air. In the next subsection, this model is used with release to soil and the results are plotted in Figure 4 to investigate the correlation between spatial range and overall persistence.

The model consists of a closed loop of 80 cells, each identical to a single level III multimedia model with soil, water and air subcompartments. In this loop with circumference $G = 40\,000$ km, the one-dimensional global circulation of chemicals is modelled. Based on macro-diffusive transport in air with $D_{eddy} = 2 \cdot 10^6 m^2/s$ (and first order degradation and phase partitioning), the model yields the steady-state concentrations for each subcompartment of the 80 cells. The atmospheric concentration distribution of perchloroethylene is shown as an example in Figure 2. For details of the model, see ref *(3)*.

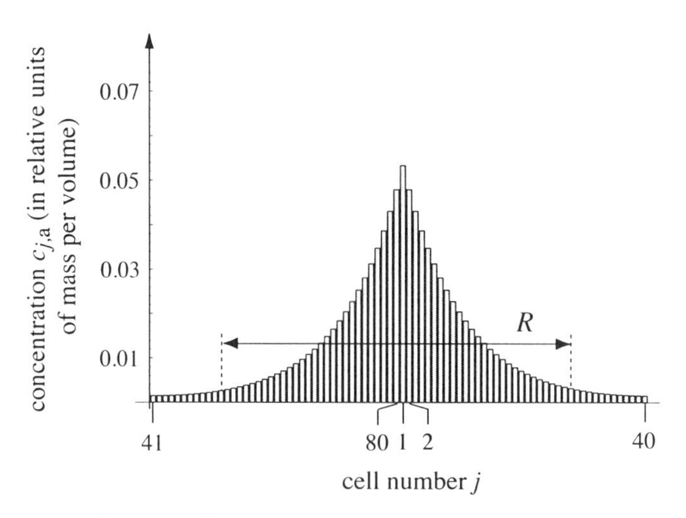

Figure 2. Atmospheric steady-state concentrations $c_{j,a}$ of perchloroethylene vs. cell number j in the circular global model. The distribution $\{c_{j,a}\}_{j=1,\ldots,80}$ is normalized to 1. The chemical is released to air at $j = 1$. In this graphic representation of the circular system, the adjacent cells 40 and 41 appear separated. The spatial range is $R = 68.5\%$, see dot "perchloroethylene" in Figure 1.

The atmospheric residence time after emission to air at cell $j = 1$ was calculated according to

$$\tau_a = \frac{1}{Q} \sum_{j=1}^{80} c_{j,a} \cdot v_j = \frac{M_{air}^{stst}}{Q} \qquad (10)$$

with the continuous source term Q (in kg/s), the atmospheric steady state concentrations $c_{j,a}$ in the cells j with volume v_j, and the steady-state mass in air M_{air}^{stst} (in kg). The spatial range was derived as the 95% interquantile range from the spatial concentration distributions, see Figure 2 and eq 3.

For chemicals with $\tau_a < 100$ d, the model results are described by the analytical expression $R = 6.00 \sqrt{D_{eddy}} \cdot \sqrt{\tau_a}$. Perchloroethylene with $\tau_a = 127$ d shows a slightly lower spatial range than predicted by this analytical expression. This is due to the fact that the model assumes a closed circular world with an upper limit for the spatial range of $G = 40\,000$ km while the analytical expression is valid for an "open" world and predicts unlimited spatial ranges for increasing τ_a. Chemicals with even higher atmospheric residence time such as tetrachloromethane and CFCs do not follow the analytical expression but show a constant spatial range of 95% of the global circumference. In Figure 3, the model results for an expanded set of chemicals, including very persistent ones, are shown on a logarithmic scale for the atmospheric residence time.

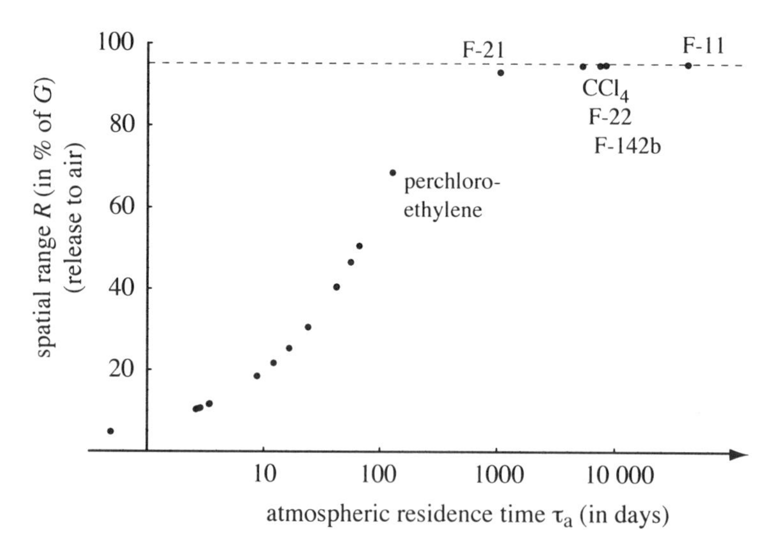

Figure 3. Model results for atmospheric spatial range R and atmospheric residence time τ_a for an expanded set of chemicals. The chemicals are released to air. τ_a is given on a logarithmic scale. The chemicals at the lower end of the scale are the same as in Figure 1. The dashed horizontal line indicates the upper limit of $R = 95\%$ of G.

Relationship between Spatial Range and Overall Persistence

The analytical relationships plotted in Figure 1 demonstrate that R and L can be calculated directly from the atmospheric residence time τ_a. On the one hand, this is useful but on the other hand, τ_a is a compartment specific residence time which does not provide the most comprehensive information on a chemical's environmental lifetime because it is determined by both degradation *and* transfer to other compartments *(15)*. Efficient removal through deposition processes can lead to a low atmospheric residence time for a chemical that may be persistent in water or soil. Therefore, it is more informative to combine the spatial range with the *overall* persistence of the chemicals in one diagram so that information on the chemicals' actual lifetime and their phase partitioning is included. However, the relation between overall persistence (denoted by τ) and spatial range cannot be described by an analytical expression. Figure 4 shows the plot of R versus $\log \tau$ for the same set of chemicals plotted in Figure 3 (dots) and, in addition, for the 12 POPs and lindane (lines) *(4)*. As can be seen in the figure, there is not a

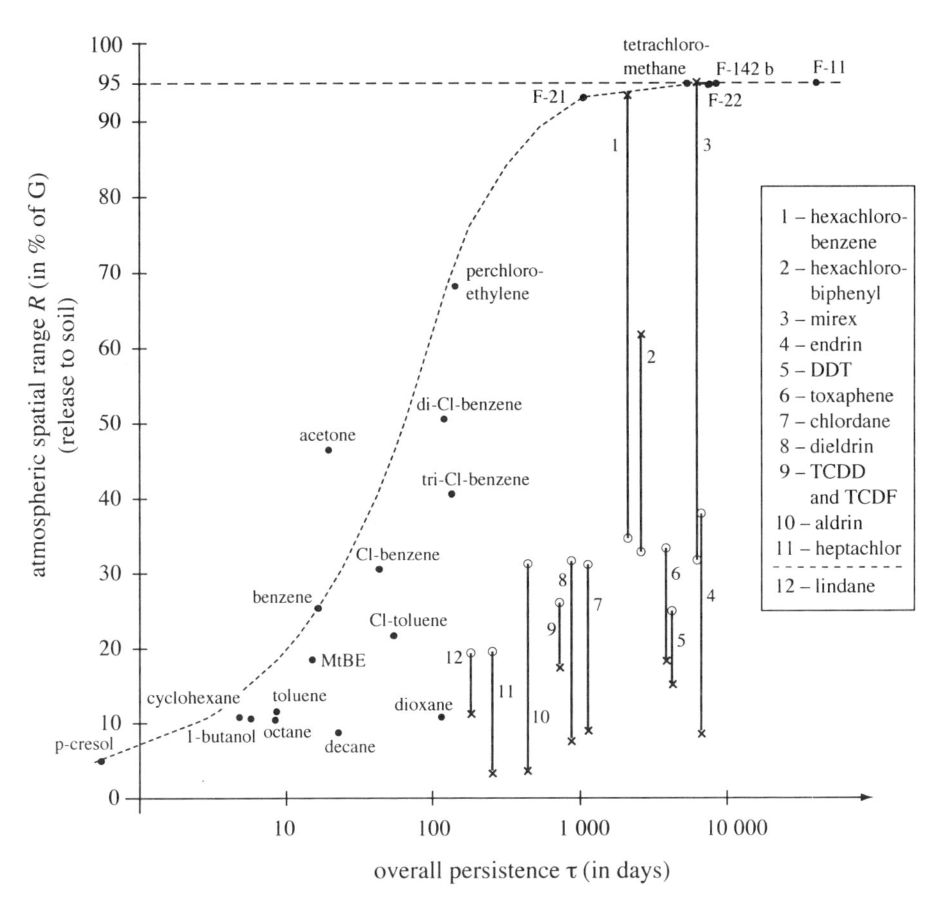

Figure 4. Model results for atmospheric spatial range R and overall persistence τ for the chemicals from Figure 3 and for the 12 POPs and lindane. All chemicals are released to the soil. The persistence values are given on a logarithmic scale. The dashed curve is the interpolated $R(\tau_a)$ relation of Figure 3. The dashed horizontal line indicates the upper limit of $R = 95\%$. The vertical lines represent spatial range intervals which are obtained if the adsorption of the POPs to aerosol particles is varied between 0% ($\times$) and 100% ($\circ$).

well defined relationship between these two measures. All chemicals were released to the soil, i.e. the immobile compartment, so that they have to volatilize before they are transported with the macro-diffusive air flow. However, the dynamics of the transport in air is independent of the size of the airborne fraction of the chemical. This means that the numerical values of the atmospheric spatial range R for release to soil (Figure 4) are almost the same as for release to air (Figure 3).

The overall persistence τ was calculated according to

$$\tau = \frac{1}{Q} \sum_{i=\mathrm{s,w,a}} \sum_{j=1}^{80} c_{i,j} \cdot v_{i,j} = \frac{M_{\mathrm{tot}}^{\mathrm{stst}}}{Q}, \tag{11}$$

compare eq 10. $M_{\mathrm{tot}}^{\mathrm{stst}}$ is the amount of chemical that is contained in the entire model system in the steady state.

The dashed curve in Figure 4 is the interpolated $R(\tau_{\mathrm{a}})$ relation of Figure 3. Most of the chemicals lie on the right of this line. This is due to the fact that their degradation rate constants in soil are lower than the degradation rate constants in air so that the fraction in soil is degraded more slowly than the fraction in air. Under this conditions, the overall persistence τ is higher than the atmospheric residence time τ_{a} while the spatial range is nearly the same as for release to the air. This effect is very pronounced for dioxane and the the semivolatile organochlorines at the lower ends of the lines.

There are some chemicals whose overall persistence τ is very close to the atmospheric residence time τ_{a} (benzene, perchloroethylene, the CFCs). This is observed if the degradation rate constants in soil, water and air are (almost) the same or if the chemical is so volatile that the overall persistence is essentially independent of the degradation rate constants in water and soil.

Finally, acetone has an atmospheric degradation rate constant which is *lower* than that in the soil. In this case, release to the soil leads to an overall persistence τ which is lower than the atmospheric residence time τ_{a}. For this reason, acetone is on the left of the $R(\tau_{\mathrm{a}})$ line in Figure 4.

The spatial ranges of the semivolatile organochlorines are subject to considerable uncertainty because the atmospheric degradation and transport of these chemicals is strongly influenced by adsorption to aerosol particles *(16, 17, 18)*. The vertical lines shown in Figure 4 represent the entire range from 0% adsorption (indicated by $\times$) to 100% adsorption (indicated by $\circ$); for details, see *(4)*. Although these extremes do not reflect realistic conditions for all of the chemicals (rather volatile compounds such as hexachlorobenzene are not completely adsorbed to particles under most environmental conditions), they are shown here to illustrate the possible influence of the aerosol particles and the sensitivity of the spatial range to this model parameter.

Influence of Phase Partitioning and Path of Release

As apparent from Figure 4, many chemicals that have a large overall persistence do not necessarily have a large spatial range. It is desirable to try to understand the relationship between the spatial range and the overall persistence in the environment. To this end, we calculated the ratio of the atmospheric spatial range after release to the soil (the R values of the dots in Figure 4) and the atmospheric spatial range given by the $R(\tau_a)$ line for the given τ value. Thus, this ratio, which is denoted by r, quantifies the vertical deviation of the dots in Figure 4 from the dashed $R(\tau_{\mathrm{a}})$ line in Figure 4. The ratio r is lower for chemicals that tend to partition out of the atmosphere, such that there is a positive correlation to vapor pressure and Henry's law constant (H) and a negative correlation to the octanol-water and octanol-air (K_{oa}) partitioning coefficients. The comparisons revealed that $\log K_{oa}$ is actually the best descriptor for this trend. In Figure 5, we plot r versus $\log K_{oa}$ for the chemicals from Figure 4. The plot shows a continuous trend to lower r values if $\log K_{oa}$ increases. The chemicals with $r \approx 1$ are the CFCs, perchloroethylene, benzene, and p-cresol; they lie on the $R(\tau_a)$ line in Figure 4. Acetone is the only chemical with $r > 1$; dioxane has $r = 0.15$ and $\log K_{oa} = 0.036$. For the POPs, the arithmetic means of the R intervals of Figure 4 were used in this analysis.

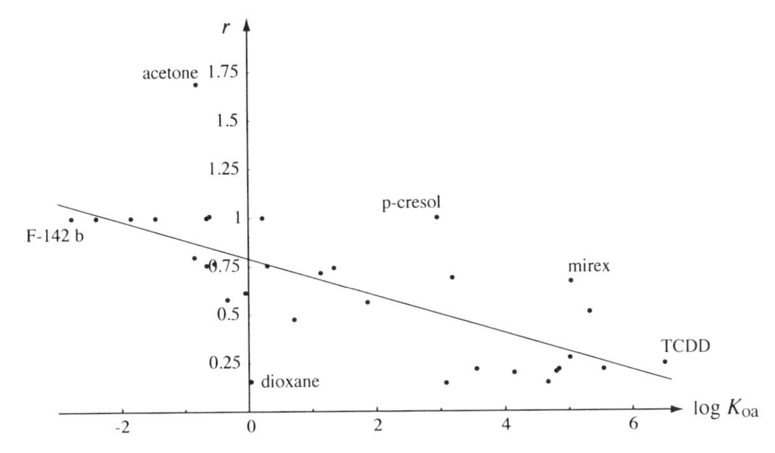

Figure 5. Correlation of $\log K_{oa}$ *with the ratio r of the calculated spatial range indicated by the dots in Figure 4 and the spatial range determined by the curve* $R(\tau_a)$ *in Figure 4. The chemicals are released to soil.* K_{oa} *equals* K_{ow}/H *with H in Pa m*^{3}*/mol. The line is obtained by linear regression.*

It is important to note that there is not a perfect correlation between r and $\log K_{oa}$ in Figure 5 because a chemical's partitioning between the environmental media in a steady-state condition is not influenced by the partitioning coefficients alone, but also by advective exchange processes such as wet deposition and by the degradation rates in the environmental media, which can be very different for the same $\log K_{oa}$ value.

Another factor influencing the spatial range relative to the overall persistence is whether a chemical is released to the air or to the soil. Many chemicals are degraded more rapidly in the air than in the soil. Thus, if a chemical is released to the air, the overall persistence is likely to be lower than if the chemical is released to the soil. Figure 6 shows the r-$\log K_{oa}$ correlation for this situation. It can be seen that all non-adsorbing chemicals (dots in Figure 4) except dioxane have r values close to 1. There is a sharp decrease in r at $\log K_{oa} \approx 3$, leading to r values below 0.3 for most of the POPs. These results indicate that the overall persistence of the volatile chemicals is dominated by the atmospheric degradation rate constant while the degradation rate constants in water and soil and the partitioning coefficients do not have a significant influence on τ. Thus, the overall persistence τ is close to the atmospheric residence time τ_a and much lower than for the "release to soil" scenario. The POPs, on the other hand, are transferred to the soil by diffusion and deposition, where they have low degradation rate constants. Thus their overall persistence is not as strongly influenced by the degradation rate constant in air and their overall persistences are higher than their atmospheric residence times.

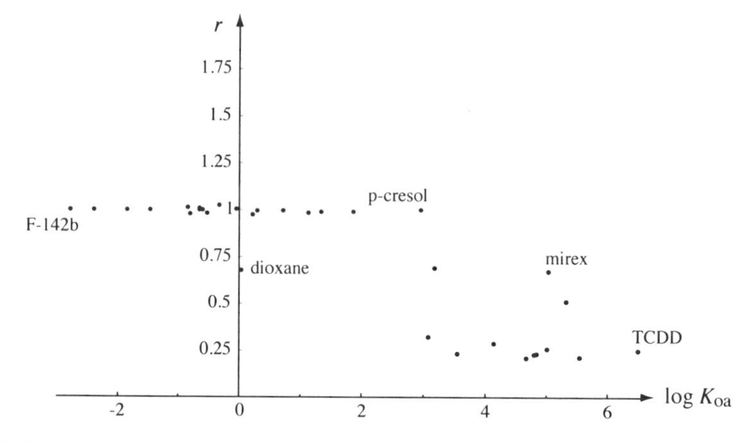

Figure 6. Correlation of log K_{oa} *with the ratio r of the calculated spatial range and the spatial range determined by the curve* $R(\tau_a)$. *In this case, the chemicals are released to air. A plot such as in Figure 4 is not shown for this case because most of the* (τ, R) *points would be very close to the* $R(\tau_a)$ *curve.*

These results illustrate that modeling a release to air is suitable only if we want to calculate the atmospheric residence time or if the chemical is actually released to the air. In general, this scenario is less interesting than release to the soil because there is often less multimedia partitioning for a release to air than to soil, and thus the different degradation rate constants are not as influential if the release is to air. If one is not sure what medium the chemical will be released to, model calculations with release to the air should be complemented by calculations with release to the soil which provide more insight into the chemicals' phase partitioning and overall degradability.

Conclusions

In conclusion, R, τ-diagrams such as in Figure 4 are useful for the screening of the exposure behavior of larger sets of chemicals. High overall persistence and high spatial range are unwanted properties which indicate a chemical's potential to cause problematic environmental exposure.

In addition to the screening of a variety of chemicals, R, τ-diagrams can be used to visualize details of the complex environmental fate of specific chemicals such as the POPs. In Figure 4, the sensitivity of the atmospheric spatial range of the POPs to the extent of adsorption to particles is shown. The modeling results of such a particular focus can be aggregated and communicated by means of a R, τ-diagram.

Both aspects, exposure screening of a larger set of chemicals and analyzing details of the transport and degradation behavior of selected chemicals, are relevant for the future debate on chemicals regulation and for the assessment of possible POPs candidates.

Acknowledgments

We thank K. Fenner for helpful discussions and P. Funck for providing a LaTeX style file for this document. D. H. Bennett and T. E. McKone were supported by the EPA Contract DW-998-38190-01-0.

References

1. Scheringer, M., Berg, M., Müller-Herold, U. In *Was ist ein Schaden?*; Berg, M. et al., Eds.; Verlag der Fachvereine: Zürich, 1994, pp 115–146.
2. Scheringer, M., Berg, M. *Fresenius Environ. Bull.* **1994**, *3*, 493–498.
3. Scheringer, M. *Environ. Sci. Technol.* **1996**, *30*, 1652–1659.
4. Scheringer, M. *Environ. Sci. Technol.* **1997** *31*, 2891–2897.
5. Müller-Herold, U., Nickel, G. *Tübinger Berichte zur Funktionalanalysis*; Department of Mathematics, University of Tübingen: Tübingen, 1998, Vol. 7, pp 181–194.
6. Bennett, D. H., McKone, T. E., Matthies, M., Kastenberg, W. E. *Environ. Sci. Technol.* **1998** *32*, 4023–4030.
7. van Pul, W. A. J., de Leeuw, F. A. A. M., van Jaarsveld, J. A., van der Gaag, M. A., Sliggers, C. J. *Chemosphere* **1998** *37*, 113–141.
8. *Evaluation of Persistence and Long-Range Transport of Chemicals in the Environment: Summary of a SETAC Pellston Workshop*; Klecka, G.; Mackay, D., Eds.; SETAC-Press: Pensacola FL, 1999.
9. Scheringer, M., Hungerbühler, K. In ECO-INFORMA *97*; Alef, K. et al., Eds.; Eco-Informa; Eco-Informa Press: Bayreuth, 1997, Volume 12, pp 173–178.
10. Scheringer, M. *Persistenz und Reichweite von Umweltchemikalien*; Wiley-VCH: Weinheim, 1999.
11. Wania, F., Mackay, D. *Sci. Total Environ.* **1995**, *160/161*, 211–232.
12. Wania, F., Mackay, D., Li, Y. F., Bidleman, T. F., Strand, A. *Environ. Toxicol. Chem.* **1999**, *18*, 1390–1399.
13. Fenner, K., Wegmann, F., Scheringer, M., Hungerbühler, K. *Proceedings of the 9th SETAC-Europe Annual Meeting*; Leipzig, 1999, p 14.
14. Czeplak, G., Junge, C. E. *Adv. Geophys.* **1974**, *18 B*, 57–72.
15. Webster, E., Mackay, D., Wania, F. *Environ. Toxicol. Chem.* **1998** *17*, 2148–2158.
16. Pankow, J. F. *Atmos. Environ.* **1988**, *22*, 1405–1409.
17. Koester C. J., Hites, R. A. *Environ. Sci. Technol.* **1992**, *24*, 502–507.
18. Harrad, S. J. In *Atmospheric Particles*; Harrison, R. H., van Grieken, R. E., Eds.; Wiley: Chichester, 1998, pp 233–251.

Chapter 6

Application of Negligible Depletion Solid-Phase Extraction (nd-SPE) for Estimating Bioavailability and Bioaccumulation of Individual Chemicals and Mixtures

Joop L. M. Hermens, Andreas P. Freidig, Enaut Urrestarazu Ramos, Wouter H. J. Vaes, Willem M. G. M. van Loon, Eric M. J. Verbruggen, and Henk J. M. Verhaar

Research Institute of Toxicology, Utrecht University, P.O. Box 80176, 3508 TD Utrecht, The Netherlands

Negligible depletion solid phase (micro) extraction (nd-SPE or nd-SPME) is a new analytical tool for measuring freely dissolved concentrations and partition coefficients. This approach can be applied in identifying bioaccumulative compounds in environmental samples. In this paper, a brief description of the procedure, as well as an overview of several published applications, are presented. The approach is illustrated with two examples: (i) the measurement of humic acid partition coefficients, and (ii) the sreening for bioaccumulative compounds in the environment.

Introduction: bioconcentration and bioavailability

Bioconcentration as well as bioavailability are two important features in risk assessment. The degree of bioconcentration determines the internal concentration in biota and bioavailability remains a crucial topic in exposure modeling because of its influence on accumulation, biodegradation and other processes in the environment (1-7). Also in toxicology, bioavailability is an important aspect of the potency of chemicals (8), because in principle it is only the freely dissolved molecules that can interact with a receptor site. The definition of bioavailability is not always straightforward. Some consider the bioavailable fraction as the one that is accessible to a target site or biological receptor, while others will consider the concentration that is freely dissolved in the aqueous phase as available, for example for uptake via fish gills.

Bioconcentration and bioavailability can be approached in several ways:

1. Both phenomena can be studied experimentally by measuring concentrations of a chemical in the organism or in some specific tissue or target site. The measurement of critical body residues has been a first attempt to relate effects to internal concentrations or body residues. The advantage of body residues is that they circumvent confounding factors associated with bioavailability and multiple uptake routes (9). Until now, the concept of body residues has successfully been applied to non-specific acting chemicals, because for these chemicals, body residues which induce narcosis are rather constant (10-12). For other types of compounds, the concept seems less applicable, not only because the concept of constant body burdens is not valid for specific modes of action, but also because whole body burdens do not always reflect the concentrations at the target site (13-15).
2. Physiologically Based Pharmacokinetic (PBPK) models (16), environmental fate models (17) or food chain models (18, 19) are powerful tools for predicting freely dissolved concentrations and internal concentrations in specific tissues or environmental compartments. PBPK models require measurements of body residues for validation.
3. As a third approach, the uptake in biota can be simulated by some surrogate material to mimic the characteristics of biological material. The objective of the work presented in this overview paper is focused on this third approach.

Södergren (20) as well as Huckins et al. (21) have applied semipermeable membrane devices to mimic and predict the bioconcentration and bioavailability of hydrophobic chemicals in natural waters. Other methods to measure freely dissolved concentrations or bioavailability include dialysis (22), fast elution on solid phase columns (23), solubility enhancement (24), measurement of headspace concentration (25) or fluorescence quenching (26). In our laboratory, "negligible depletion solid phase extraction" (nd-SPE) has been developed and applied to simulate accumulation (27-30), to estimate the freely dissolved concentrations and to measure partition coefficients (8, 31-33). A similar approach has been applied by Kopinke et al. and Poerschman et al. (34-36).

Although freely dissolved concentrations and bioavailability are not one and the same topic, precise information on freely dissolved concentration is a basic requirement to understand complex phenomena such as bioavailability.

Negligible depletion solid phase extractions (nd-SPE) for measuring freely dissolved concentrations and to simulate bioconcentration

Most solid phase extractions are performed under conditions of high extraction efficiency in order to achieve complete extraction and obtain maximum sensitivity of the analysis. Solid phase extraction techniques, however, can also be performed as a partition extraction and SPME is an example of such a solvent free partition extraction that has been developed by Pawliszyn and co-workers (37-39).

The most important characteristic of the nd-SPE or SPME is that the amount extracted is negligibly small in comparison with the total amount that is freely dissolved in the aqueous phase. As a consequence, existing equilibria between the freely dissolved and the matrix bound chemical remains virtually undisturbed during the extraction. A negligible depletion extraction can be performed by: (i) keeping the volume ratio of aqueous phase to hydrophobic phase very high (see Figure 1b) and/or (ii) to use short exposure times of the hydrophobic phase. In (ii) the difference attributable to hydrophobicity may not apply because equilibrium is not approached. Under these conditions, the freely dissolved concentration is not affected and the extraction will also be hydrophobicity dependent. In other words, at a similar aqueous concentration, the concentration on the solid phase will be higher for a more hydrophobic chemical than for a less hydrophobic one (see Figure 1a). Verhaar et al. (27) who used this type of extraction for estimating the accumulation of complex mixtures, introduced the term "biomimetic extraction", to indicate that the extraction mimics the biological uptake. Advantage of such an extraction technique is, as suggested by Kopinke et al. (34) and as has been developed by Vaes et al. (8, 31) and Poerschman et al. (35), that it also represents a procedure for measuring freely dissolved concentrations and partition or distribution behavior of chemicals in the environment. Because the aqueous concentration hardly decreases during the extraction, equilibria with a matrix (M), for example DOC, in the aqueous phase will not be disturbed and the concentration on the hydrophobic solid phase reflects the freely dissolved aqueous concentration.

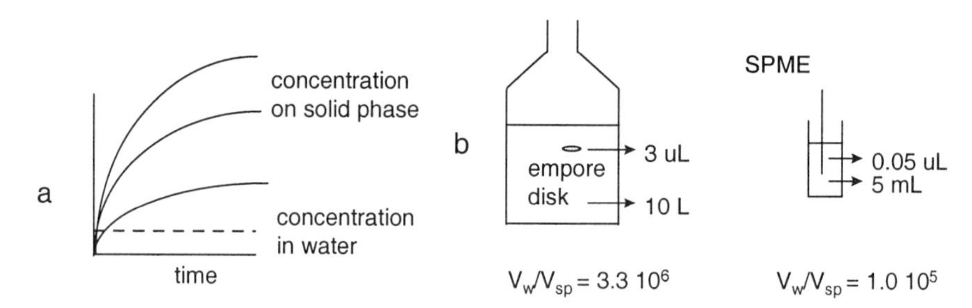

Figure 1. Principle of a biomimetic and negligible depletion extraction. (a) the aqueous concentration remains constant and the concentration on the hydrophobic solid phase (sp) depends on the hydrophobicity; (b) this is achieved by a large ratio of $V_{water} / V_{solid\ phase}$.

Extraction disks containing C18 particles (Empore disks) or solid-phase micro-extraction (SPME) devices coated with, for example, polyacrylate are suitable for performing a negligible depletion or biomimetic extraction because they contain a small amount of hydrophobic material.

The biomimetic approach for concentrating chemicals on a hydrophobic phase resembles extraction techniques using semi-permeable membranes or similar devices, which have been used by Södergren et al. (20) and Huckins and co-workers (21, 40). The advantages of the use of solid phases such as empore disks or SPME are that no clean up is needed and that the amount of hydrophobic phase can be kept low, which makes it possible to use relatively small water volumes and to perform the extraction at laboratory scales. Moreover, exposure times are shorter because the kinetics are relatively fast. In addition, the hydrophobic material in empore disk as well as SPME fibers is defined and partition coefficients and the kinetics of the partition process are relatively easy to control if stirring or agitating is kept reproducible (41, 42).

Details about the negligible depletion solid phase extractions are presented in several publications (8, 27-29, 31-33). Here, we only briefly describe the principles of the method (see Figure 2 in which the SPME fiber is chosen as solid phase). A chemical in a pure aqucous solution with a concentration $[X_{aq}]$ will partition to the SPME fiber reaching a concentration $[X_{SPME}]$. In the presence of a matrix M, the freely dissolved concentration in the aqueous phase may decrease to $[X'_{aq}]$. In a negligible depletion extraction, the equilibrium between X and M will remain undisturbed because $[X'_{aq}]$ remains almost constant and the concentration on the fiber $[X'_{SPME}]$ is directly related to the freely dissolved concentration $[X'_{aq}]$.

The concentration of the freely dissolved chemical X in a matrix can be determined from the concentration on the fiber, using a measured partition coefficient (K_{SPME}) of the chemical in pure water, or by a calibration curve established in pure water at a fixed exposure time or at equilibrium (see Figure 2).

Some examples of the negligible depletion solid phase extraction

This negligible depletion extraction technique has been applied in a number of studies, with the objectives:

1. to estimate the accumulation of complex mixtures in the environment (27-30).
2. to measure freely dissolved concentrations in *in vitro* test systems and partition coefficients with biomembranes (8, 31).
3. to measure freely dissolved and bioavailable concentrations in the presence of humic acids and to determine partition coefficients with humic acids (33-35).
4. to identify bioaccumulative chemicals in the environment (29, 30).

A few examples of these studies are presented in this overview paper.

Measurement of humic acid - water partition coefficients

Urrestarazu Ramos et al. (33) applied the negligible depletion SPME technique for estimating humic acid - water partition coefficients of pentachlorobenzene, hexachlorobenzene, PCB#77 and DDT. Experimental conditions were as follows: phosphate buffer pH 7.0 / matrix: Aldrich humic acid (HA) / final HA concentration: 0.1-100 mg/l / SPME fiber: PDMS 7μm / exposure time of the fiber was 1 min.

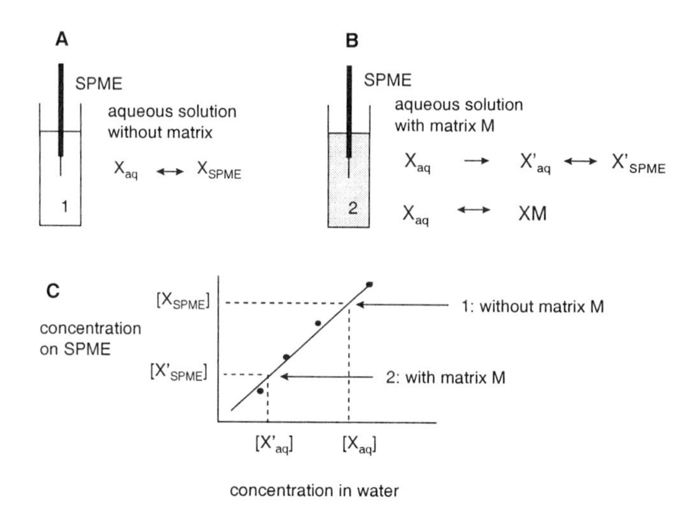

Figure 2. Principle of measurement of freely dissolved concentrations. A,B: equilibrium distribution in vials with and without matrix. C: calibration plot of concentration on SPME ($[X_{SPME}]$) versus aqueous concentration ([X]).

Concentrations on the fiber were measured on a GC with ECD detection (33). The decrease of the free concentration with increasing concentration of humic acid is given in Figure 3 and these data were fitted to the following expression:

$$f = 1 / (1 + K_{DOC} \cdot C_{DOC})$$

where f is the freely dissolved fraction (defined as the ratio between the free and total concentrations), K_{DOC} is the HA/water partition coefficient of the chemical, and C_{DOC} is the aqueous concentration of humic acids.

The K_{DOC} obtained are given in Table I and these values correspond well with literature data from studies that have used other experimental methods (see ref. (33)).

Table I. Humic acid-water partition coefficients determined in this study (from Urrestarazu Ramos et al. (33)).

Chemical	log K_{DOC}
Pentachlorobenzene	4.50
Hexachlorobenzene	4.97
PCB#77	5.97
DDT	5.57

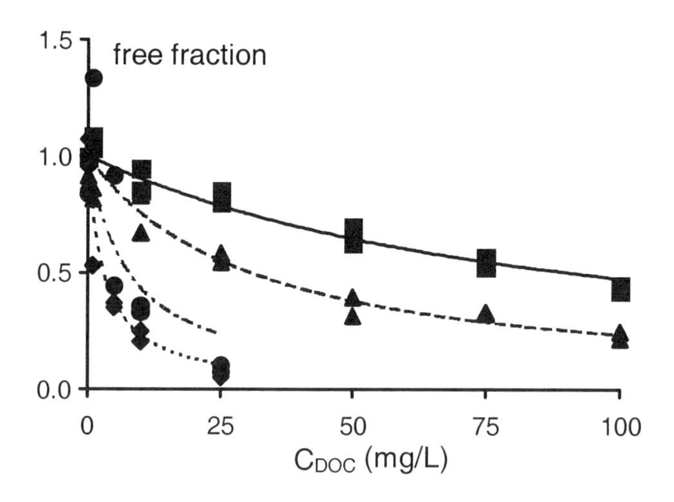

Figure 3. Free fraction determined using SPME *vs.* the humic acids concentration (C_{DOC}) in water. The lines represent the fitted curve to the experimental values. (■ —— pentachlorobenzene; ▲ – – – hexachlorobenzene; ◆ -------- PCB#77, ● – - – - – DDT). Figure taken from Urrestarazu Ramos et al. (33).

Simulation of bioconcentration

It is well known that bioconcentration of organic compounds increases with hydrophobicity. In a negigible depletion or biomimetic extraction the concentration on the hydrophobic phase is a reflection of the hydrophobicity of compounds (see Figure 1). The concentration of an individual chemical on the hydrophobic phase (C_h) will be equal to C_a.K, in which: C_a is the concentration in the aqueous phase and K is the partition coefficient of the chemical between the hydrophobic phase and water. This equation is similar to the accumulation in biota, where the concentration in biota (C_b) is equal to C_a.BCF, in which BCF is the bioconcentration factor. In this sense, the extraction mimics the bioconcentration in biota. This characteristic can be applied to screen for bioaccumulative compounds in environmental water samples. Compounds that are almost absent in standard exhaustive extracts, can become apparent in the biomimetic extracts (29, 30). An interesting illustration of the use of the biomimetic extraction to identify bioaccumulative compounds in environmental samples is given in Figure 4. Experimental details about this application of solid phase extraction for identifying hydrophobic chemicals are given by Verbruggen et al. (30). Briefly: a small piece of an empore disk is inserted in 10 L of an aqueous sample. Empore disk contain C18 and the amount of disk in this study corresponds with 11 μL C18. The aqueous solution is stirred for 14 days and most chemicals have reached steady state within this time frame, except for the more hydrophobic compounds of log K_{ow} above 6, for which about 50 % of the steady state is reached. Because the volume of the hydropobic phase is small, the aqueous solution is hardly depleted by the absorption to the empore disk. The observed two polycyclic musk compounds (HHCB and AHTN) are not or hardly visible in exhaustive extracts but become major components in chromatograms obtained with the biomimetic extraction procedure. The biomimetic extraction is able to focus on compounds with a high bioconcentration potential and is therefore a powerful tool to search and identify these potential hazardous compounds in the environment. As such, the approach may be useful in monitoring and screening for persistent and bioaccumulative, toxic chemicals (PBT's). These compounds often occur at low concentrations in the aqueous phase but may have a major contribution to the internal body residues.

A limitation of the procedure is that it does not take into account biotransformation or additional uptake routes and these phenomena may result in lower (43, 44) or higher internal concentrations (19, 45) than expected based on equilibrium partitioning.

Estimation of Total Body Residues

The biomimetic procedure can also be applied in a more quantitative manner. This approach was used by Verhaar et al. (27) and van Loon et al. (28, 29) to estimate total body residues (TBR) by measuring total molar concentrations in the biomimetic extracts. Total molar concentrations were determined using vapor pressure osmometry or GC-MS. TBR's were estimated from the total molar concentrations on the empore or GC-MS. TBR's were estimated from the total molar concentrations on the empore disk and a correlation between disk-water partition coefficients and bioconcentration factors. TBR's are useful to estimate the accumulation of mixtures, but can also be used to estimate the contribution of compounds with narcosis type of activity. Narcosis occurs at a constant Critical Body Residue (CBR) of 50-100 mMol/kg-lipid and this CBR is valid for individual compounds as well as mixtures (11, 27), because the mechanism of narcosis is completely concentration additive (46, 47). Many chemicals in the environment act by narcosis and also specific acting compounds may contribute to narcosis if they are present at concentration levels below effect concentrations for their specific mode of action (48). The estimated TBR includes those chemicals that are usually not detected because they are below their detection limit, or because they are not analysed as priority pollutants.

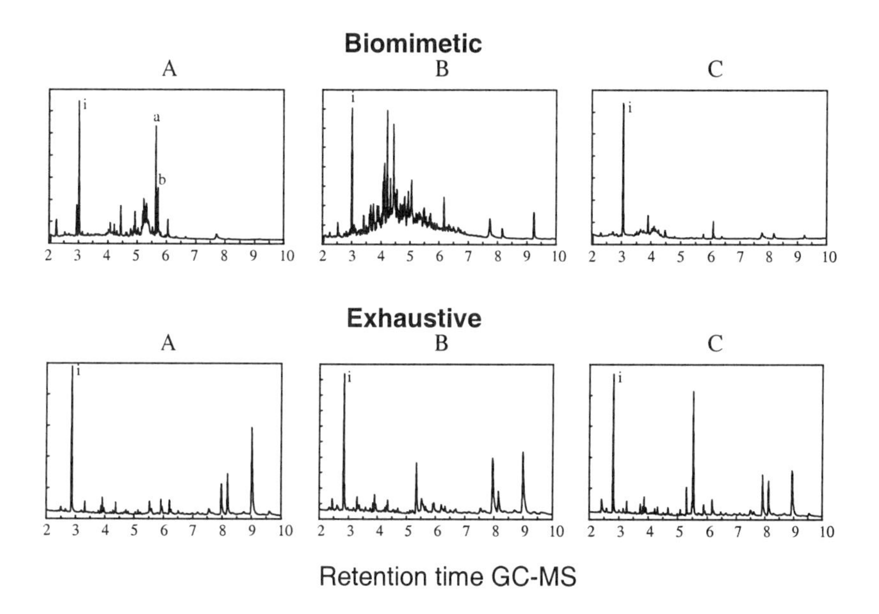

Figure 4. Biomimetic extract vs. exhaustive extracts. A: effluent of sewage treatment plant 3, B and C: industrial effluents, i: internal standard (2,4,5-trichlorotoluene), a: HHCB and b: AHTN. Note the absence of the two peaks of HHCB and AHTN from the exhaustive extract of the sewage treatment plant sample (reproduced form Verbruggen et al. (30).
HHCB: 1,3,4,6,7,8-hexahydro-4,6,6,7,8,8-hexamethylcyclopenta-gamma-2-benzopyran
AHTN: 6-acetyl-1,1,2,4,4,7-hexamethyltetraline
(Reproduced with permission from reference 30. Copyright 1999, American Chemical Society).

Conclusion

Negligible depletion solid phase extractions using empore disks or SPME nd-SPME) minimizes perturbation of equilibria in a matrix. Because of this, the extraction simulates bioconcentration, and can be applied to (measure the freely-dissolved (bioavailable) concentrations. Despite these advantages, one should be alert for pitfalls such as interfering binding of matrix to the solid phase or disturbing effects in the diffusion layer (49). Interfering binding of humic acids to empore disk has not been observed (28). In addition, both Vaes (31) and Urrestarazu Ramos (33) compared their results with dialysis experiments and did not observe binding of matrix to the SPME fiber material.

We believe that this technique can play an important role in identifying persistent bioaccumulative and toxic compounds in the environment and determining how they partition to natural matrices such as DOC as well as biota.

References

1.McCarthy, J.F.; Jimenez, B.D. *Environ. Toxicol. Chem.* **1985**, *4*, 511-521.

2.Day, K.E. *Environ. Toxicol. Chem.* **1991**, *10*, 91-101.

3.Kullberg, A.; Bishop, K.H.; Hargeby, A.; Jansson, M.; Petersen, R.C.J. *AMBIO* **1993**, *22*, 331-337.

4.Larsson, P.; Collvin, L.; Okla, L.; Meyer, G. *Environ. Sci. Technol.* **1992**, *26*, 346-352.

5.Alexandar, M. *Environ. Sci. Technol.* **1995**, *29*, 2713-217.

6.Kukkonen, J. *Wat. Res.* **1992**, *26*, 1523-1532.

7.Naes, K.; Axelman, J.; Naf, C.; Broman-D. *Environ. Sci. Technol.* **1998**, *32*, 1786-1792.

8.Vaes, W.H.J.; Urrestarazu Ramos, E.; Hamwijk, C.; van Holsteijn, I.; Blaauboer, B.J.; Seinen, W.; Verhaar, H.J.M.; Hermens, J.L.M. *Chem. Res. Toxicol.* **1997**, *10*, 1067-1072.

9.Shephard, B.K. In *National Sediment Bioaccumulation Conference Proceedings*; U.S. Environmental Protection Agency, Office of Water: Washington, D.C., 1998, pp 2-31 to 2-52.

10.McCarty, L.S.; Mackay, D.; Smith, A.D.; Ozburn, G.W.; Dixon., D.G. *Sci. Total Environ.* **1991**, *109/110*, 515-525.

11.McCarty, L.S.; Mackay, D. *Environ. Sci. Technol.* **1993**, *27*, 1719-1728.

12.Van Wezel, A.P.; Opperhuizen, A. *Crit. Rev. Toxicol.* **1995**, *25*, 255-279.

13.Hermens, J.L.M. In *Handbook of Environmental Chemistry*; Hutzinger, O., Ed.; Springer Verlag: Berlin, 1989; Vol. 2E, pp 111-162.

14.Legierse, K.C.H.M.; Verhaar, H.J.M.; Vaes, W.H.J.; De Bruijn, J.H.M.; Hermens, J.L.M. *Environ. Sci. Technol.* **1999**, *33*, 917-925.

15.Freidig, A.P.; Verhaar, H.J.M.; Hermens, J.L., M. *Environ. Sci. Technol.* **1999**, *accepted*,

16.Clewell, H.J., III *Toxicology Letters* **1995**, *79*, 207-217.
17.Mackay, D.; Paterson, S. *Environ. Sci. Technol.* **1991**, *25*, 427-436.
18.Broman, D.; Naf, C.; Rolff, C.; Zebuhr, Y.; Fry, B.; Hobbie, J. *Environ. Toxicol. Chem.* **1992**, *11*, 331-345.
19.Gobas, F.A.P.C.; Zhang, X.; Wells., R. *Environ. Sci. Technol.* **1993**, *27*, 2855-2863.
20.Södergren, A. *Environ. Sci. Technol.* **1987**, *21*, 855-859.
21.Huckins, J.N.; Tubergen, M.W.; Manuweera, G.K. *Chemosphere* **1990**, *20*, 533-552.
22.Escher, B.; Schwarzenbach, R.P. *Environ. Sci. Technol.* **1996**, *30*, 26-270.
23.Landrum, P.F.; Nihart, S.R.; Eadis, B.J.; Gardner, W.S. *Environ. Sci. Technol.* **1984**, *18*, 187-192.
24.Chiou; C.T.; Brinton, T.I.; Malcom, R.L.; Kile, D.E. *Environ. Sci. Technol.* **1986**, *20*, 502-508.
25.Resendes, J.; Shiu, W.Y.; Mackay, D. *Environ. Sci. Technol.* **1992**, *26*, 2381-2387.
26.Backhus, D.A.; Gschwend, P.M. *Environ. Sci. Technol.* **1990**, *24*, 1214-1223.
27.Verhaar, H.J.M.; Busser, F.; Hermens, J.L.M. *Environ. Sci. Technol.* **1995**, *29*, 726-734.
28.Van Loon, W.M.G.M.; Wijnker, F.G.; Verwoerd, M.E.; Hermens, J.L.M. *Anal. Chem.* **1996**, *68*, 2916-2926.
29.Van Loon, W.M.G.M.; Verwoerd, M.E.; Wijnker, F.G.; van Leeuwen, C.J.; van Duyn, P.; van de Guchte, C.; Hermens, J.L.M. *Environ. Toxicol. Chem.* **1997**, *16*, 1358-1365.
30.Verbruggen, E.M.J.; van Loon, W.M.G.M.; Tonkes, M.; van Duijn, P.; Seinen, W.; Hermens, J.L.M. *Environ. Sci. Technol.* **1999**, *33*, 801-806.
31.Vaes, W.H.J.; Urrestarazu Ramos, E.; Verhaar, H.J.M.; Seinen, W.; Hermens, J.L.M. *Anal. Chem.* **1996**, *68*, 4463-4467.
32.Freidig, A.P.; Artola Garicano, E.; Busser, F.J.M.; Hermens, J.L.M. *Environ. Toxicol. Chem.* **1998**, *17*, 998-1004.
33.Urrestarazu-Ramos, E.; Meijer, S.N.; Vaes, W.H.J.; Verhaar, H.J.M.; Hermens, J.L.M. *Environ. Sci. Technol.* **1998**, *32*, 3430-3435.
34.Kopinke, F.-D.; Pörschmann, J.; Remmler, M. *Naturwissenschaften* **1995**, *82*, 28-30.
35.Poerschmann, J.; Zhangh, Z.; Kopinke, F.-D.; Pawliszyn, J. *Anal. Chem* **1997**, *69*, 579-600.
36.Poerschmann, J.; Kopinke, F.-D.; Pawliszyn, J. *Environ. Sci. Technol.* **1997**, *31*, 3629-3636.
37.Arthur, C.L.; Pawliszyn, J. *Anal. Chem.* **1990**, *62*, 2145-2148.
38.Boyd-Boland, A.A.; Chai, M.; Luo, Y.Z.; Zhang, Z.; Yang, M.J.; Pawliszyn, J.B.; Gorecki, T. *Eviron. Sci. Technol.* **1994**, *28 (13)*, 569 A-574 A.
39.Pawliszyn, J. *Trends Anal. Chem.* **1995**, *14*, 113-122.
40.Ellis, G.S.; Huckins, J.N.; Rostad, C.E.; Smitt, C.J.; Petty, D.D.; MacCarthy, P. *Environ. Toxicol. Chem.* **1995**, *14*, 875-1884.
41.Louch, D.; Motlagh, S.; Pawliszyn, J. *Anal. Chem.* **1992**, *64*, 1187-1199.

42. Vaes, W.H.J.; Hamwijk, C.; Urrestarazu Ramos, E.; Verhaar, H.J.M.; Hermens, J.L.M. *Anal. Chem.* **1996**, *68*, 4458-4462.
43. De Wolf, W.; de Bruijn, J.H.M.; Seinen, W.; Hermens, J.L.M. *Environ. Sci. Technol.* **1992**, *26*, 1197-1201.
44. Sijm, D.T.H.M.; Wever, H.; Opperhuizen., A. *Chemosphere* **1989**, *19*, 475-480.
45. Thomann, R.V.; Connolly, J.P.; Parkerton., T.F. *Environ. Toxicol. Chem.* **1992**, *11*, 615-629.
46. Könemann, H. *Toxicology* **1981**, *19*, 229-238.
47. Hermens, J.L.M.; Könemann, J.H.; Leeuwangh, P.; Musch, A. *Environ. Toxicol. Chem.* **1985**, *4*, 273-279.
48. Hermens, J.L.M.; Leeuwangh, P. *Ecotoxicol. Environ. Safety* **1982**, *6*, 302-310.
49. Jeannot, M.A.; Cantwell, F.F. *Anal. Chem.* **1997**, *69*, 2935-2940.

Chapter 7

Modeling Historical Emissions and Environmental Fate of PCBs in the United Kingdom

Andrew J. Sweetman and Kevin C. Jones

Department of Environmental Science, Institute of Environmental and Natural Sciences, Lancaster University, Lancaster, LA1 4YQ, United Kingdom (email: a.sweetman@lancaster.ac.uk; phone: +44 1524 593300; fax: +44 1524 593985)

A primary emission driven fugacity model of the historical fate, behavior and distribution of PCBs in the UK environment is described. The model attempts to re-create the temporal release trend of PCBs over the last 40 years and to replicate the observed historical trends in soils and sediments. The releases of PCBs to the UK atmosphere are modeled using emission curves calculated from production and use data and emission factors. Life-spans of end uses, such as capacitors and transformers, are included, resulting in the removal or reduction of potential sources. As a result of release to the atmosphere from primary sources and the advection of contaminated air into the atmosphere, the UK environment has become contaminated, with the soil accounting for most of the burden. The model predictions agree reasonably well with measured data from archived soils and fresh water sediment cores, both in terms of temporal trends and predicted concentrations. The use of soil-air fugacity ratios suggests that soil changed from being a net sink through the 1950's until the mid 1980's into a net source during the 1990's. Current measured and predicted ratios suggest that near equilibrium conditions exist. A sensitivity analysis of the model is also included and discussed along with recommendations as to possible future improvements to models of this type.

Introduction

The use of environmental fate and behavior models has become increasingly widespread in recent years as they have proven to be useful in the absence of measured environmental data. At their simplest level they can be used to provide fate profiles based on few physicochemical data and a 'unit world' model construction. This approach can be used to provide environmental fate and behavior information for new and existing chemicals, such as the potential for long range atmospheric transport and to identify which compartment(s) are likely to be important repositories. However, there is an increasing demand to adapt these models to simulate site specific scenarios, to provide answers to more specific questions. This paper describes a fugacity based Mackay type Level IV 'dynamic' model used to simulate the historical release of PCBs in the UK. Hence the purpose of this model is to attempt to simulate the long term environmental fate of PCBs as a result of their production and release. Further, the model provides an insight into inter-compartmental exchange processes and their relative importance to the overall fate of PCBs in the UK.

Compartment and model construction

The model was constructed using the established Mackay Level III design (1) with four bulk compartments, namely air, soil, water and sediment, with each compartment sub-divided into constituent parts, for example soil pore water and interstitial air (see Figure 1). Compartment volumes and characteristics for the UK were taken from the Countryside Information System, a land cover database using satellite and national field surveys commissioned by the UK Department of the Environment, Transport and the Regions (2) and are included in Table 1. In the absence of measured data or alternatives, default values for compartment dimensions, for example atmospheric height, were taken from Mackay (1). Soil organic carbon data were taken from a study of 46 UK soils carried out at our laboratory (3).

Models such as this are constructed from a large range of input parameters such as contaminant physicochemical properties, emission estimates and transport parameters. Rather than choose single ‘preferred’ we have examined the literature and selected a more realistic range of values. By using a range rather than a single value, a more realistic output from the model was expected. Physicochemical property data, for example octanol-water partition coefficients (Kow), vapor pressure and aqueous solubility were taken from Mackay *et al.* (4), whilst half life data for the atmospheric compartment were taken from Anderson and Hites (5) and adjusted for temperature as appropriate using the supplied algorithms. Half life estimates for soil, water and sediment compartments were taken from Mackay *et al.* (4) with mean values selected. However, we recognized that the soil compartment was likely to be the major repository for PCBs, so for the purposes of a sensitivity analysis this parameter was assigned a range of values. The physicochemical property data reported in the literature were log transformed (except for Kow) to reduce skewness and 95% percentiles taken. The chosen temperature range was typical for the UK but kept above 5°C in order to avoid problems associated with air-ice partitioning (maximum 30°C).

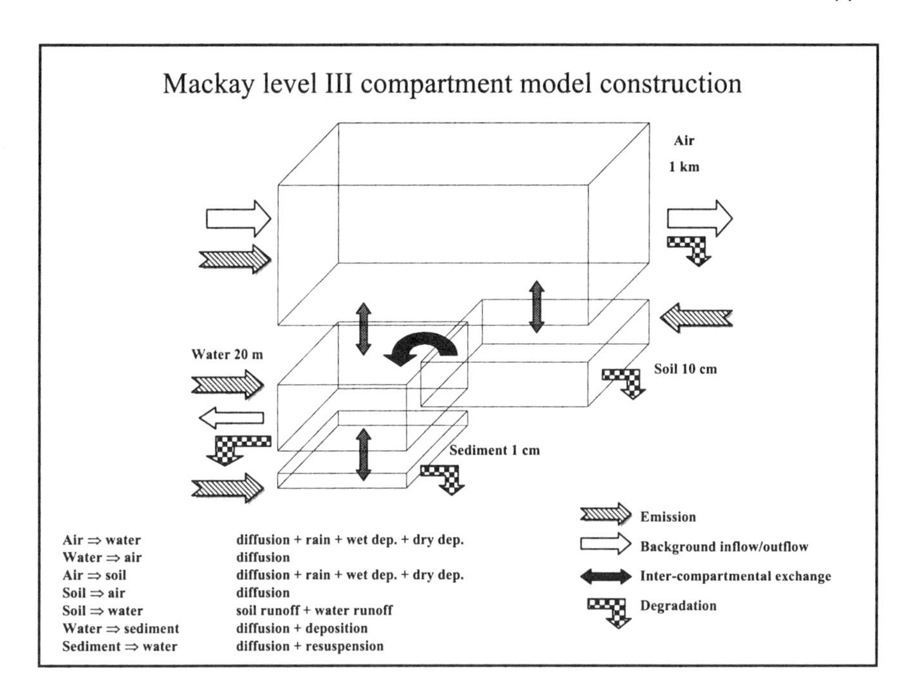

Figure 1 - Model construction

Precipitation rates were taken from UK Environment Agency statistics, and scavenging ratios and deposition velocities were taken from Swackhamer *et al.* (6). A normally distributed random number function was used to construct a matrix of a 1000 input values for each parameter using the prescribed ranges. As a result, this matrix was then fed into the model providing a range of simulation results. The model's sensitivity to these input properties was also examined individually, to determine which have the greatest influence on the predicted output and hence are likely to control contaminant fate in the environment. The model was run varying one parameter at a time, whilst maintaining the other parameters at mean values.

The model is driven by primary emission into the UK atmosphere calculated from production and use data and the application of emission factors for each designated end use. PCB production data, based on Monsanto sales for the UK, were taken from OECD statistics (7). Apart from production destined for transformers and capacitors, a large proportion of production that remained in the UK was designated under 'other' uses, which would have included end uses such as plasticizers, hydraulic fluids, lubricants, carbonless copy paper production, heat transfer fluids and petroleum additives. In the absence of available data for the UK, the proportions of each of these PCB containing products was derived using USEPA data (8). It is realized that this represents an estimation, but the proportion of production being used in each end use is likely to be similar in most of the OECD countries.

Table 1 UK model parameter values used in calculations

	Compartment properties		Source
Air	Bulk volume	2.34E+14 m^3	Ref. (2)
	Atmospheric height	1000 m	Ref. (1)
	Volume fraction aerosol	1.3E-11 (≡ 20μg.m^{-3})	Ref. (2)
Soil	Bulk volume	2.25E+10 m^3	Ref. (2)
	Depth	0.1 m	Ref. (1)
	Fraction organic carbon	0.1 (w/w)	Ref. (3)
	Volume fractions		
	air	0.2	Ref. (4)
	water	0.3	Ref. (4)
Water	Bulk volume	1.88E+11 m^3	Ref. (2)
	Depth	20 m	Ref. (4)
	Volume fractions		
	suspended sediment	5E-6	Ref. (4)
	biota	1E-6	Ref. (4)
Sediment	Bulk volume	9.40E+07 m^3	Ref. (2)
	Depth	0.01 m	Ref. (2)
	Fraction organic carbon	0.1	Ref. (2)
	Volume fraction water	0.63	Ref. (4)

It was recognized that each of these end use categories has a finite life-span as the equipment becomes obsolete and hence is recycled or disposed to landfill. In order to take this into account a proportion of each end use category was removed from the 'pool' after expiration of its life expectancy. For example, for transformers this period is 30 years, for large capacitors 25 years and for small capacitors 10 years. Without data on the amounts of PCB containing equipment or materials disposed of over the last 30 years available in the literature, we have assumed that 75% of each PCB 'pool' was disposed after expiration of its life-span. For open source categories we have assumed a life-span of 15 years.

In order to calculate emission estimates for the UK, production figures for each category were adjusted for disposal and then multiplied by an emission factor appropriate to the potential for release. Emission factors for 'closed' systems such as transformers and capacitors were calculated from surveys of appliance failure and leakage in combination with evaporation studies (9). Semi-open uses such as plasticizers and heat transfer fluids were assigned an emission factor based on evaporation tests using plasticizer products (10). Although disposal of PCB equipment has been included in the calculations, emissions from disposal processes, for example, scrap metal recycling, incineration and landfill, have not been included. However, these process may be important for certain source categories and hence will require further investigation. For example, high temperature incineration has been used to dispose of PCBs in the UK although there is some uncertainty as to the quantities involved.

The emission values for 'total PCB' were converted to congener specific data by assuming an Aroclor production ratio of 10:5:2 for 1242, 1254 and 1260 respectively (11). These calculations suggest that as a result of their manufacture and release in the UK, direct emission into the UK atmosphere began in the mid to late 1950s, increased rapidly during the 1960s and peaked in the early 1970s. After restrictions were introduced production decreased rapidly until a ban was imposed in 1977.

However, this is not the complete picture as historical concentration data from sediment cores (12) and archived soil data (13) has shown that PCBs were present in the UK environment prior to their production and release within the UK. Also, the historical profiles suggest that peak emission occurred in the late 1960's rather than the early 1970's as the emission calculations suggest. In order to take these factors into account the emission curves were adjusted to include the advection of contaminated air into the UK prior to 1955. From these calculations peak emission of PCBs into the UK atmosphere resulted in a concentration of 140, 130 and 90 $pg.m^{-3}$ for congeners 52, 101 and 153, respectively. Measured ambient levels in the 1990s for congeners 52, 101 and 153 have been reported at 18, 7 and 2 $pg.m^{-3}$, respectively (14).

An important question in models of this type is how rapidly the atmospheric box is ventilated and what will be the contaminant concentration in the advected inflow air that replaces it. A residence time of 13 hours has been calculated for the UK using compartment volume and mean wind speed taken over a 30 year period. This results in a rapid exchange of air over the UK surface. In order to calculate the advected inflow air burden a combination of wind direction distribution frequencies for 6 sectors and typical contemporary PCB concentrations for each sector has been used. These calculations show that as a result of primary and secondary emissions the UK exports approximately 25% more PCBs than it receives from its neighboring regions. These data are, of course, for contemporary air, but the assumption is made that this ratio has remained constant over time.

Model output

The output from the model provides a time integrated fugacity profile for each of the four compartments and values for each of the inter-compartmental fluxes and advective flows. The fugacity values were then converted into concentrations. Typical concentration profiles for PCB-153 are shown in Figure 2. The air concentrations, not surprisingly, closely follow the primary emission curve with peak emission and peak concentration occurring during the late 1960s. This is a reflection of the high atmospheric advective flow term and hence short residence time. A predicted peak in water compartment concentrations occurred 2 to 3 years later around 1970. A peak in the sediment concentration occurs a further 2 to 3 years later. The soil compartment as a result of its large capacity for PCBs is the slowest compartment to respond, peaking during the mid 1970s. As a result of a cessation of production and an increase in the amount of PCB containing equipment and materials being disposed, the atmospheric burden rapidly decreases during the late 1970s and 1980s.

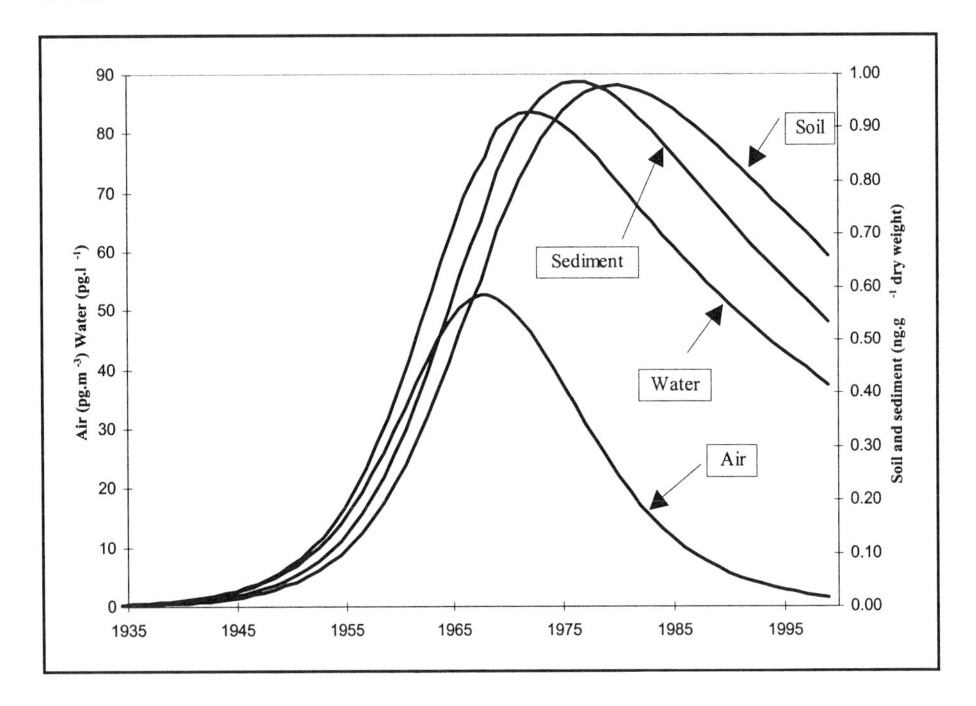

Figure 2 Example model output for PCB-153

The water compartment also shows a rapid decrease of its burden, but the soil and sediment are slower to respond as a result of their higher fugacity capacities and degradation half lives.

Direct comparison of predicted concentration data for each compartment is extremely difficult as measurement data is subject to spatial and temporal variability. However, the predicted fugacity values have been converted to concentration units to allow easier comparison of data with more familiar units. Predicted concentration data for congeners 52, 101 and 153 during peak and contemporary emission scenarios along with measured data are contained in Table 2. Although there are no data for direct historical measurements of air concentrations (e.g. in the 1950-80s) with which to compare model predictions, the model values for contemporary air concentrations

Table 2 - Model predictions (range represents 95 percentiles after 1000 simulations)

		PCB-52		PCB-101		PCB-153	
		predicted	measured	predicted	measured	predicted	measured
Air	Peak	140-150	na	110-150	na	25-80	na
	1993	5	18	5-8	6.5	2-5	1.7
Soil	Peak	0.08-0.22	5.2	0.3-2.4	1.55	0.6-1.4	0.77
	1993	0.02-0.08	0.06-0.2	0.25-1.8	0.04-0.26	0.4-0.9	0.13-0.55
Sediment	Peak	0.1-0.28	na	1-3.6	2.11	0.6-1.4	na
	1993	0.01-0.07	na	0.2-2	na	0.3-0.8	na
Water	Peak	9-12	na	35-75	na	50-110	na
	1993	1-2	na	5-30	na	20-60	na

Note; na not available, measured air data from Ref. 18 and soil data from Ref. 8, units air pgm^{-3}, water pgl^{-1}, soil and sediment ngg^{-1} dry weight for PCB-52 and PCB-101 were close, i.e. within a factor of 3 of the measured values (14). The predicted soil concentrations for both peak and contemporary periods were close to the measured values, although the peak concentration for PCB-52 was underestimated by a factor of 50. The poorly predicted soil peaks for PCB-52 and PCB-101 probably resulted from inaccuracies in the emission calculations. However, there were also some problems reported with the analysis of the archived soils which led to sample contamination by laboratory air. The measured sediment data taken from a fresh water lake in Northern England (12) only provided concentration data for PCB-101. However, for this congener the model predictions for peak emission were close to the measured values.

Sensitivity analysis

As the soil compartment is the largest repository for PCBs a sensitivity analysis was carried out using predicted soil concentrations as the indicator of variability. Parameters incorporated into the sensitivity analysis included all of the physicochemical properties, temperature, atmospheric deposition parameters and soil organic carbon fraction and degradation half life. The results for PCB-101 show that degradation half life had the greatest impact on model output. Varying the half life from 3 to 15 years resulted in a maximum residual standard deviation (RSD) of 70% of the mean peak concentration and a shift of the year of peak concentration from 1968-70 to 1980-82. Precipitation rate and scavenging ratio also had a significant affect on model output with maximum RSD values of 66% and 57%, respectively. This is not surprising as these properties control deposition to the soil surface and have been assigned a wide range of values. Varying the deposition velocity, Henry's law constant and environmental temperature resulted in a 10% RSD, whilst variation in Kow and soil organic carbon content resulted in 1-2% RSDs.

Calculated emission profiles for the PCB congeners will obviously have a great impact on model predictions. Factors such as the amount of PCBs associated with the various end uses, along with emission factors chosen for this study will strongly affect output. In order to investigate this further, a separate sensitivity analysis was carried out to examine the influence of the assumptions used in calculating emission on the model predictions. The emission factors used for each end use were varied by a factor of ±2 and the proportion of each end use category removed by recycling/disposal was varied between 25 and 75%. As a result of varying the emission input, the predicted air concentrations varied by a factor of 1.8 and the year of peak concentration ranged from 1967 to 1975. Not surprisingly the predicted soil concentrations also varied by a factor of 1.8 with the peak year of concentration ranging from 1975 to 1985.

Historical profiles

As a test of the predictive abilities of the model, historical trends provided by the model have been compared to those available for UK soils and sediments, Figure 3 contains data for PCB-101 as an example. The two solid lines represent the 95 percentiles from replicate model runs. A comparison of the general shape of the predicted historical trends and the measured data suggests that the model is providing a reasonable approximation. The peak sediment concentrations occur around 1970-75, which coincides with model predictions. This suggests that the inter-compartmental transfer terms in the model were providing a good approximation. Measured historical soil patterns are less clear but would appear to peak around 1970 (13) whereas the model suggests that the peak would occur around the mid-1970's.

As soil is the most important repository for PCBs, accounting for 98% of the total emitted, the flux between the soil and air has been (and is) one of the most important fate processes. The historical relationship between soil and air is explored in Figure 4, which shows how the soil turned from being a repository to a secondary source as PCB production increased and then declined. The model predicts that during the 1950s and 1960s deposition to the UK soil surface increased as a result of emission peaking at approximately 420 kg per annum, equivalent to 2μgPCB-101m^{-2}. The soil-air fugacity ratio during this phase is constant at around 0.06. Emissions decreased after 1970, resulting in a rapid decrease in deposition and zero net transfer by 1976-77. After this period soil becomes a source of PCB-101 which is reflected in the fugacity ratio which increases steadily. Contemporary soil-air fugacity ratios have been calculated for the UK by Cousins and Jones (15) based on measured soil and air data. Assuming a mean temperature of 13°C and soil organic carbon content of 8% fugacity ratios of 0.58, 0.77 and 2.5 were calculated for PCBs 52, 101 and 153, respectively, suggesting that near equilibrium conditions exist. The fugacity ratios provided by the model for contemporary emission scenarios were 2, 0.35 and 1.4 for the same congeners also suggesting near equilibrium conditions.

Future trends

The model predictions can be extrapolated, with some caution, to provide estimates of future trends for PCBs in the UK environment. Using PCB-101 as an example, the model suggests that soil, sediment and water concentrations should continue decreasing with half lives of approximately 6 to 7 years, while air concentrations should continue decreasing with a half life of approximately 4 years. Hillery *et al.* (16) reviewed half life data for PCBs in a range of matrices at various locations, and the findings broadly agree with the model predictions. However, there is some discrepancy about the decrease in atmospheric concentrations. Levels of PCBs at some sites e.g. Lake Michigan and Lake Erie have shown a decrease with a half life of about 6 years, but at other sites e.g. Lake Superior and Bermuda this decrease was not observed. This suggests that contemporary sources and soil out-gassing may still be maintaining levels in the atmosphere.

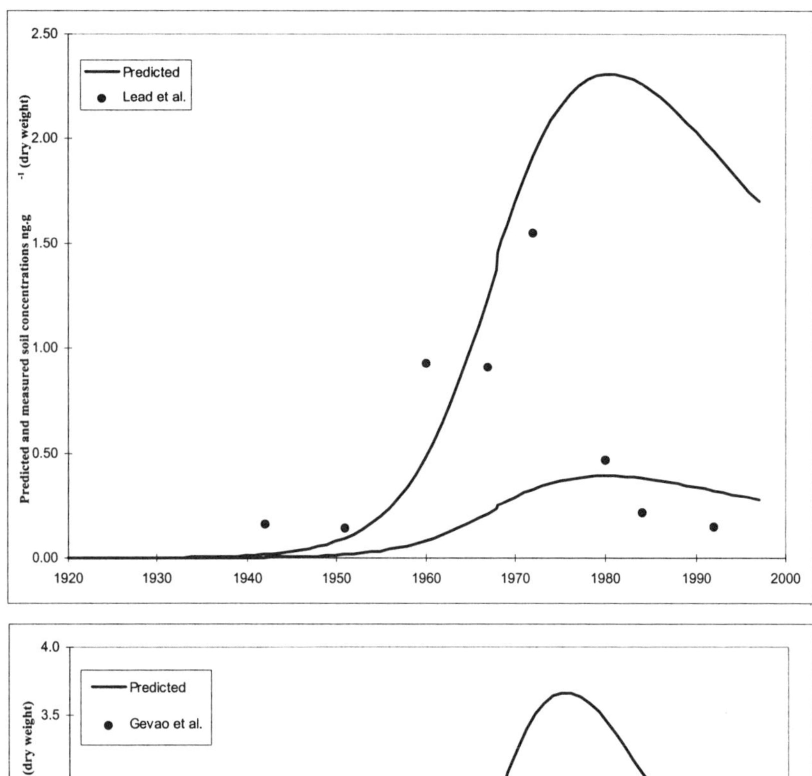

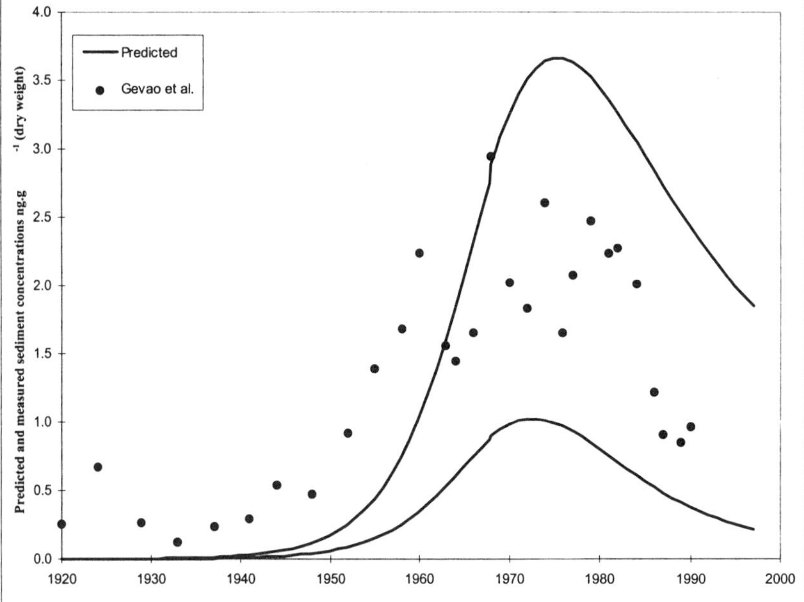

Figure 3 *Comparison of model output and measured historical concentration profiles for UK soils and sediments for PCB-101*

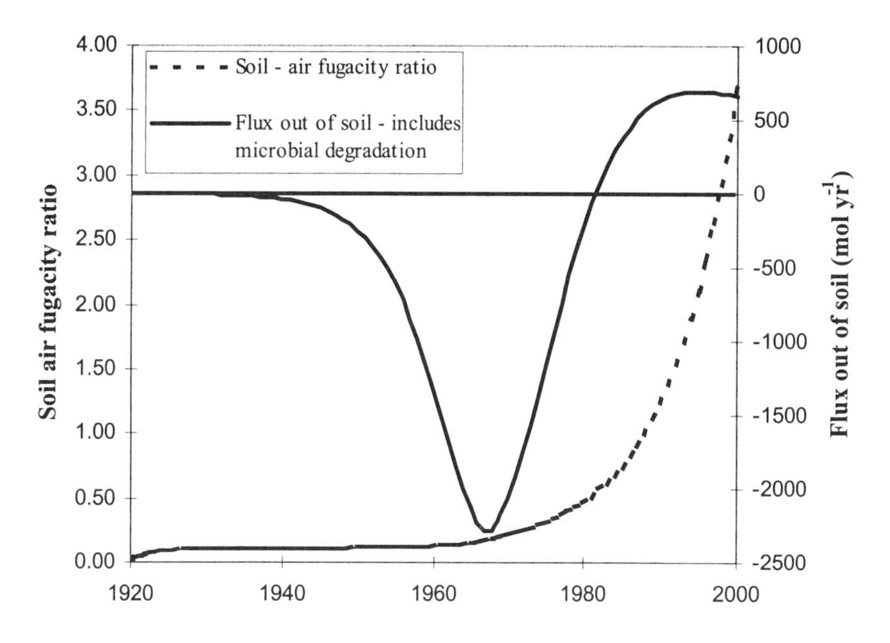

Figure 4 - Soil to air flux and soil-air fugacity ratio for PCB-101

To investigate whether there is any evidence for a decrease of PCBs in the UK environment, as the model suggests, we examined PCB air concentration data measured at a meteorological site in north west England since 1992. This data set, comprising of over 200 data points, suggests that PCB levels are decreasing with average congener specific half lives ranging from approximately 2 to 6 years. With the exception of congener 52, which shows the steepest decline, the slopes of other ICES congeners included in this study (i.e. 28,101,118,153,138) were not found to be significantly different from each other. Figure 5 shows a plot of fugacity against time for each congener after correction for differences in ambient temperature at time of collection. As mentioned previously, both the model and contemporary soil-air fugacity ratio calculations suggest that deposition of PCBs from the UK atmosphere is currently approximately balanced by volatilisation from soil. However, uncertainties remain with the importance of degradative processes such as atmospheric OH^- radical reaction. It is important to point out that whilst soil in polluted/urban areas could still be providing a source, pristine/remote soils could also be providing a repository. Remaining primary emissions (or those resulting from recycling) appear to be still supporting the UK atmospheric burden, but in the long term, with the continual reduction of these sources, it will be fate processes within soil that will most likely control the long term fate of PCBs, and their supply to the atmosphere, and 'driving' their regional and global cycling.

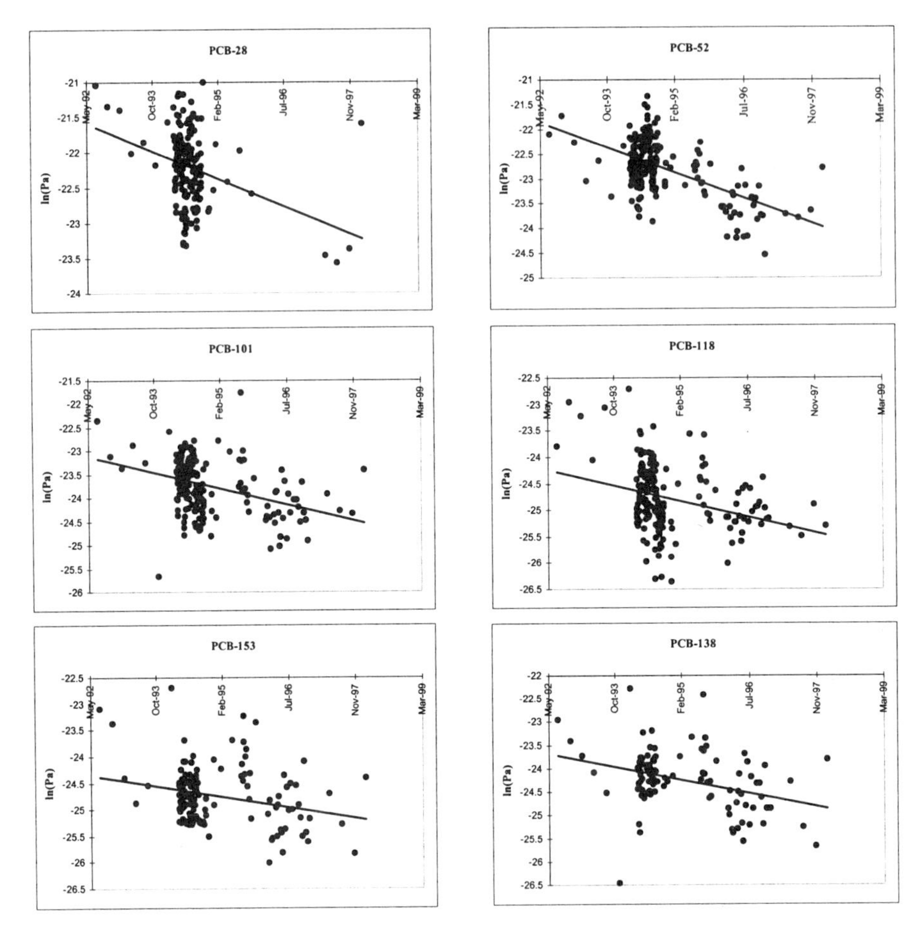

Figure 5 Temporal variation of atmospheric PCB fugacities in UK air (1992-1998)

Model improvements

The model presented in this paper has provided an encouraging reproduction of the observed historical trends for the UK situation. This suggests that the existing structure of the model provides a reasonable estimation of the real environment and that increasing the level of detail may not be warranted. Models of this type can be calibrated using measurement data for a range of PCB congeners and, if successful, can then be applied to others for which data is sparse.

However, it is important to stress that there are many areas which could be improved.

- The most important part of a model of this type is the emission calculations. Many assumptions have been made in order to calculate emission curves. Subtle changes in these calculations, such as altering the emission factors, can result in significant changes in the model predictions. The rôle of emission from disposal processes, such as incineration and outgassing from landfill operations, also needs to be examined.

- Output from the sensitivity analysis has shown that some input parameters have a strong influence on model predictions. In the case of PCBs, factors affecting atmospheric deposition and soil degradation have the greatest influence. For example, there are very few measurements of soil biodegradation half life and those there are do not adequately differentiate biodegradation from biodegradation and volatilisation. By making improvements to the measurement of these factors the overall predictive capabilities of the model can be improved.

- The output from the model compared favorably with the measured historical data for soils and sediments and provided a reasonable comparison with contemporary levels measured in the UK environment. However, more historical data e.g. from sediments would provide further assessment of the model's usefulness.

- Finally, there is a great deal of interest in predicting the rate at which contaminants such as PCBs are decreasing in the environment and their 'clearance' in response to management options. Hopefully, models of this type with continual improvement can help to provide answers to questions of this type.

The approach used in this study could usefully be applied to other persistent, bioaccumulative and toxic chemicals (PBTs) and other locations to build up an improved predictive capability. Thus it will contribute to better management of PBTs at national and ultimately global scales.

Acknowledgement

We are grateful to the UK Department of the Environment, Transport and the Regions for funding our research on environmental fate modeling of persistent organic contaminants. We would also like to acknowledge Professor Donald Mackay for his useful and constructive comments during the development of this model.

References

1. Mackay (1991) Multimedia Environmental Models. Lewis Publishers/CRC Press, Boca Raton, Fl.
2. Countryside Information System (1994), Produced for the UK Department of the Environment, Transport and the Regions by the Institute of Terrestrial Ecology.
3. Lead, WA., Steinnes, E., Bacon, JR. and Jones, KC. Sci. Total Environ., **1997**,193(3), 229- 236
4. Mackay, D., Shui, W-Y. and Ma, K-C. (1992) Illustrated handbook of physical-chemical properties and environmental fate for organic chemicals. Volume 1, Monoaromatic hydrocarbons, chlorobenzenes and PCBs. Lewis Publishers/CRC Press, Boca Raton, Fl.
5. Anderson, P.N and Hites, R.A. Environ. Sci. Technol., **1996**, 30, 1756-1763
6. Swackhamer, D., McVeety, B and Hites, R. Environ. Sci. Technol., **1988**, 22 (6), 664-672
7. Bletchly, J.D. (1983) Polychlorinated biphenyls: Production, current use and possible rates of future disposal in OECD member countries. In, Proceedings of PCB Seminar, Scheveningen, Netherlands. Editors. Barros, M.C., Könemann, H. and Visser, R.
8. Environmental Protection Agency. Management of Polychlorinated Biphenyls in the United States (1997). Office of Pollution Prevention and Toxics, Washington.
9. Annema, J., Beurskens, J.E.M. and Bodar, C.W.M. (1995) Evaluation of PCB fluxes in the environment. RIVM report No 601014011
10. Duiser, J.A. and Velt, C. (1989) Emissions into the atmosphere of polyaromatic hydrocarbons, polychlorinated biphenyls, lindane and hexachlorobenzene in Europe. TNO, Apeldoorn.
11. Harrad, S.J., Sewart, A.P., Alcock, R.E., Boumphrey, R., Burnett, V., Duarte-Davidson, R. Halsall, C. Sanders, G, Waterhouse, K., Wild, S. and Jones, K.C. Environ. Pollut., **1994**, 85, 131-146.
12. Gevao, B., Hamilton-Taylor, J., Murdoch, C., Kelly, M., Tabner, B. and Jones, K.C. Environ. Sci. and Technol., **1997**, 31, 3274-3280
13. Alcock, R., McGrath, S. and Jones, K. Environ. Toxicol. Chem., **1995**, 14(4), 553-560
14. Lee, R.G.M. and Jones, K.C. (1998) The influence of meteorology and air masses on daily atmospheric PCB and PAH concentrations at a UK location. Environ. Sci. Technol., **1999**, 33, 705-712
15. Cousins, I.T. and Jones, K.C. Environ. Pollut., **1998**, 101, 1-14
16. Hillery, B.R., Basu, I., Sweet, C. and Hites, R.A. Environ. Sci. Technol., **1997**, 31, 1811-1815

Chapter 8

Estimating Accumulation and Baseline Toxicity of Complex Mixtures or Organic Chemicals to Aquatic Organisms

The Use of Hydrophobicity Dependent Analytical Methods

Eric M. J. Verbruggen and Joop L. M. Hermens

Research Institute of Toxicology, Utrecht University, P.O. Box 80176, 3508 TD Utrecht, The Netherlands

Many chemical products, such as oils, are complex mixtures. Also in the environment itself chemicals always occur as complex mixtures. Physicochemical parameters however are mostly single-compound properties. One such a parameter that has a crucial role in many environmental processes is the hydrophobicity of organic chemicals. Two methods are presented for measuring the hydrophobicity of complex mixtures: a) hydrophobicity-based fractionation (hydrophobicity distribution profile) and b) hydrophobicity dependent extraction (biomimetic extraction). For two petroleum products the estimated toxicity by these two approaches was in good agreement with toxicity to *Daphnia magna*.

Introduction: risk assessment of complex mixtures

Risk assessment is usually based on the evaluation of the fate and toxic effects of single chemicals. However, many chemical products are not pure compounds but complex mixtures, of which the precise composition is unknown. Typical examples of these mixtures are petroleum derived products (*1*). Once emitted into the environment, all chemicals in effluents and surface waters are part of complex mixtures, of which the composition even after chemical analysis is largely unknown (*2*).

Physicochemical properties that are defined for single compounds are often used in quantitative structure activity relationships (QSARs) to predict the behavior in the environment. A very important compound property for organic chemicals is hydrophobicity. The parameters that are mostly used to reflect differences in hydrophobicity are the *n*-octanol-water partition coefficient (K_{ow}), aqueous solubility (S) or retention factors in reversed-phase high-performance liquid chromatography (k).

The use of these hydrophobicity parameters in QSARs to estimate partition behavior is manifold. Some examples are the partition coefficient for sorption to soil (*3*), the bioconcentration factor (*BCF*) (*4*), and narcosis or baseline-toxicity (*5, 6*). Baseline toxicity is related to the concentration of compounds in the cell membranes (*7*) and it has been shown that the internal effect concentrations are almost constant. This level is often called the critical body residue (*CBR*) (*8, 9*). Consequently external effect concentration directly reflect the differences in bioconcentration factors. Another important feature of narcosis is that toxicity exerted by different chemicals is completely concentration additive (*10-12*). Because of the additivity of narcotic effects and the constant internal effect concentrations, analytical methods to estimate bioconcentration of complex organic mixtures would be very useful. It is not necessary to identify and quantify all individual compounds, but to have information about total molar concentrations. In both approaches in this paper, the hydrophobicity distribution profile and the biomimetic extraction, total molar concentrations are estimated using either vapor pressure osmometry (VPO) (*13-15*) or gas chromatography-mass spectrometry (GC-MS) (*15*).

Approaches to measure hydrophobicity and baseline toxicity of complex mixtures

Hydrophobicity distribution profiles (HDP)

In order to determine a hydrophobicity distribution profile (HDP) of a complex mixture, it is first separated on a reversed-phase high performance liquid chromatography (RP-HPLC) column (*14, 16*). The retention time in gradient elution RP-HPLC (*16*) is a good measure for separation of a complex organic mixture according to hydrophobicity, even for micropollutants with very diverse chemical structures (see Figure 1). Individual compounds are grouped together in fractions with a small range of hydrophobicity. In each fraction the total molar concentrations is determined.

The resulting hydrophobicity distribution profile is presented as a limited number of blocks that can be considered to be homogeneous with respect to hydrophobicity. An example of such a profile for kerosene is given in Figure 2. The HDP approach is very similar to the hydrocarbon block approach by CONCAWE (*18*). However, here hydrophobicity is the discriminating factor, while the hydrocarbon block approach distinguishes for example between aromatic and aliphatic hydrocarbons and on carbon chain length. The results of the hydrophobicity distribution profile can be used in QSARs on basis of hydrophobicity, to estimate the fate of the mixture in the aquatic environment and baseline toxicity.

The hydrophobicity parameters K_{ow} and solubility of liquid compounds are strongly correlated. Both aromatic and aliphatic liquid compounds from petroleum derived substances, show almost the same relationship between K_{ow} and solubility (*19*). The HDP determined for such complex mixtures can thus be seen as a (subcooled) liquid solubility distribution as well as a K_{ow} distribution. Also other types of compounds show almost the same correlation between solubility and K_{ow} (*20*).

If a mixture does not dissolve completely, for example if a layer of oil is equilibrated with water, the data from a HDP can also be used to calculate to what extent the compounds in each fraction will be dissolved. For nearly ideal mixtures of organic compounds, the ratio of the equilibrium concentration of a compound in the aqueous phase and the aqueous solubility of the pure compound approaches the mole fraction of that compound in the organic phase of the mixture (*21*). For the partitioning of some fuel components between gasolines and water, this assumption of ideal behavior gave very good results (*22*). From this equation it follows that for a mixture that is just above its solubility in water, the compounds with the lowest (subcooled) liquid solubility will aggregate first. The aqueous concentration of the most hydrophobic fractions is strongly reduced, if a mixture is not completely dissolved.

Biomimetic extraction

The biomimetic extraction procedure (*13*) is a method to determine the total amount of chemicals that will accumulate in organisms after exposure to an aqueous sample. In this procedure the bioconcentration process is simulated by a surrogate hydrophobic phase with similar partition behavior. The procedure is based on an extraction with negligible depletion and selective for compounds that will concentrate in biota. After that the total amount of organic chemicals on the hydrophobic phase is determined. The procedure was first developed making use of extraction disks coated with octadecylsilica (*13, 15, 23*) or solid phase microextraction (SPME) fibers (*24*). Polyacrylate SPME fibers also appear to be

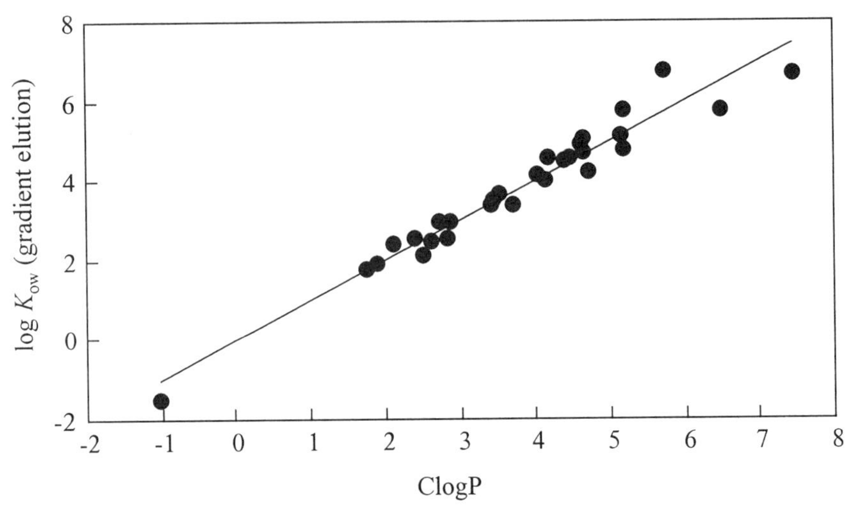

Figure 1: log K_{ow} estimated from RP-HPLC gradient elution retention times (16) versus their log K_{ow} values; log K_{ow} star values from (17).

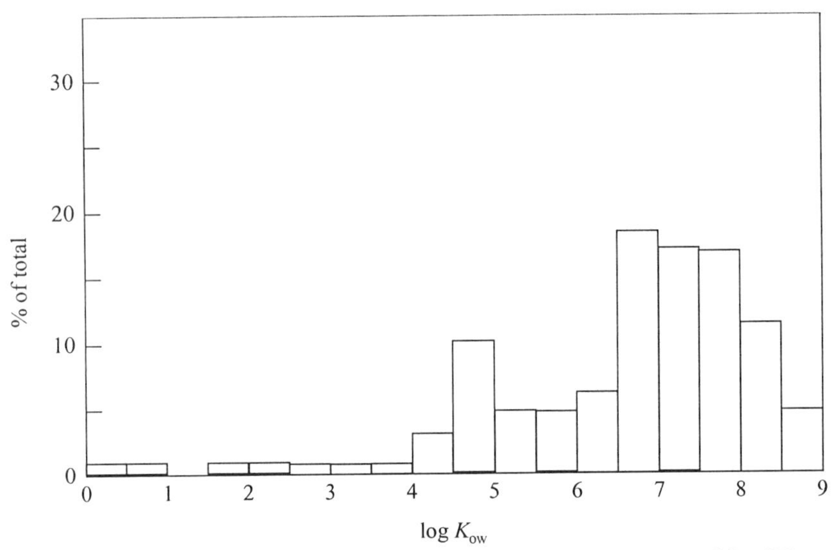

Figure 2: Hydrophobicity distribution profile of kerosene determined by GC-MS.

suitable for biomimetic extraction purposes (*25*). The partition coefficients to these fibers (K_{SPME}) are equally dependent on hydrophobicity as the partition coefficients to artificial membranes (K_m) (see Figure 3), which make them a very good surrogate to simulate partitioning to the target site for narcosis, i.e. the cell membranes. Other advantages of the SPME procedure compared to the procedure with the extraction disks are the much shorter period to obtain equilibrium (one day instead of two weeks) and the much smaller volumes to achieve negligible depletion (250 mL instead of 10 L).

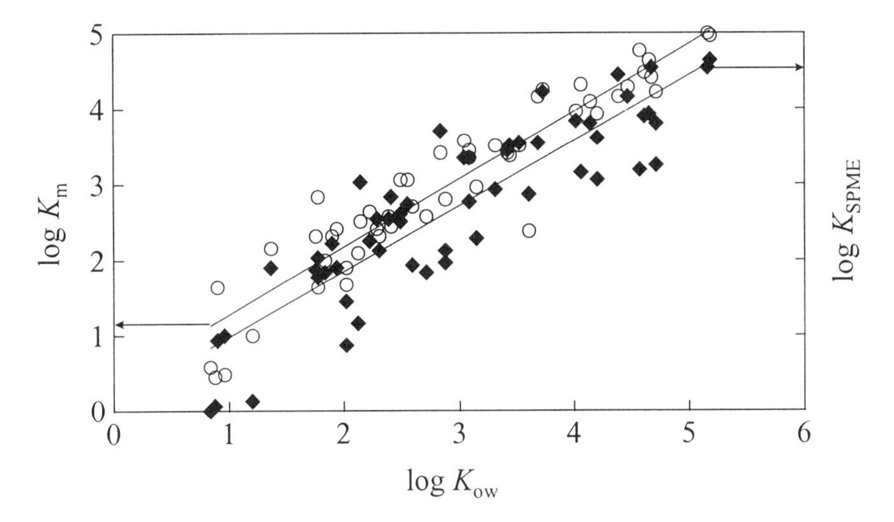

Figure 3: Polyacrylate (SPME)-water (◆) and membrane-water (○) partition coefficients as a function of log K_{ow}; data from (25).

Deriving internal concentrations from HDP and biomimetic extraction

Bioconcentration factors of compounds that are not strongly metabolized, are strongly correlated with their hydrophobicity (*4, 26-29*). However, at high hydrophobicity (log K_{ow} > 5.5) constant or even decreasing *BCF*s have been observed (*30-32*). Gobas *et al.* (*33*) showed the same effect for the partition coefficient between water and artificial membranes, which is confirmed by additional data from Dulfer and Govers (*34*) for compounds with high log K_{ow} and data from Vaes *et al.* for compounds of moderate hydrophobicity (*35, 36*). The following relationship between K_m and K_{ow} can be established:

$$\log K_m = -0.017 \cdot \log K_{ow}^3 + 0.131 \cdot \log K_{ow}^2 + 0.797 \cdot \log K_{ow} - 0.119$$

$$r^2 = 0.961, \qquad s.e. = 0.437\ (n = 86)$$

A possible explanation for this effect is that for large solutes the solubility in membranes decreases as much as or faster than the solubility in water with increasing molecular volume. This can be explained by a relative fast increase in the activity coefficient of the highly structured membrane bilayers in comparison with the activity coefficients in the bulk phases water and octanol for compounds with large molecular volumes (*29, 37*). Also membrane permeation, which is important for the uptake of compounds into an organism e.g. via the gills of fish, is affected by the size of the molecule (*38, 39*). However, for compounds that are not biotransformed, slower membrane permeation will only influence the kinetics but not the equilibrium constant (*38*).

To estimate the toxicity of a mixture due to narcosis, the total concentration at the target site, i.e. the membranes, has to be calculated from the HDP. Based on the considerations above, it is probably more accurate to use a relationship between log K_m and log K_{ow} that is not linear at high hydrophobicity, especially for complex mixtures, containing many very hydrophobic compounds such as petroleum products.

In the case of the biomimetic extraction, the membrane concentration is directly estimated from the concentration on the hydrophobic phase. The partitioning to this hydrophobic phase has to be similar to the membrane-water partitioning. The few available data on very hydrophobic compounds indicate that partitioning to polyacrylate SPME fibers is also not linear with hydrophobicity in the range above log K_{ow} 5.5 (*25*).

Application of HDP and biomimetic extraction to predict baseline toxicity of a petroleum products

Petroleum products mainly consist of aliphatic and aromatic hydrocarbons (*18*). For this type of compounds acute toxicity can probably be attributed to narcosis, e.g. (*40, 41*). Therefore, this type of complex mixtures is most suitable to test whether the two methods do provide good estimates for baseline toxicity. The biomimetic extraction procedure will give a direct estimate of the internal concentration, which is compared to an internal effect concentration (Figure 4). As internal effect concentrations, a critical target concentration (*CTC*, i.e. concentration in the membranes) of 125 mM can be used for median lethality due to narcosis (*25*).

To estimate toxicity from the hydrophobicity distribution profile, aqueous concentrations, resulting from dosing in the toxicity test have to be calculated first (Figure 4). For this purpose, the hydrophobicity of the different fractions in the HDP can be used to estimate the aqueous solubility of each fraction. For nearly ideal mixtures of organic compounds, the equilibrium concentration of a compound in the aqueous phase can be estimated as the product of its aqueous solubility and the mole fraction in the organic phase of the mixture (*21*). Because the HDP fractions are considered homogeneous, all compounds in it have the same aqueous solubility and can be treated as one. Then, the different fractions are the components, to which the ideal mixture theory can be applied, using a mass balance that accounts for the part of each fraction in the organic phase of the mixture and the part that is dissolved in the aqueous phase. From the aqueous concentrations of each fraction and its hydrophobicity, the contribution to the total internal concentration can be calculated.

The hydrophobicity distribution profile is needed to estimate the toxicity after emission or dosing to water for complex organic chemical products, such as petroleum products. The biomimetic extraction procedure is suitable to estimate toxicity from a water sample that contains a complex mixture of organic chemicals, which is common for environmental water samples, for example water contaminated by oil (Figure 4).

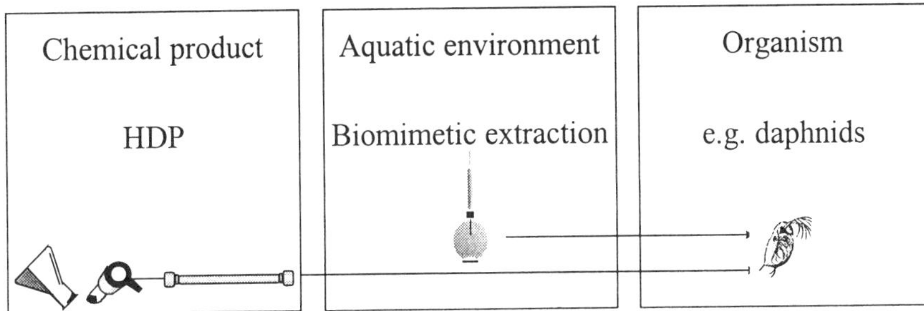

Figure 4: Overview of analytical methods to assess mixture toxicity.

Comparison of the analytical methods with toxicity for two petroleum products

For two petroleum products (a gas oil and kerosene), the two approaches are applied and compared with a 48-h toxicity test with water fleas (*Daphnia magna*). Details are described in (*19*). The observed median lethal loadings (LL_{50}) are displayed in Table 1. The somewhat lower lethal loading of kerosene can probably be explained from the relative higher amount of compounds with log K_{ow} 4-6 in comparison with the heavier gas oil. These compounds are relatively soluble and probably have almost equal membrane-water partition coefficients as the highly hydrophobic compounds.

From the biomimetic extractions with polyacrylate SPME fibers the internal concentrations (total target concentrations = concentration in the cell membranes) were estimated. By comparing these *TTC* values with the critical target concentration of 125 mM for lethality, median lethal loadings of the petroleum products could be estimated (Table 1). By expressing the percentage lethality that was observed in the toxicity experiment as a dose-response function of the estimated *TTC* values, the critical target concentration can also be estimated (Table 1).

For the calculations of lethal loadings from the hydrophobicity distribution profiles, the internal concentrations at equilibrium are calculated from the membrane-water partition coefficients and the aqueous concentrations after dosing. By combining with the results of the toxicity experiment, the *CTC* values are derived (Table 1). Kinetic behavior of the uptake process can be estimated from a QSAR on basis of log K_{ow} for the first order elimination rate constant for daphnids (*42*), log k_2. However, with the assumption of ideal mixture behavior with regard to solubility and membrane-water partition coefficients that level off at high hydrophobicity, non-equilibrium plays a minor role in a 48-h toxicity test with water fleas.

Table 1: Lethal loading rates (*LL_{50}*) and estimated critical target concentration (*$CTC_{est.}$*) determined by a 48-h toxicity test with *Daphnia magna*, the SPME biomimetic extraction procedure or hydrophobicity distribution profiles (HDP).

Parameter	Kerosene	Gas oil
LL_{50} [µl/L]		
Toxicity test Daphnia magna	5.1	1.7
Biomimetic extraction	3.6	0.8
HDP	2.7	5.7
$CTC_{est.}$ [mM]		
Biomimetic extraction	166	206
HDP	149	78

In the calculations, it is assumed that the petroleum products are ideal mixtures with respect to mixture solubility. Further, it is supposed that membrane-water partitioning can be estimated from hydrophobicity in the case of the HDPs or can be mimicked by polyacrylate SPME fibers in the case of the biomimetic extractions. These assumptions will lead to small differences between the methods. Still, both methods, biomimetic extraction and hydrophobicity distribution profile, provide good estimates of the toxicity of these mixtures. It can be concluded that

1. Acute toxicity of complex petroleum products can be explained entirely by narcosis.

2. The SPME biomimetic extraction procedure is a very useful tool in estimating the total internal concentrations in aquatic organisms after exposure to complex mixtures of organic chemicals that are present in the aquatic environment.

3. The hydrophobicity distribution profiles provide useful information for the risk assessment of chemical products that are complex mixtures of organic chemicals, not only for predicting environmental concentrations but also for estimation of baseline toxicity.

The advantage of both methods is that they provide information that can be expressed directly in terms of concentrations and toxic effects. Moreover, in both methods it is not necessary to identify individual compounds. Still, with GC-MS as molar detection technique it is possible to identify major compounds.

References

(1) CONCAWE Ecology group *The chemistry and grouping of petroleum-derived substances. Paper for the DG XI/CONCAWE workshop on the environmental risk assessment of petroleum substances to be held at the EU Joint Research Centre, Ispra, on 6/7 December 1994*; CONCAWE: Brussels, Belgium, 1994.

(2) Hendriks, A. J.; Maas-Diepeveen, J. L.; Noordsij, A.; van der Gaag, M. A. *Water Res.* **1994**, *28*, 581-598.

(3) Karickhoff, S. W.; Brown, D. S.; Scott, T. A. *Water Res.* **1979**, *13*, 241-248.

(4) Neely, W. B.; Branson, D. R.; Blau, G. E. *Environ. Sci. Technol.* **1974**, *8*, 1113-1115.

(5) Könemann, H. *Toxicology* **1981**, *19*, 209-221.

(6) Veith, G. D.; Call, D. J.; Brooke, L. T. *Can. J. Fish. Aquat. Sci.* **1983**, *40*, 743-748.

(7) van Wezel, A. P.; Opperhuizen, A. *Crit. Rev. Toxicol.* **1995**, *25*, 255-279.

(8) McCarty, L. S.; Mackay, D.; Smith, A. D.; Ozburn, G. W.; Dixon, D. G. *Sci. Total Environ.* **1993**, *109/110*, 515-525.

(9) McCarty, L. S.; Mackay, D. *Environ. Sci. Technol.* **1993**, *27*, 1719-1728.

(10) Könemann, H. *Toxicology* **1981**, *19*, 229-238.

(11) Hermens, J.; Canton, H.; Janssen, P.; de Jong, P. *Aquat. Toxicol.* **1984**, *5*, 143-154.

(12) Broderius, S.; Kahl, M. *Aquat. Toxicol.* **1985**, *6*, 307-322.

(13) Verhaar, H. J. M.; Busser, F. J. M.; Hermens, J. L. M. *Environ. Sci. Technol.* **1995**, *29*, 726-734.

(14) Verbruggen, E. M. J.; van Loon, W. M. G. M.; Hermens, J. L. M. *Environ. Sci. Poll. Res. Int.* **1996**, *3*, 163-168.

(15) van Loon, W. M. G. M.; Wijnker, F. G.; Verwoerd, M. E.; Hermens, J. L. M. *Anal. Chem.* **1996**, *68*, 2916-2926.

(16) Verbruggen, E. M. J.; Klamer, H. J. C.; Villerius, L.; Brinkman, U. A. T.; Hermens, J. L. M. *J. Chromatogr. A* **1999**, *835*, 19-27.

(17) Biobyte Corp.; MedChem: Claremont, CA, USA, 1994.

(18) CONCAWE Ecology group *Environmental risk assessment of petroleum substances: the hydrocarbon block approach*; CONCAWE: Brussels, Belgium, 1996.
(19) Verbruggen, E. M. J. PhD, Utrecht University, Utrecht, 1999.
(20) Schwarzenbach, R. P.; Gschwend, P. M.; Imboden, D. M. *Environmental organic chemistry*; John Wiley & Sons: New York, 1993.
(21) Banerjee, S. *Environ. Sci. Technol.* **1984**, *18*, 587-591.
(22) Cline, P. V.; Delfino, J. J.; Rao, P. S. C. *Environ. Sci. Technol.* **1991**, *25*, 914-920.
(23) van Loon, W. M. G. M.; Verwoerd, M. E.; Wijnker, F. G.; van Leeuwen, C. J.; van Duyn, P.; van de Guchte, C.; Hermens, J. L. M. *Environ. Toxicol. Chem.* **1997**, *16*, 1358-1365.
(24) Parkerton, T. F.; Stone, M. A. *SETAC 17th annual meeting*, Washington, DC, USA, 17-21 November 1996; Abstract book, SETAC Press; 150.
(25) Verbruggen, E. M. J.; Vaes, W. H. J.; Parkerton, T. F.; Hermens, J. L. M. *Environ. Sci. Technol.* **2000**, *34*, 324-331.
(26) Kenaga, E. E. *Environ. Sci. Technol.* **1980**, *14*, 553-556.
(27) Mackay, D. *Environ. Sci. Technol.* **1982**, *16*, 274-278.
(28) Chiou, C. T. *Environ. Sci. Technol.* **1985**, *19*, 57-62.
(29) Banerjee, S.; Baughman, G. L. *Environ. Sci. Technol.* **1991**, *25*, 536-539.
(30) Opperhuizen, A.; van de Velde, E. W.; Gobas, F. A. P. C.; Liem, D. A. K.; van de Steen, J. M. D. *Chemosphere* **1985**, *14*, 1871-1896.
(31) Anliker, R.; Moser, P. *Ecotoxicol. Environ. Saf.* **1987**, *13*, 43-52.
(32) Oliver, B. G.; Niimi, A. J. *Environ. Sci. Technol.* **1985**, *19*, 842-849.
(33) Gobas, F. A. P. C.; Lahittete, J. M.; Garofalo, G.; Shiu, W. Y.; Mackay, D. *J. Pharm. Sci.* **1988**, *77*, 265-272.
(34) Dulfer, W. J.; Govers, H. A. J. *Environ. Sci. Technol.* **1995**, *29*, 2548-2554.
(35) Vaes, W. H. J.; Urrestarazu Ramos, E.; Hamwijk, C.; van Holsteijn, I.; Blaauboer, B. J.; Seinen, W.; Verhaar, H. J. M.; Hermens, J. L. M. *Chem. Res. Toxicol.* **1997**, *10*, 1067-1072.
(36) Vaes, W. H. J.; Urrestarazu Ramos, E.; Verhaar, H. J. M.; Cramer, C. J.; Hermens, J. L. M. *Chem. Res. Toxicol.* **1998**, *11*, 847-854.
(37) Gobas, F. A. P. C.; Shiu, W. Y.; Mackay, D. In *QSAR in Environmental Toxicology – II*; Kaiser, K. L. E., Ed.; D. Reidel Publishing Company: Dordrecht, The Netherlands, 1987, pp 107-123.
(38) Gobas, F. A. P. C.; Opperhuizen, A.; Hutzinger, O. *Environ. Toxicol. Chem.* **1986**, *5*, 637-646.
(39) Opperhuizen, A.; Sijm, D. T. H. M. *Environ. Toxicol. Chem.* **1990**, *9*, 175-186.
(40) Hutchinson, T. C.; Hellebust, J. A.; Tam, D.; Mackay, D.; Mascarenhas, R. A.; Shiu, W. Y. In *Hydrocarbons and halogenated hydrocarbons in the aquatic environment*; Afghan, B. K.; Mackay, D., Eds.; Plenum Press: New York, NY, USA, 1980, pp 577-586.
(41) Bobra, A. M.; Shiu, W. Y.; Mackay, D. *Chemosphere* **1983**, *12*, 1137-1149.
(42) Hawker, D. W.; Connell, D. W. *Ecotoxicol. Environ. Saf.* **1986**, *11*, 184-187.

Assessment

Chapter 9

Selection and Evaluation of Persistent, Bioaccumulative, and Toxic Criteria for Hazard Assessment

Han Blok[1], Froukje Balk[1], Peter Okkerman[2], and Dick Sijm[3,4]

[1]Haskoning, Consulting Engineers and Architects, P.O. Box 151, 6500 AD Nijmegen, The Netherlands
[2]BKH Consulting Engineers, P.O. Box 5094, 2600 GB Delft, The Netherlands
[3]National Institute of Public Health and the Environment, P.O. Box 1, 3720 BA Bilthoven, The Netherlands
[4]Corresponding author

Criteria for persistence (P), bioaccumulation (B) and toxicity (T) were selected from national and international lists of priority pollutants. The PBT-criteria were used to search for PBT-substances in two databases, starting with empirical information on mammalian toxicity and on ecotoxicity, and complemented with estimated values for bioaccumulation and persistence. Biodegradation as measure of persistence is further discussed.
The search thus identified 301 potential PBT-substances. Of these candidates, 97 already were found on one or more existing lists and 204 substances were not earlier selected. The availability of data on ecotoxicity, mammalian toxicity, and environmental properties is poor, which calls for estimated properties. Since estimates for toxicity are not considered of much value, the search for PBT-substances from a large database may start with estimations on bioaccumulation and persistence, and followed by experimental toxicity testing for the remaining substances. The identification of PBT-substances can be used as a tool in environmental policy, such as for hazard assessment, prioritization or selecting substances.

Introduction

For most of the chemicals in use, insufficient toxicity and ecotoxicity data is publicly available for "minimal" risk assessment under OECD guidelines. With regard to the approximately 2,700 High Production Volume Chemicals (HPVCs) in

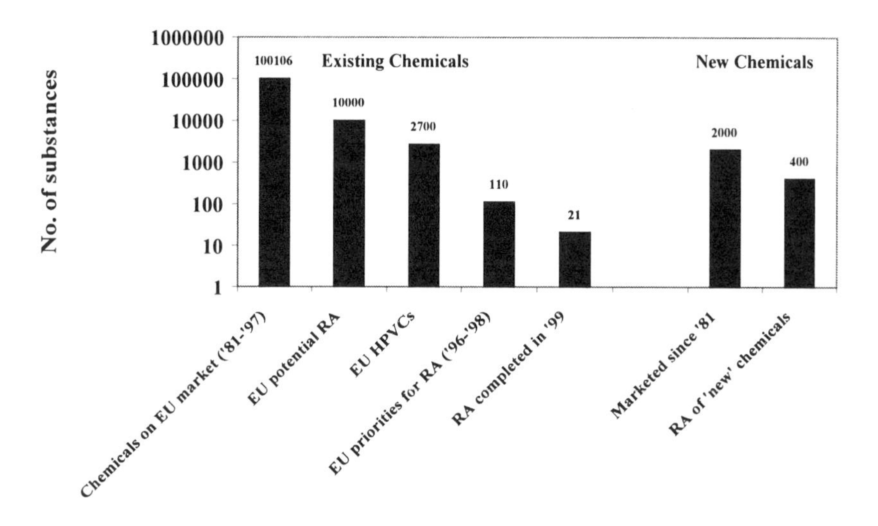

Figure 1. Summary of number of substances on the European market before 1981 ("existing chemicals"); number of existing chemicals for which potentially a risk assessment (RA) needs to be performed, the number of High Production Volume Chemicals (HPVCs), the number of substances that are on the priority list, for which it is agreed that RA needs to be performed, and the number of substances for which RA has been completed in 1999. In addition, the number of chemicals that are notified since 1981, and the number of those "new chemicals" for which RA has been completed. (Sources: 1-3)

use in the European Union, there is insufficient (eco) toxicity data for 70% of these chemicals (1-4). Figure 1 illustrates that for only a few "existing chemicals" risk assessment, including hazard assessment, has been completed. "Existing chemicals" are those that were marketed on the European market before September 18, 1981. For the "new chemicals", the chemicals that are marketed after that date, we currently have more, but still insufficient information on hazard assessment (Figure 1). Thus, for the vast majority of substances in use or in possible use, there is insufficient information on their hazardous properties.

Because of the poor data availability on hazardous properties, the Dutch government developed a new approach. In the third National Environmental Policy Plan (NMP3) in The Netherlands (5) several actions are formulated which include the definition, identification and formulation of restrictive measures for persistent toxic substances. The policy plans include also actions concerning the contribution to international activities of UNEP, UN-ECE, EC and OSPAR on emission reduction for a selected group of substances. These actions are an implementation of Chapter 19 on toxic substances in Agenda 21 of the Declaration of Rio (6). The NMP3 calls for a policy to replace toxic, bioaccumulating, and persistent (PBT) substances by less hazardous alternatives, to further reduce emissions of environmentally hazardous chemicals as agreed during the Fourth Ministers Conference on the North Sea and thus to develop a new approach for the 'non-assessed chemicals'.

The implementation of these actions requires in the first place a systematic listing of the hazardous substances. In the past the awareness of environmental hazards of substances was built in particular on the effects of persistent bioaccumulating substances and many hazardous substances, which use was to be reduced or banned, were put on lists ('black lists'). However, these lists were composed mainly by expert judgement and the selection criteria for inclusion of substances remained mostly undefined. Many of the substances occur on various lists and they represent only a very small selection of the 'chemical universe'.

To identify the number of potential hazardous substances that are in use, the Dutch Ministry of Housing, Spatial Planning and the Environment (Ministry of VROM) sponsored a series of desk studies. These studies were to identify those chemical substances that can be considered hazardous because they possess a combination of inherent substance properties, viz. persistence (P), bioaccumulation (B), and toxicity (T).

The first study was initiated in 1995 and reviewed criteria for selecting priority substances which were used by several national and international bodies and resulted in the recommendation for a set of criteria to select PBTs. A combination of BKH Consulting Engineers and HASKONING Consulting Engineers and Architects has been contracted to select PBTs from databases for chemical substance properties according to these criteria. At first the selection was driven by empirical ecotoxicity data and later on an extension was made to empirical data on mammalian toxicity. In each part the first automatic "black box" selection was followed by a more in-depth evaluation of the selected substances. Thus, the studies only selected substances with known toxicity, for their persistency and liability to bioaccumulate. The results of the project are reported in full detail in four reports (7-10) and an overview in a summary report (11).

The aim of the present study is to select PBT-criteria, to identify all chemicals that meet the PBT-criteria, and to evaluate the selected PBT-substances with respect to their occurrence on (inter) national lists.

Selection of PBT-criteria and PBT-substances

Several national and international criteria for selecting PBT-substances were reviewed to develop a set of PBT-criteria that would link with the current international opinions (7). As lists of priority substances are put together for many different reasons, wide ranges for cut-off values were thus found. Since the aim of the study was to identify all chemicals that meet specified PBT-criteria, this meant that the criteria should be applied to a large database of chemicals. First the databases will be described, which is followed by a description of the selected PBT-criteria, and the selection procedure.

Databases

Two databases were used for the PBT-selections: ISIS/Riskline and SMILES-CAS. ISIS/Riskline (version 6, 1996, HASKONING Consulting Engineers and Architects, Nijmegen, The Netherlands) is the International Substances Information System developed by HASKONING together with the National Chemicals Inspectorate in Sweden (KEMI). The database contains about 180,000 substances and Riskline bibliographic subfiles on about 6,000 substances. Starting point of ISIS was the EINECS list (European Inventory of Existing Chemical Substances), supplemented by substances in RTECS, IRPTC and around 1,000 substances from the ELINCS list (European List of Notified Chemical Substances). ISIS includes toxicity data from AQUIRE, RTECS, IRPTC and KNS (Dutch database with information on newly notified substances). The SMILES-CAS database of Syracuse (12) (version: 30-7-1997) contains SMILES notations for 102.905 chemical substances. The origin of this database is the American TOSCA inventory; later on substances are added from other sources such as an unofficial Danish Inventory.

Mammalian toxicity data are available for about 33% of the 180.000 substances in ISIS/Riskline. The database contains ecotoxicity data for about 3.600 substances. Although ISIS/Riskline includes properties like log Kow and degradation, the number of experimental data is very low. Instead, the log Kow and the probability of biodegradation are estimated from the SMILES notation.

The compilations in Syracuse and ISIS/Riskline do not match completely. Although the ISIS/Riskline and Syracuse databases contain altogether approximately 220.000 substances, the screening for PTB substances could be applied only to a set of 28,600 substances, where both (eco) toxicity data and estimated log Kow and probability of biodegradation were available.

PBT-criteria

Criteria were selected for Persistence (P), Bioaccumulation (B), and Toxicity (T), both for human health and the environment (7-11).

Persistence

Persistence is the opposite of degradability. Substances may be degraded by biotic as well as by abiotic processes. In the search for PTB-substances, biodegradation was estimated for each organic substance, and abiotic atmospheric degradation was estimated for those organic substances that distribute significantly to air. Biodegradability may be expressed in a variety of ways, reflecting the variety of available test methods. Currently the EU base-set requires substances to be tested for ready biodegradability (OECD TG 301A-F). This is a set of rather stringent[1] tests and thus covers just the opposite of the scale from ready degradability to persistence. The number of chemicals with measured data for biodegradation is limited to about 1,500, with a wide variety of test methods used. Therefore, for the selection of persistent substances from the database, biodegradability is estimated by using the biodegradation probability program, BIOWIN (Syracuse BIOWIN V3.61). In this way a uniform assessment of all substances is assured. Substances with a probability of rapid biodegradation > 0.5 are expected to degrade rapidly. For the selection of very persistent substances two classes of persistence are defined: Very persistent chemicals are those that have a probability of rapid degradation < 0.1 (linear regression model) and the time for ultimate degradation is months (class I) or more than months (class II).

For organic substances with high volatility, degradation by atmospheric oxidation may reduce their persistence in the environment. The partitioning over the environmental compartments air, soil and water is estimated by a Mackay level I fugacity model (13). In the selection of very persistent substances, those that partition to air for more than 10% and are photodegraded with a half-life < 2 days (Syracuse AOP, V1.83), are excluded. The criteria for persistence are listed in table I.

Bioaccumulation

The degree of bioaccumulation is either predicted by log Kow or is based on an experimental bioconcentration factor (BCF) value. In the present study for most of the substances, the estimated log Kow (Syracuse, KOWWIN 96/V1.54) is used. The criteria for the identification of substances with a high bioaccumulation potential are summarised in table II.

Toxicity

Since measured toxicity data are assumed to be much more reliable than estimates, the search for PBT-substances only used measured toxicity data for either mammalian toxicity or ecotoxicity. The criterion for toxicity is thus subdivided in mammalian toxicity and ecotoxicity. In both cases the criteria include acute toxicity but also some chronic toxicity data are taken into account.

The levels for the classes of mammalian toxicity are mainly derived from, but are not identical to, the classification system in Annex VI of the EC Directive 67/548/EC (14). For substances that have already been classified according to the EC system, this classification is directly used. The criteria for acute toxicity are based on the rat. When these rat toxicity data are not available, data for other mammals are extrapolated to rat. Three classes are used: Very Toxic, Toxic and Harmful. These

---

1 Stringent conditions: relatively low concentrations of test substance and inoculum, so only easily degradable substances are mineralised.

classes include the EC classes with the same name, but substances that show carcinogenicity and reproduction toxicity have also been attributed to the class Very Toxic. The three classes for ecotoxicity all belong to the EC classification Very Toxic to Aquatic Organism (R50). The criteria are summarised in table III.

Selection procedure

The selection procedure as well as the results are schematically presented in figure 2. Firstly, In ISIS/Riskline inorganic substances were excluded before the toxicity criteria were applied (both mammalian and ecotoxicity criteria). In parallel, for all substances in the SMILES-CAS database, log Kow was estimated. For the substances meeting the criteria for log Kow, the biodegradation rate was estimated, and the substances meeting the criteria for persistence and liability to bioaccumulate were selected. The overlap between the persistent and bioaccumulative substances from the Syracuse database and the toxic substances from ISIS is the first collection of PBT candidates.

In the next step, the Mackay fugacity model was applied to this first collection to identify the substances that significantly partition to air. Within this group substances with $t\frac{1}{2}$ (photodegradation) < 2 days were excluded from the collection of PBT substances. The PBT-candidates were compared with current international lists and further searches took place to provide additional identification of the substances.

Results

The results of the selection process are included in figure 2. Starting from the almost 180,000 chemical substances in ISIS/Riskline a strong reduction is obtained due to the lack of mammalian toxicity and ecotoxicity data, and the subsequent selection of tests meeting the toxicity criteria. The final number of substances meeting the criteria for mammalian toxicity is 15,527 and the final number meeting the ecotoxicity criteria is 634. From the 103,000 substances in the SMILES-CAS database, the number of substances that are estimated to bioaccumulate is 8,502, of which approximately 56% are estimated to be persistent. When the selections from both databases are combined, 268 substances are selected based on mammalian toxicity and 61 substances based on ecotoxicity. Due to some overlap and to exclusion of some errors, the total of the two lists is 301 PTB-substances. Since the overlap between the two databases is poor, the number of substances presently selected as PBT-candidates is not selected from around 220,000 substances but from a subset of approximately 65,000 chemicals.

Table I. Criteria for persistence.

Persistence criteria	*Class I*	*Class II*
Probability of rapid biodegradation[a]	< 0.1, AND	< 0.1, AND
Time for ultimate biodegradation[a]	months, AND	> months, AND
Half-life time for photo-oxidation[a,b]	> 2 days	> 2 days

[a] Estimated by using the BIOWIN estimation program (11).[b] Only for substances that partition to air for more than 10%.

Table II. Criteria for bioaccumulation.

Bioaccumulation criteria	*Class I*	*Class II*	*Class III*
Log Kow OR BCF	4 ≤ log Kow < 4.5 OR 1000 ≤ BCF < 3000	4.5 ≤ log Kow < 5 OR 3000 ≤ BCF < 5000	Log Kow ≥ 5 OR BCF ≥ 5000

Table III. Criteria for mammalian toxicity and ecotoxicity.

Toxicity criteria			
Mammalian toxicity	*Harmful Class I*	*Toxic Class II*	*Very Toxic Class III*
Acute toxicity			
- Oral	R22 OR	R25 OR	R 28 OR
(LD_{50}rat, mg/kg)	$200<LD_{50} \leq 2000$	$25<LD_{50} \leq 200$	$LD_{50} \leq 25$
- Inhalation	R20 OR	R23 OR	R 26 OR
(LC_{50}rat, mg/l/4h)	$2 <LC_{50} \leq 20$	$0.5<LC_{50} \leq 2$	$LC_{50} \leq 0.5$
- Dermal	R21 OR	R24 OR	R 27 OR
(LD_{50} rat/rabbit, mg/kg)	$400<LD_{50} \leq 2000$	$50<LD_{50} \leq 400$	$LD_{50} \leq 50$
Carcinogenicity			R45 OR R49 OR IARC 1 OR IARC 2A
Reproduction toxicity			R60 OR R61 OR R62 OR R63
Ecotoxicity criteria	*Very Toxic Class I*	*Very Toxic Class II*	*Very Toxic Class III*
Acute toxicity (mg/l)	$0.1<L(E)C_{50} \leq 1$	$0.01<L(E)C_{50} \leq 0.1$	$L(E)C_{50} \leq 0.01$
OR	OR	OR	OR
Chronic toxicity (mg/l)	0.01<NOEC≤ 0.1	0.001<NOEC ≤ 0.01	NOEC≤ 0.001

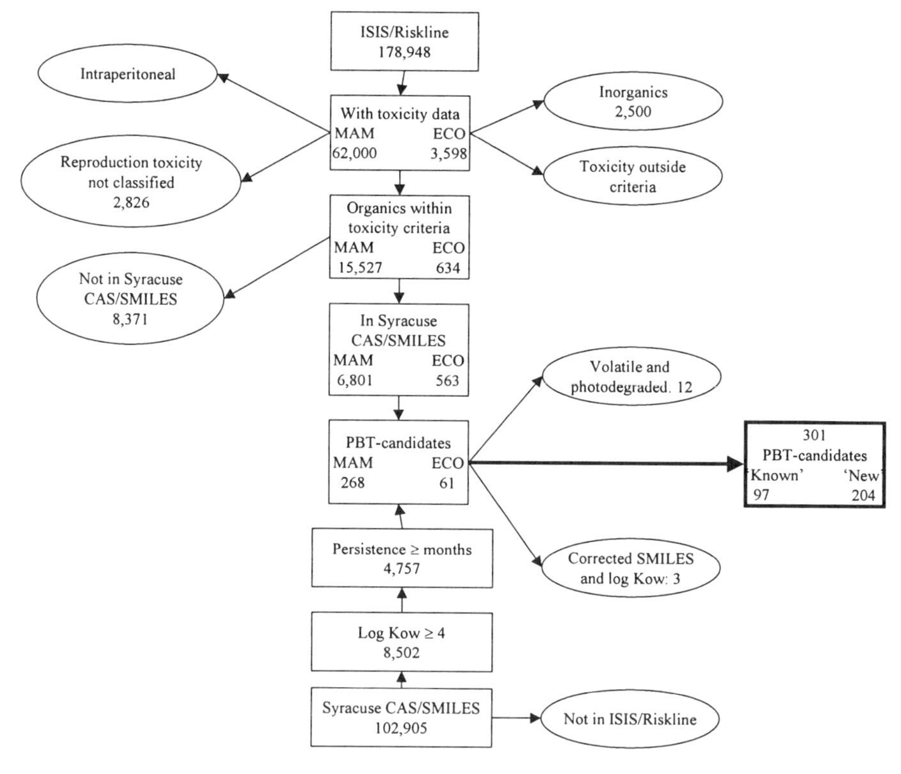

Figure 2. The BKH/HASKONING selection procedure to identify PBT-candidates from two databases (ISIS/Riskline and SMILES-CAS) starting from known ecotoxicity and mammalian toxicity data, and complemented with estimated data on liability to bioaccumulate (log Kow) and persistence. The numbers refer to the least stringent levels of the criteria.

The criteria were graded in two or three levels to study the influence of the cut off values. In table IV the influence of the different cut off values on the numbers of selected substances are presented. The fraction of substances meeting the mammalian toxicity criteria is approximately 25%. For the 15,527 substances selected on the basis of mammalian toxicity, a SMILES code was available only for 6,801 organic substances, of which 529 have a log Kow ≥ 4. For the 529 substances that show mammalian toxicity and liability to bioaccumulate, 268 are persistent. The number of substances selected by the mammalian toxicity route varies from 268 to 41 for the three toxicity classes.

The fraction of substances meeting the ecotoxicity criteria is 18%. The list of 634 substances selected on the basis of ecotoxicity data with available SMILES codes contains 282 substances with log Kow ≥ 4, of which 217 substances have a log Kow

≥ 4.5, and 167 have a log Kow ≥ 5. Persistency was estimated for the substances that were selected for toxicity and bioaccumulation. For the 282 substances that were selected based on ecotoxicity and liability to bioaccumulate, 80 substances are also persistent. The number of PBT-substances via the ecotoxicity route thus varies with the three toxicity classes from 80 to 40 and is reduced to 19 for the most stringent classes for bioaccumulation and persistency.

Only 97 of the selected substances already occur on various international lists of priority substances, while 204 have not been detected on other lists. Many PBT-candidates are halogenated organic substances. The number of substances with three or more halogens is 108. Most of the substances selected through the mammalian toxicity pathway are applied as medicinal drug (97). The second largest group (81) includes agricultural pesticides, or ingredients of them. In 35 cases of mainly pesticides, the substance is mentioned in some way in a national regulation. The group of industrial chemicals includes only 16 substances, including a cluster of polyfluorinated organic substances. For 25 substances, including two natural products, no application was found.

Discussion

The present study was initiated to develop a procedure to identify substances that are potentially hazardous because of their intrinsic properties, such as persistence, liability to bioaccumulate, and toxicity. The criteria developed for these properties are in line with levels used in internationally recognized procedures. The procedure makes a uniform and automated treatment of the substances possible. If toxicity data and SMILES codes are available, the selection of PBTs is relatively straightforward.

Since empirical data for persistence and bioaccumulation are almost completely lacking, using estimated values for these properties broadens the basis for the selection. In spite of the use of the estimated values for two of the three criteria, the selection procedure is still severely limited by the low numbers of SMILES notations for substances with toxicity data.

With respect to the criteria for bioaccumulation and persistence a discussion on the levels of the criteria is not based on science, or on sufficient data, but on policy. The estimated log Kow as a surrogate for bioaccumulation and the estimated value for persistence may be regarded as a relative selection tool that ranks substances according to these criteria. The combined application of a threshold for toxicity with the least stringent selection criteria for bioaccumulation and persistence already limits the number of chemicals that should be considered hazardous for the environment to only 0.5 to 1 % of the chemicals, with toxicity data. In view of this strong reduction in numbers, a further restriction of the selected number by applying more stringent threshold values for Kow and/or persistence does not seem to make much sense.

Discussing persistence

The criteria and levels that were selected (Tables I-III) were based on international priority and/or selection methods. The criteria for toxicity were chosen in such a way that empirical data could be used and retrieved. Since these empirical data were derived under internationally accepted guidelines, toxicity will not be discussed further here. Furthermore, since only organic substances were taken, the choice of log Kow as a worst-case estimate for bioaccumulation, seems to be justified. Therefore, also bioaccumulation will not be further discussed here. Persistence, however, will be further discussed, simply because there are only few or almost no validation studies for the estimation of persistence.

For estimating persistence, only the half-lives for biodegradation and for atmospheric oxidation were estimated using the Syracuse estimation methods. One general comment on these estimation methods is that they do not include other degradation processes, such as biotransformation, anaerobic degradation. These other processes that may have been included, however, were found not to be relevant or were found to be accompanied with poor or no data.

Biodegradability may be expressed as a probability that an adequate adaptation of microorganisms may occur. For readily degradable substances such an adaptation will easily occur and result in a high removal rate and complete disappearance within several days. For xenobiotic substances that are inherently degradable (OECD TG 302), this may take several weeks and for persistent chemicals the total disappearance may take months or even more.

Recently, estimation programs to predict the persistence on the basis of molecular fragments have been improved. Limitations still exist for relatively uncommon structures and for steric isomers. The Syracuse program BIOWIN was validated for the prediction of the outcome of the MITI-I test for ready biodegradability, showing that this model is capable of reliably predicting not-ready biodegradability. Of a total set of 894 substances with experimental data on biodegradability, it was shown that 88% of 208 not-ready biodegradable substances were predicted correctly (15, Table V). The good correlation is explained by the fact that the presence of a biodegradation-retarding fragment will prevent mineralisation. Thus, an estimation program was used because there were insufficient data on persistency. The Syracuse program, BIOWIN, was used, since it provides a uniform approach for dealing with estimating persistency, and it is considered to be a good model for that purpose.

For atmospheric oxidation as an expression of degradation, Rorije et al. (19) compared two estimation models. For 917 HPVC compounds, an estimation of the reaction rate constant for reaction with hydroxyl radical was possible using the Syracuse AOP (hydroxyl radical and ozone reaction) model, and for 864 compounds using the MOOH-method (hydroxyl radical reaction) of Klamt (20). The two models,

Table IV. The number of resulting substances that meet the various levels of the PBT-criteria.

Data source	*Number of substances*	*Class I*	*Class II*	*Class III*	*Total*
Mammalian toxicity (179,948)	62,000	Harmful 10,184	Toxic 2,969	Very Toxic 2,374	15,527
Ecotoxicity (179,948)	3,598	Very Toxic I 285	Very Toxic II 173	Very Toxic III 176	634
Bio-accumulation	102,905 (with SMILES notation)	Log Kow ≥ 4 1,128	Log Kow ≥ 4.5 1,113	Log Kow ≥ 5 6,261	8,502
Persistence	8,502 (with log Kow ≥ 4)	$T_{1/2}$ = Month 3,362	$T_{1/2}$ >Months 1,395		4,757

Table V. Comparison of models in predicting ready biodegradability. Percentages of number of correct predictions are given. For BIOWIN, 733 substances from the MITI database were taken, for OECD, ECB-model and MULTICASE, 894 substances were taken (15).

	BIOWIN	*BIOWIN Adapted*[a]	*OECD (16)*	*ECB-model (17)*	*MULTI-CASE (18)*
Readily biodegradable	68.1 %	88 %	81.9 %	88.2 %	84.8 %
Not readily biodegradable	76.0 %	88 %	88.2 %	80.0 %	72.0 %

[a] An adapted version of BIOWIN included the MITI database and resulted in an improved accuracy of estimating both 'readily biodegradable' and 'not ready biodegradable' (Loonen, unpublished results)

however, gave very different results, again without telling which model result is the best.

Other abiotic degradation processes, such as hydrolysis, oxidation and reduction in the aqueous phase, resulted in estimates for only 237 HPVC compounds. These other reactions were facing missing data or time-consuming calculations (19).

The results of Rorije et al. (19,21) thus show that estimation programs should be used with caution, but that for some classes of chemicals they may be very useful.

Dealing with data gaps

The absence of (eco) toxicity data and SMILES codes for almost 75% of the 220,000 chemicals that occur on inventories significantly limits the identification of PBT-substances. Furthermore, filling the data gaps on (eco) toxicity for about 165,000 chemicals by performing toxicity tests seems not feasible. Instead of starting with experimental data on (eco) toxicity, the cheapest part of the information, the supply of the SMILES notation and the estimation of bioaccumulation (log Kow) and persistence by computer programs should be performed first for the largest number of chemicals. This selection will probably identify about 4% of the 165,000 chemicals that require information on toxicity. If in analogy of the present data selection, toxicity data will already be available for about 25% of them[2], an estimated 5,000 chemicals remain to be tested for (eco) toxicity.

Conclusions

- The described approach to identify and to select PBT-substances using the PBT-criteria seems to be a powerful and relatively easy method to identify substances that are potentially hazardous to man and the environment.
- The current approaches to generate data for each of the PTB-criteria probably present the 'state-of-the-art'. Of the PTB-criteria, that for toxicity is based on empirical data, which are relatively sound, that for bioaccumulation has also a strong basis, but is not applicable for all substances, while that for persistence seems to be the best available.
- The databases that were used do not match completely. Significant numbers of (eco) toxicity data are not available. Almost no empirical data on persistence and bioaccumulation are available. In addition, the estimation of persistency and bioaccumulation potential is considerably hampered by the lack of readily available SMILES notations for many substances.

---

[2] Under the assumption that the availability of toxicity data is distributed randomly over all substances.

- A significant part of the selected substances (97) is already included on one or more priority lists and the majority of the others (201) is applied as medicinal drugs or pesticides.
- The identification of PBT-substances can be used as a tool in environmental policy, such as for prioritization, selecting substances or hazard assessment. The availability of ecotoxicity, toxicity and environmental data, however, is very poor. This either calls for further testing or the (further) use of estimation methods. Further testing to obtain data on the selected PTB-criteria would add to enormous amounts of money for the (hundreds of) thousands substances for which data are lacking. On the other hand, the estimation of the missing data on the selected PTB-criteria would go hand in hand with a high degree of uncertainty.
- When first persistence and bioaccumulation properties of substances are estimated, the number of substances thus selected that requires additional data on toxicity will probably be limited and manageable.

Acknowledgements

The Dutch Ministry of Housing, Spatial Planning and the Environment (Ministry of VROM) is thanked for their financial and other support of this work.

References

1. Chemicals in the European environment: low doses, high stakes? The EEA and UNEP Annual Message 2 on the State of Europe's Environment (1998).
2. Verhaar, H. J. M.; van Leeuwen, C. J.; Bol, J.; Hermens, J. L. M. *SAR and QSAR in Environmental Research* **1994,** *2*, 39-58.
3. Ahlers, J. *Environ. Sci. & Pollut. Res.* **1999,** *6*, 127-129.
4. Allanou, R.; Hansen, B.G.; Van der Bilt, Y. Public availability of data on EU High Production Volume Chemicals. *European Chemical Bureau Report EUR 18966 EN, Ispra, Italy,* **1999**.
5. Ministries of Housing, Physical Planning and the Environment, Economic Affairs, Agriculture, Nature Management and Fisheries, Transport, Public Works and Water Management, Finance, Foreign Affairs. *Nationaal Milieubeleidsplan 3 (National Environmental Policy Plan),* **1998.**
6. United Nations Conference on Environment and Development UNCED, Rio de Janeiro, Chapter 19 of Agenda 21, June **1992**.
7. BKH *Criteria voor zeer persistente toxische stoffen. Phase 1 of PTB project.* Report to the Ministry of VROM, DGM/SVS. BKH/R0216076/SR0, **1995**
8. BKH *Selection of Toxic, Persistent and Bioaccumulative Substances based on ecotoxicity data. Phase 2/eco of PTB project.* Report to the Ministry of VROM, DGM/SVS. BKH/R0216007/2471P, **1998**.

9. BKH *Selection of Toxic, Persistent and Bioaccumulative Substances based on mammalian toxicity data. Phase 2/mam of PTB project.* Report to the Ministry of VROM, DGM/SVS. BKH/R0216007/2319P, **1998**.
10. BKH/HASKONING *Detailed evaluation of automatically selected PTB candidate substances. Phase 3 of PTB project.* Report to the Ministry of VROM, DGM/SVS. HASKONING/G1260.A0/R003/FBA/AS, **1999**.
11. BKH/HASKONING *Identification of Persistent, Toxic and Bioaccumulating substances (PTBs). Summary report.* Report to the Ministry of VROM, DGM/SVS. BKH/HASKONING/M0216008, **1999**.
12. SRC. Syracuse Research Corporation. Syracuse, NY, USA.
13. Mackay, D. *Multimedia environmental models. The fugacity approach.* Lewis Publishers, Inc., Chelsea, Michigan, USA, **1993.**
14. EC Directive EU *Guidance document for the implementation of Annex VII D of council Directive 67/548/EEC (Directive 93/105/EEC) and Requirements for spectral data, **1993**.*
15. Rorije, E.; Loonen, H.; Müller, M.; Klopman, G.; Peijnenburg, W. J. G. M. *Chemosphere* **1999,** *38*, 1409-1417.
16. Degner, P.; Müller, M.; Nendza, M.; Klein, W. *Structure-activity relationships for biodegradation.* OECD Environment monographs No. 68, Paris, France, **1993**.
17. Loonen, H.; Lindgren, F.; Hansen, B. G.; Karcher, W. *Prediction of biodegradation from chemical structure: Use of MITI data, structural fragments and multivariate analysis for the estimation of ready and not-ready biodegradability.* In Peijnenburg, W. J. G. M.; Damborsky, J., Eds, *Biodegradability Prediction*, Kluwer Academic Publishers, Dordrecht, The Netherlands, 1996, pp. 105-114.
18. Klopman, G.; Zhang, Z.; Balthasar, D.M.; Rosenkranz, H.S. *Environ. Toxicol. Chem.* **1995,** *14*, 395-403.
19. Rorije, E.; Müller, M.; Peijnenburg, W. J. G. M. *Prediction of environmental degradation rates for High Production Volume Chemicals (HPVC) using Quantitative Structure-Activity Relationships.* National Institute of Public Health and the Environment, Report No. 719101030, Bilthoven, The Netherlands, **1997**.
20. Klamt, A. *Chemosphere* **1993,** *32*, 717-726.
21. EU-DG-XII *Overview of structure-activity relationships for environmental endpoints. Part 1: General outline and procedure.* Report of the EU-DG-XII project QSAR for predicting fate and effects of chemicals in the environment. Contract # EV5V-CT92-0211, **1995**.

Chapter 10

An Environmental Protection Agency Multimedia Strategy for Priority Persistent, Bioaccumulative, and Toxic Pollutants

Sam Sasnett, Thomas Murray, Sheila Canavan, John Alter, Kathy Davey, and Paul Matthai

Office of Pollution Prevention and Toxics, U.S. Environmental Protection Agency, 401 M Street, SW, Washington, DC 20460

The goal of this strategy is to reduce risks to human health and the environment from existing and future exposures to priority persistent, bioaccumulative and toxic (PBT) pollutants. These pollutants pose risks because they are toxic, persist in ecosystems, accumulate in fish and up the food chain. The remaining PBT challenges stem from the ability of these pollutants to travel long distances, to transfer rather easily among air, water, and land, and to linger for generations, making EPA's traditional single-statute approaches less than the full solution to reducing these risks. Due to a number o f adverse health and ecological effects linked to PBT pollutants – especially mercury, PCBs and dioxin – it is key for EPA to aim for further reductions in PBT risks. The fetus and child are especially vulnerable, putting pregnant women and young children at particular risk. EPA is committing, through this Strategy, to work with its partners at all levels of government, in the private sector, with interested non-governmental organizations, and in the international arena to integrate and coordinate all the tools available to address the cross-media issues associated with priority PBT pollutants.

The Problem

Persistent, bioaccumulative and toxic pollutants are highly toxic, long-lasting substances that can build up in the food chain to levels that are harmful to human and ecosystem health. They are associated with a range of adverse human health effects, including effects on the nervous system, reproductive and developmental problems, cancer and genetic impacts. The populations at risk, especially to PBTs such as mercury, dioxins and Polychlorinated Biphenyls (PCBs), are children and the developing fetus. EPA's challenge in reducing risks from PBTs stems from the pollutants' ability to travel long distances, to transfer rather easily among air, water, and land and to linger for generations in people and the environment.

Although, over the years much work has been done to reduce the risk associated with these chemicals, the nation still finds them in its fish supply. The total number of advisories in the United States increased by 80% from 1993 to 1997 and the number of waterbodies under advisory increased from 1,278 to 2,299. Only 17 States and territories have stayed at the same level or have had a decrease in the number of advisories since 1993. In the other 38 states, advisories to restrict or avoid eating the fish have increased. Six states have increased advisories more than 30% and 13 states had added statewide advisories applying to all fresh water, all coastal waters, or both (*1*). All of the substances that are causing the advisories are PBTs. While some may argue that a part of this increase may be due to the fact that the states are doing a better job of monitoring and setting protective levels, the facts are clear that we have much work ahead of us to reduce the risks of these PBT chemicals.

Groups outside EPA recognize the need for a cross-program, multimedia approach to environmental problems like PBTs. The 1998 Natural Resources Defense Council Report, *Contaminated Catch -- The Public Health Threat From Toxics in Fish* (*2*) challenges the nation to be more aggressive in preventing persistent pollution by controlling pollutants that cross media. The National Academy of Public Administration's 1995 Report, *Setting Priorities, Getting Results -- A New Direction for EPA* (*3*) emphasizes the need to do a better job of integrating the Agency's efforts across environmental media programs and statutes that support them. The Organization for Economic Cooperation and Development's 1996 report, *Environmental Performance Review of the United States* (*4*) calls for better coordination between EPA's chemical programs and its media programs. Finally, in his Earth Day message of 1998, Vice President Gore announced a new Chemical-right-to-know program with an emphasis on PBT chemicals.

The Strategy

This Strategy was developed by the Persistent, Bioaccumulative and Toxics Plenary Group. This group is comprised of program and technical experts from the following EPA Program Offices: The Office of Air and Radiation; the Office of Enforcement and Compliance Assurance; the Office of International Activities; the Office of Prevention, Pesticides and Toxic Substances; the Office of Research and Development; the Office of Solid Waste and Emergency Response; the Office of Water; the Great Lakes National Program Office and the ten EPA Regional Offices. The Office of Prevention, Pesticides and Toxic Substances chairs the group. This developmental activity was overseen by The Office Directors' Multimedia and Pollution Prevention (M2P2) Forum. This is a group of senior level management officials charged with examining and coordinating a variety of multimedia and pollution prevention issues. The PBT strategy is a central focus of the M2P2 Forum. The Office of Pollution Prevention and Toxics and the Office of Water currently co-chair the Forum.

On November 16, 1998 the Environmental Protection Agency released for public comment a draft Multimedia Strategy on Priority Persistent Bioaccumulative and Toxic Pollutants and a draft national Mercury Action Plan. The Strategy is a blueprint to link together all of EPA's work on priority PBTs. It calls on EPA to use all of its tools, across all media -- voluntary, regulatory, compliance and enforcement, international, and research -- in a coordinated and targeted way. For example, EPA will better integrate pollution prevention in its air rules and better consider the water impacts of waste disposal and pesticide use. This Strategy also fulfills a promise made under the 1998 Clean Water Action Plan (*5*) released by President Clinton and Vice President Gore to speed restoration of the nation's waterways. In that plan, the Agency committed to write a strategy by the end of 1998 that would outline a plan for reducing the risks associated with PBT substances.

Purpose And Goal

The goal of this Strategy is to further reduce risks to human health and the environment from existing and future exposure to priority persistent, bioaccumulative, and toxic pollutants.

The U.S. Environmental Protection Agency has developed this draft national strategy to overcome the difficult challenges associated with priority PBT pollutants. These pollutants pose risks because they are toxic, persist in ecosystems, and accumulate in fish and other organisms all along the food chain. The PBT challenges remaining stem from the pollutants' ability to travel long distances, to transfer rather easily among air, water, and land, and to linger for generations, making EPA's traditional single-statute approaches less than the full solution to reducing risks from PBTs. Due to a number of adverse health and ecological effects linked to PBT pollutants it is key for EPA to aim for further reductions in PBT risks. The fetus and child are especially vulnerable. EPA is committing, through this strategy, to create an enduring cross-office system that will address the cross-media issues associated with priority PBT pollutants.

Four Main Elements Of EPA's Strategy

1. Develop and Implement National Action Plans for Priority PBT Pollutants

EPA is affirming the priority given by the United States and Canada to the Level 1 substances under the 1997 Canada-U.S. Binational Toxics Strategy (*6*), making these substances the first focus for action.

The Level 1 substances are:

- aldrin/dieldrin
- alkyl-lead
- benzo(a)pyrene
- chlordane
- DDT/DDD/ DDE
- dioxins and furans
- hexachlorobenzene
- mercury and compounds
- mirex
- octachlorostyrene
- PCBs
- toxaphene

EPA is developing action plans that will use the full range of its tools to prevent and reduce releases of these 12 (and later other) PBTs. EPA will analyze PBT pollutant sources and reduction options as bases for grouping pollutants, activities, and sectors to maximize efficiencies in achieving reductions. EPA will integrate and sequence actions within and across action plans, and will seek to leverage these actions on international and industry-sector bases.

2. Screen and Select More Priority PBT Pollutants for Action.

Beyond the Level 1 substances, EPA will select additional PBT pollutants for action. EPA will apply selection criteria in consultation with a technical panel. Candidate chemicals will be those highly scored by EPA's Waste Minimization Prioritization Tool (*7*) and other chemicals of high-priority to EPA offices. EPA will seek internal and external comment on the proposed substances and the selection methodology.

3. Prevent Introduction of New PBTs

EPA is acting to prevent new PBT chemicals from entering commerce by: (a) proposing criteria for requiring testing/restrictions on new PBT chemicals; (b) developing a rule to control attempts to re-introduce out-of-use PBT chemicals into commerce; (c) developing incentives to reward the development of lower-risk chemicals as alternatives to PBTs; and (d) documenting how PBT-related screening criteria are taken into account for approval of new pesticides and re-registration of old pesticides.

4. Measure Progress
EPA is defining measurable objectives to assess progress. EPA will use direct and indirect progress measures, including: (a) human health or environmental indicators (such as National Health and Nutritional Examination Surveys (*8*) and a national study of chemical residues in fish); (b) chemical release, waste generation or use indicators (such as enhancing the Toxics Release Inventory (*9*) and using other release reporting and monitoring mechanisms); and, program activity measures (such as EPA compliance/enforcement data).

How EPA Will Make This Strategy Work

This strategy is built on a strong foundation fortifying existing EPA commitments related to priority PBTs, such as the 1997 Canada – U.S. Binational Toxics Strategy, the North American Agreement on Environmental Cooperation, and the Clean Water Action Plan released in 1998. EPA is forging a new approach to reduce risks from and exposures to priority PBT pollutants through increased coordination among EPA national and regional programs. This approach also requires the significant involvement of stakeholders, including international, state, local, and tribal organizations, the regulated community, environmental groups, and private citizens.

Recent Actions
EPA's strategy outlines a number of actions the Agency will take to reduce exposures to and uses of PBTs. Some of the near-term actions include:

- Preventing the introduction of new PBTs into commerce that may pose unreasonable risk to human health and the environment, and to require testing to confirm a chemical's PBT status. Please refer to Toxic Substances Control Act (TSCA) New Chemicals *Federal Register* Notice dated October 5, 1998 (*12*).
- Encouraging voluntary reductions of priority PBTs in hazardous waste. The EPA's Office of Solid Waste has challenged industry to voluntarily target priority PBTs found in hazardous waste for waste minimization activities. EPA has proposed a list of 53 PBT substances for this purpose in the draft Resource Conservation and Recovery Act (RCRA) PBT list in the *Federal Register* Notice dated November 9, 1998 (*13*).
- Giving the public information on mercury emissions from utilities. The EPA will require utilities to conduct coal and emissions sampling for mercury in order to analyze the link between mercury emissions and sources.
- Increasing the public's right-to-know about local sources of PBT emissions. EPA published a proposed rule on January 5, 1999 (*14*) that will add certain PBTs to the Toxic Chemical Release Inventory (TRI) and lower reporting thresholds for PBTs already on the inventory. The purpose of this rule amendment is to obtain more complete reporting on PBTs, particularly because

these types of substances may be produced or used at volumes significantly below the standard TRI reporting thresholds (e.g. manufacture of 25,000 pounds per year

- Evaluating fish in United States water bodies for PBT contamination. EPA's Office of Water will conduct a comprehensive study of PBT contamination in fish tissue as an indication of PBT contamination in our nation's water bodies.

Emphasis on prevention

All of the agency's tools play a role in the Strategy, especially for those situations such as risks from contaminated sediments that are less amenable to prevention-based approaches. However, there is a preference in the Strategy for pollution prevention and cooperative solutions. EPA believes that preventing pollution is the best way to safeguard the future well being of public health and the environment.

Emphasis on partnerships

EPA cannot accomplish the goals of this Strategy alone. It will rely on close cooperation with its regulatory partners, the States, to carry out these shared priorities. EPA will need their input to ensure that local and regional PBT problems are adequately addressed. Additionally, EPA will be engaging in partnerships with industry, environmental groups, and the public and will strive to fully involve stakeholders. Long-term success will be based on cooperative efforts that are mutually beneficial. The following partnerships exemplify the spirit of EPA's strategy:

- The American Hospitals Association, Healthcare Without Harm, and the EPA reached a landmark agreement with the goal of virtually eliminating mercury-containing waste from hospital waste streams by the year 2005 (*10*).
- Three Indiana steel facilities - Bethlehem Steel Burns Harbor, Ispat Inland Incorporated Indiana Harbor Works and United States Steel Gary Works - signed an agreement to reduce the use of mercury at their facilities through pollution prevention (*11*).
- The Chlor-alkali sector of the chemical industry has committed to reduce mercury use by 50 percent by 2005.

Mercury -- An Action Plan Example

This strategy is a living document that supports the development and implementation of action plans on priority PBTs. Attached to the Strategy is EPA's draft Mercury Action Plan. It illustrates an action plan that is national and even international in scope, and describes the kinds of actions EPA may take to reduce risks posed by other priority PBT pollutants. Each substance or group of substances will present its own set of action opportunities.

Mercury was selected for the first action plan because it is a well-documented threat to public health. Scientists know that mercury causes harmful changes in brain functions and slows fetal and child development in ways that may last a lifetime. Mercury moves easily among air, water, and land, so that coordinated actions are needed. Also, mercury is the leading and fastest growing cause of fish consumption advisories in this country. For example during the period from 1994 through 1997, thirty-four States had increases in fish consumption advisories due to mercury. Fourteen of these States adopted advisories statewide for their fresh waters or coastal waters.

Why Is Mercury A Problem?

Mercury is well known and long established as a neurotoxin that slows fetal and child development and cause irreversible deficits in brain function. Scientific debate is ongoing to more precisely determine the level of mercury exposure at which effects begin to occur. Several, but not all, existing studies show adverse human health effects at the level at which many Americans are exposed today from fish consumption. Tens of thousands of babies are born each year after being exposed in the womb to levels of mercury at which some studies have shown adverse health effects. This is the same uncertainty the Agency faced with respect to lead, decades ago. Like lead, mercury poses threats to our children that we must address now.

Key Mercury Action Items

The draft Mercury Action Plan contains a list of reasonable actions underway and planned across the Agency to address this potent neurotoxin. These include control of air emissions, seeking better treatment methods for mercury-containing wastes, reduced uses of mercury-containing products, further limited discharges to water, community right-to-know efforts, better estimating the amount of mercury emitted by power plants and revising the Agency's human health water quality criterion for mercury to better reflect bioaccumulation and the amount of fish typically eaten by people. Finally, research is needed on product substitution, control technologies, and improving the Agency's understanding of the link between air deposition and water pollution.

Public Reaction To The PBT Strategy / Mercury Action Plan Proposal

EPA provided a 90-day comment public period and held three public meetings across the country that concluded in early March, 1999. The Agency received over

300 comments from a broad range of interested parties, including industry, environmental and other public interest groups, State government, and many individual citizens. Overall, comments indicated support for the Strategy; in particular it's emphasis on the need for cross-media, cross-programmatic approaches to reducing risk.

Public Interest Groups and Citizens Comments

The Public Interest Group sector and individual citizens indicated that the goal of the Strategy needs to be stronger and adopt the principles of virtual elimination, which to which the Agency ascribes to under the Canada/U.S. Binational Toxics Strategy. Regarding the identification of additional substances for national action plans these comments emphasized that the Agency should use the precautionary principal and not presume that it must prove risk before acting. These comments also cautioned the Agency to not rely solely on pollution prevention and the development of voluntary partnerships. They urge the Agency to make regulatory control options and enforcement strong components of any action plan because they provide powerful incentives for release and risk reductions. Certain comments also emphasize the need to promote risk reduction through public education programs.

Industry Comments

Comments from the Industry sector emphasized that EPA should be consistent in what it defines as a PBT and that such substances should exhibit all three properties – that is, it must be persistent and bioaccumulative and toxic. For purposes of selecting additional substances for national action plans these comments favored a bioaccumulation factor of 5000 and a media specific half life of two days for air or six months for soil, water or sediment. They contend that these levels are consistent with levels agreed to in international conventions and that EPA should adopt them for purposes of the Strategy. EPA should also prioritize its choices based upon relative risk and in any event the list of additional substances should be kept small for the purpose of all parties focusing limited resources. Many comments addressed metals and indicated that they should not be evaluated in the same manner as organic substances and that bioavailability was the more pertinent issue regarding metal compounds. Comments also indicated that most progress in reducing PBTs would come through voluntary measures – that emphasis on control actions would thwart partnership development.

State Comments

A number of States indicated that they are also developing similar strategies for PBTs. Some indicated that they are stressing actions to reduce mercury emission and further contamination problems. Some, such as the New England States have also developed approaches with neighboring provinces of Canada to address mercury. Many welcomed the federal level of attention and stressed the need for more resources and better data on emissions and presecence of PBTs in the environment. Comments also stressed that States need flexibility in addressing PBTs. PBT problems and risk reduction options may vary considerably depending by region because of varying factors such as climate, types and numbers of PBT pollutant sources, and the contribution from global sources.

Comments On the Mercury Action Plan

Comments on the draft Mercury Action Plan were also generally favorable but many stressed the need for the plan to be more strategic. These comments stressed that the Plan needed to be more specific and say what it plans to do and by when. Also certain comments stressed the need to prioritize actions based upon relative risk contribution. Other comments indicated that EPA should pursue virtual elimination of mercury and take action to phase out non-essential uses of mercury.

Finding Out More

For copies of the EPA draft PBT Strategy and other related documents, please call the Pollution Prevention Information Clearinghouse at (202) 260-1023. Documents are also available on the World Wide Web at http://www.epa.gov/pbt.

References

1. USEPA National Listing of Fish and Wildlife Consumption Advisories, 1997, EPA-823-F-98-009
2. Natural Resources Defense Council. 1998. Contaminated Catch -- The Public Health Threat From Toxics in Fish.
3. National Academy of Public Administration, Washington D.C. 1995 Setting Priorities, Getting Results -- A new Direction for EPA.
4. Organization for Economic Cooperation and Development 1996. Environmental Performance Review of the United States. OECD Publications Center

5. USEPA and US Department of Agriculture 1998. Clean Water Action Plan: Restoring and Protecting America's Waters. EPA-840-R-98-001.
6. The Great Lakes Binational Toxics Strategy: Canada - US Strategy for the Virtual Elimination of Persistent Toxic Substances in the Great Lakes 1997.
7. USEPA, Office of Solid Waste, Waste Minimization Prioritization Tool (WMPT) (Beta test version 1.0): Users Guide and System Documentation (EPA 530-R-97-019) June 23, 1997. www.epa.gov/wastemin.
8. National Health and Nutrition Examination Surveys (NHANES), http://www.cdc.gov/nchswww/about/major/nhanes/nhanes.htm
9. USEPA, EPCRA Section 313 Toxic Chemical Release Inventory, EPA 745-B-97-008, Office of Pollution Prevention and Toxics, Washington D.C., November 1997. http:// www.epa.gov/tri/
10. 10. Memorandum of Understanding between the United States Environmental Protection Agency and the American Hospital Association, June 24, 1998.
11. Mercury Pollution Prevention Initiative, Voluntary Agreement between the Lake Michigan Forum, Indiana Department of Environmental Management, U.S. Environmental Protection Agency and Bethlehem Steel Burns Harbor, Ispat Inland Inc. Indiana Harbor Works and U.S. Steel Gary Works. September 25, 1998.
12. USEPA, Proposed Category for Persistent, Bioaccumulative and Toxic Chemical Substances, Federal Register OPPTS-53171; FRL-5771-6. October 5, 1998.
13. USEPA, Notice of Availability of Draft RCRA Waste Minimization PBT Chemical List, November 9, 1998, www.epa.gov/wastemin.
14. USEPA, Persistent Bioaccumulative Toxic (PBT) Chemicals; Proposed Rule Federal Register: January 5, 1999 (64 FR 687) www.epa.gov/tri

Chapter 11

Dioxin Pollution Prevention Inventory for the San Francisco Bay

Greg Karras

Communities for a Better Environment (CBE), 1611 Telegraph Avenue, Suite 450, Oakland, CA 94612

The first comprehensive inventory of root causes contributing to dioxin exposure in San Francisco Bay shows that actions to zero out dioxin at its industrial sources can be identified, prioritized, measured, and verified with existing scientific tools. Sources of chlorine in dioxin-forming reactions can be identified. Options to block these reactions in processes and production systems can be defined and verified. Existing evidence on sources that contribute to cumulative exposure, and preventable root causes of dioxin formation, provides an objective basis for identification of action priorities. Monitoring of food resources and highly exposed humans can track progress and identify needs for new actions.

A food resource for indigenous people and a major commercial fishing center in the past, the San Francisco Bay is fished by subsistence anglers today *(1,2,3,4,5,6)*. This food resource is contaminated with dioxin compounds and other persistent, bioaccumulative chemicals at levels which may be toxic, despite decreasing dioxin exposure elsewhere, especially for children of S.F. Bay anglers who are exposed to dioxin in utero and via breast milk *(3,7,8,9,10)*. In early 1999 the City of Oakland and the City and County of San Francisco adopted a policy to eliminate dioxin production and exposure from all sources in this region *(11,12)*. The new policy requires a new approach, to measure the steps toward zero dioxin at all sources in a region.

This inventory is conducted to assess whether existing data support a measurable, verifiable approach to organize and effect the goal of zero dioxin creation and exposure in San Francisco Bay and environs. Its scope is limited to preventable root causes of dioxin creation and exposure in S.F. Bay and areas within approximately 60 kilometers of the bay. Chemicals included are the polychlorinated dibenzodioxin and dibenzofuran compounds (PCDD/Fs) and the dioxin-like coplanar polychlorinated biphenyls (dioxin-like PCBs), also referred to as dioxin compounds or dioxin.

Data, Limitations and Methods

Despite recent advances in sampling and analysis, environmental concentrations of dioxin compounds are at the extreme low end of those measured successfully *(13,14, 15)*. In light of this measurement challenge, existing data are analyzed to assess the support for linking exposure to dioxin compounds in the bay food chain back to regional sources, root causes of dioxin creation in these sources, and specific pollution prevention solutions. Data gathered from many research efforts are discussed in this section. The following sections analyze these data to measure differences in exposure, trace exposure to root sources where reactions that form dioxin occur, define steps to block these reactions, and assess priorities for action.

Data from Exposure Measurements

Dioxin has not been measured directly in S.F. Bay anglers. However, more than 90 percent of total dioxin exposure in the general population is from food *(16,17,18)*.

PCDD/Fs and some dioxin-like PCBs were analyzed in 29 composite samples of more than 160 fish collected from 16 sites in the bay during May and June, 1994 and in June and July, 1997 *(3,8,9)*. Fifteen of the samples, which were sorted by site and species, measured fillets with skin from 105 white croaker *(Genyonemus lineatus)* collected at 11 S.F. Bay sites. The samples were collected and prepared by the State Department of Fish and Game and analyzed by the State Hazardous Materials Laboratory according to HML Method 880. These data are limited by what was not measured: fish organs eaten by some anglers; seasonal changes in contamination; dioxin levels in other years; many 'hot spot' locations in the bay; some dioxin-like PCB compounds; and many other species, such as humans who eat the fish *(3)*.

Three multilingual surveys with more than 1,000 S.F. Bay anglers were reported *(3,4,5,6)*. These data support a lower-bound estimate of the number of people who fish regularly for food, a maximum level of fish consumption, and the abundance of white croaker in subsistence anglers' catch and diet, but do not support estimates of the total number of people eating any specific amount of bay fish because many fishing sites and anglers are not included in the surveys *(3,4,5,6)*.

Dioxin exposure in S.F. Bay fish was measured as a range of possible mean concentration in white croaker fillets with skin. The low end of this range assumes that all data flagged by the laboratory as below analytical detection limits, below quantitation limits, or with the compound detected in the blank have a value of zero; the high end assumes all data have the reported value. High-end human exposure from fishing the bay daily for food was measured from these data and the range of daily consumption values reported in the surveys. This estimate is expressed as toxicity equivalence ("TEQ"); the sum of the concentrations of the various PCDD/Fs and dioxin-like PCBs multiplied by potency factors that describe the relative toxicity of each compound *(8,17,19,20)*. All data reported in this inventory as TEQ use the I-TEQ scheme used by the bay surveys *(8,9)* and by U.S. EPA in 1998.

Source Identification Data

Data from worldwide process, emission, and environmental measurements over time provide evidence that reactions involving chlorine in various industries are an important cause of increased dioxin in the environment *(3,13,17,21)*. Although distant sources cause substantial dioxin fallout in some areas hundreds of km. downwind, *(21,22,23,24,25)*, this is not shown in S.F. Bay, which is downwind of an ocean with few dioxin sources. There is less dioxin fallout from ocean versus continental air masses *(22,24)*, and less dioxin in lakes near the Pacific versus other U.S. lakes *(26)*.

Fish collected from a rural lake approximately 160 km north of S.F. Bay and from a reservoir adjacent to the bay were analyzed by the same laboratory and methods as were used for S.F. Bay croaker *(3,27)*. Prevailing winds bring air masses from the Pacific Ocean that do not pass by Bay Area sources en route to the rural lake. Largemouth bass and catfish were each more contaminated in the lake near the bay than in the distant lake. Largemouth bass data from the rural lake were compared with bay croaker data because these species occupy similar trophic levels. These results indicate that differences between freshwater and estuarine fish alone would not explain lower dioxin levels in the rural lake.

PCDD/Fs were analyzed using EPA method 1613 in 50 storm water runoff samples collected on 17 dates from December, 1995 through December 1998 from 23 Bay Area sites *(28,29)*. Duplicate analyses of five samples indicated good reproducibility. All PCDD/Fs were detected at least once, but only 40% were detected across all 50 samples, due in part to the small, roughly one-liter samples that were collected and prepared for analysis *(14,28,29,30)*. Each sample is only a snapshot of the dioxin entrained in the runoff stream at a moment in a highly variable, dynamic storm. Suspended solids can vary a thousand-fold between samples *(28)*. In a small study using different methods, dioxin concentration changed by more than a hundred-fold during one runoff event and peaked at different times in different outfalls *(31)*. Data on suspended solids, and precise sample collection details were not reported for some samples. Many storms, water-years, and sites were not sampled. These data bracket ranges of dioxin concentration in repeated samples, but do not measure its variability well enough to calculate mean levels with any confidence.

Limited PCDD/F emission data were found for 211 Bay Area samples, including waste water, process water, sludge and ash analyzed by EPA methods 1613 or 8280, and stack tests analyzed by EPA Method 23 or CARB Method 428. Process and chlorine source data were found for these sources. All these data were from reports to government agencies required by state and federal laws. PCBs sources were not analyzed for dioxin-like PCBs. At least one release route (effluent, stack, or product) was not tested at 96 percent of sources. Most PCDD/Fs were below detection limits in effluent. There were too few stack tests to quantify emission rates. Process data were limited, especially for chlorine sources of dioxin emission from vehicles and wood fires. To aid in assessment of the Bay Area sources, data from elsewhere in the world were gathered from the general literature. A full report on the inventory describes these data in detail *(3)*. Sources of dioxin production located from the data were compared with the fish and runoff data described above.

Data on Preventable Root Causes of Dioxin

Source identification data confirmed dioxin-forming reactions in 94 percent of the plants known to be releasing dioxin to the bay, and all these plants already used relatively advanced end-of-pipe treatment *(3)*. Three observations indicated that understanding the dioxin-forming reactions requires understanding how chlorine becomes available to them:

- Global trends in dioxin exposure and industrial chlorine use match *(3,13,21)*;
- Chlorinated dioxin compounds cannot be formed without chlorine present; and
- Blocking chlorine input has eliminated dioxin in some processes *(25)*.

Data were gathered on chlorine sources in dioxin-forming reactions, on options for altering processes and production systems to block these reactions, and on verification of zero dioxin results. All data are detailed in the inventory report *(3)*. These data were assessed for all sources to identify questions for further analysis.

Catalytic cracking provides an example of chlorine source analysis. The inventory identified this newly-discovered dioxin source in the petroleum refining industry from source test and process data *(3,32,33,34,42)*. In this process, aromatic carbon buildup is burned off a silica-alumina catalyst. Chlorine was measured in refined products and intermediates *(33,34)*. However, more test data were needed to quantify chlorine inputs to cracking units from incomplete crude oil desalting, and from intentional chlorine addition to interconnected refinery processes *(3)*.

Chlorine is added in reforming catalyst regeneration, a refinery process that removes carbon buildup from a platinum/rhenium catalyst by incineration, and creates dioxin: Here data were limited to characterize options that maintain catalyst activity without using chlorine, and thus block dioxin formation *(3,35,36,37,38,39)*. Options for process change were also assessed in multi-plant systems, such as incineration of wastes from hospitals using disposable polyvinyl chloride produced elsewhere. Work at pulp paper plants that replaced chlorine with peroxide processes provides examples of measurements which verify dioxin elimination. Data from other areas were used to assess the process measurements in many cases *(13,21,25, 33,34,35,37,39,40,41,42,43,44,45,46)*. These examples name some of the Bay Area processes. All identified sources were compared with respect to the data available to answer these chlorine source, process alteration and verification questions.

Some remaining limitations in the data were analyzed. The inventory addresses whether fireplaces and motor vehicles are "sources" by asking if other preventable causes add to wood contamination via atmospheric dispersion of chlorine compounds *(43)*, and refinery inputs of chlorine and dioxin to diesel fuel and motor oil *(34)*. Data were too limited to ensure identification of all sources: Better surveillance might answer this question. Data were too limited to prioritize sources based on harm, and increased monitoring alone cannot solve this problem: Existing tools cannot predict all future biological effects of a dioxin release today, particularly in future generations. Developing children are exposed to dioxin originally ingested by their parents in utero and via breast milk.

Measuring Dioxin Exposure in Subsistence Anglers

Many hundreds of people fish from the shorelines and piers of S.F. Bay. Most of the anglers eat their catch or share it with family and friends. Most species caught are consumed, and white croaker, also known as kingfish or tomcod, is caught and eaten most often, especially by subsistence anglers. Participants in three different surveys reported eating bay fish daily in one, two or three meals per day. Individuals who ate the most fish ate from 100 to 450 grams per day, with the higher value reported by different people and surveys *(3,4,6)*. This is consistent with eating a four to eight ounce (112-224 g.) portion in one to three meals per day.

Concentrations of dioxin compounds in the largest comparable bay-wide sample of fish eaten regularly by subsistence anglers are summarized in Table I. Most of the 22 compounds analyzed in 105 fish were detected, including 90 to 100 percent of those that contributed most to dioxin-like toxicity (TEQ). Analytes below detection or quantitation were present between zero and the detection limit or reported value, respectively. Thus, mean total TEQ was 9.5 to 10 picograms per gram.

Table I. Dioxin in Fillets with Skin from 105 S.F. Bay White Croaker

Parameter	*Detection rate*	*pg/g detected*	*Mean pg/g TEQ*
2,3,7,8-TCDD	93%	0.1 to 0.5	0.1 to 0.3
1,2,3,7,8-PeCDD	40%	0.4 to 0.7	0.09 to 0.2
2,3,7,8-TCDF	100%	0.5 to 3.7	0.18
2,3,4,7,8-PeCDF	93%	0.4 to 1.5	0.42 to 0.45
3,3',4,4',5-PeCB	100%	22.0 to 130.0	5.4
Seven PCDDs	40%	—	0.2 to 0.5
Ten PCDFs	64%	—	0.66 to 0.72
Five PCBs	96%	—	8.6
22 dioxin compounds	83%	—	9.5 to 10.0

NOTES: Units are pg/g wet wt. Means shown with analytes below detection or quantitation at values of zero, and at values equal to detection limits or reported values.
SOURCE: Reprinted with permission from reference 3. Copyright 1998.

A 70 kilogram person who eats 100 grams of these fish per day ingests 14 picograms of TEQ per kilogram of body weight per day, and one who eats 450 g/d ingests 63 pg/kg/day, from bay fish alone. However, measurements in the general population indicate that human exposure is approximately two to six pg/kg/day, and varies little between U.S. regions *(17,20)*. This comparison suggests that high-end S.F. Bay angler exposures are as much as 10 to 30 times those in the general population. Anglers receive additional dioxin exposure from dairy and meat products, and eat other fish with less dioxin content than croaker, but these factors cause smaller exposure differences, especially for high-end anglers who eat mainly croaker. Significant exposure differences are measurable with existing food chain data.

Tracing Root Sources of Dioxin in the Region

Mean dioxin TEQ in S.F. Bay fish is nearly thirty times greater than that in similar trophic level fish from Black Butte Lake. See Figure 1A. This difference exists across PCDD, PCDF and dioxin-like PCB compounds and is not due to differences between freshwater and estuarine systems alone, as there is more dioxin in the same species at a lake near the bay than at Black Butte Lake *(3,8,9,27)*.

Prevailing winds bring air masses from the Northeastern Pacific Ocean over S.F. Bay, Black Butte Lake, and other U.S. west coast areas. However, Black Butte is a rural lake approximately 160 km. north of the bay which receives winds that have not passed over the Bay Area. Source emissions in the Pacific Ocean are few, and measurements elsewhere suggest less PCDD/F *(22)* and PCB *(24)* fallout from air that has passed over oceans versus air that passed over land. PCDD/F and dioxin-like PCB sediment concentrations in other U.S. lakes decrease with proximity to the Pacific *(26)*. The greater dioxin levels in Bay Area versus Black Butte fish cannot be explained by fallout from distant sources, suggesting important nearby sources.

Twenty seven dioxin-producing sources that were identified from emission and process data are shown in Figure 1B. These include six incineration processes, six petroleum refineries, six furnaces that recover electronics and aluminum scrap, three chemical drum furnaces, two gray iron foundries, a cement kiln, a utility burning petroleum coke, a hazardous waste facility, and an organic chemicals plant *(3)*. PCDD/F releases to the bay are confirmed from at least 17 of these facilities *(3)*. This concentration of sources corresponds to elevated dioxin levels in S.F. Bay fish.

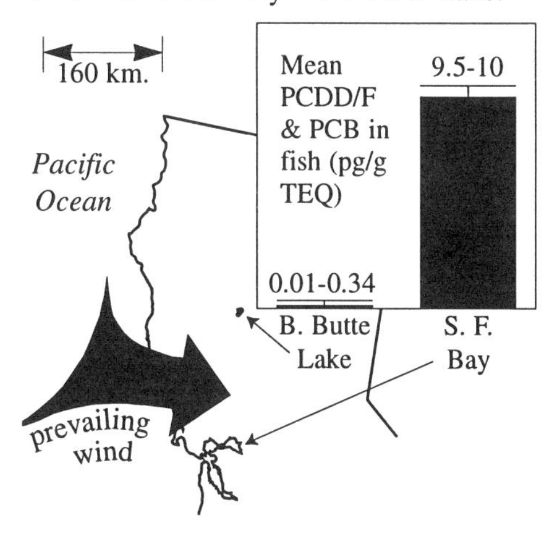

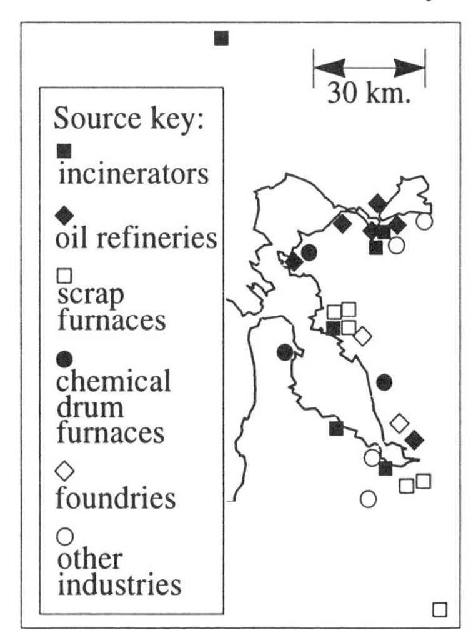

Figure 1. Comparison of Dioxin in Fish from San Francisco Bay and a Rural Lake, and Locations of Bay Area Dioxin Sources. (Reproduced with permission from reference 3. Copyright 1998.)

The gradient of increasing dioxin with proximity to Bay Area sources suggested by the fish tissue analysis is confirmed by extending the analysis to storm water runoff. Fifty runoff tests at 23 sites can be compared with distance from the sources in Figure 1B *(3,28,29)*. The number and breadth of samples add to the power of the comparison, and support estimates of runoff dioxin concentration ranges (minima to maxima), despite limitations in sampling of the highly dynamic storm events.

As shown in Figure 2, differences in runoff contamination are apparent when the data are grouped by distance from the runoff site tested to dioxin sources. PCDD/F TEQ in the most contaminated runoff sample within one block (<100 m.) of a source is nearly 200 times greater than any measured in six samples taken 16 to 25 km from sources. Variable levels over time also suggest nearby industrial sources: TEQ in runoff nearest to a source ranges a thousand-fold between samples; while TEQ in runoff 16 to 25 km from sources ranges less than ten-fold between samples. Maxima and ranges of dioxin contamination measured in runoff 4 to 10 km from sources, and in runoff several blocks away from dioxin-producing processes on petroleum refinery land, are consistent with this pollution gradient.

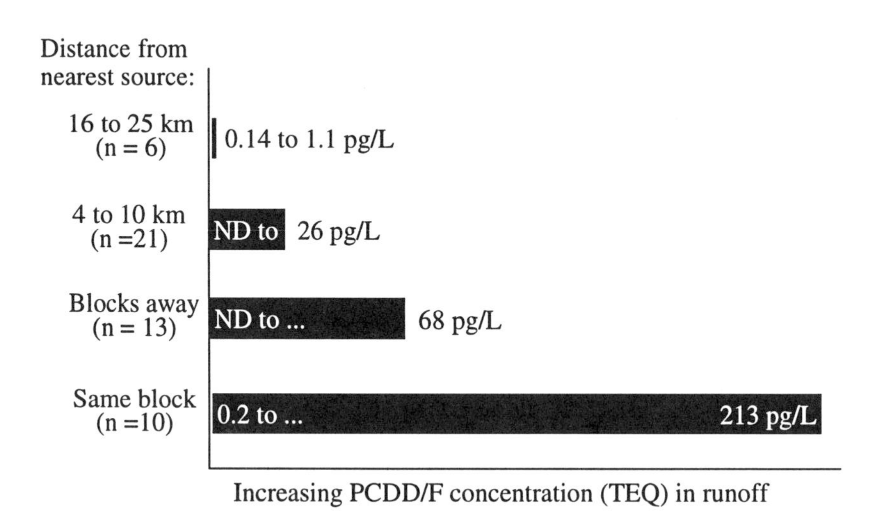

Figure 2. Ranges of PCDD/F TEQ Detected in 50 Samples of S.F. Bay Area Storm Water Runoff. (Data from references 3, 28 and 29.)

In addition to the stack-to-food chain pathway reflected in this gradient, many sources emit dioxin via effluent *(3)*, and products such as motor oil and fuels *(34)*.

The creation of dioxin from chlorine and organic compounds is confirmed at 16 of the 17 sources in Figure 1B that are known to release dioxin to the bay *(3)*. These 16 sources use relatively advanced emission controls *(3)*. Thus, cumulative pollution from them is not caused by a lack of end-of-pipe treatment. Therefore, industrial processes involving chlorine can be identified as root causes of dioxin exposure.

Measuring Preventable Dioxin Sources

Technically answerable questions can be defined about chlorine sources for dioxin creation in the petroleum refining catalytic cracking process. See Figure 3. Inputs to a catalyst incineration process where dioxin is detected in this newly discovered source include chlorine addition to various interconnected refinery processes, and incomplete chlorine removal at desalting units *(3,33,42)*. These inputs can be traced, as chlorine can be measured in refinery intermediates and products *(33,34)*.

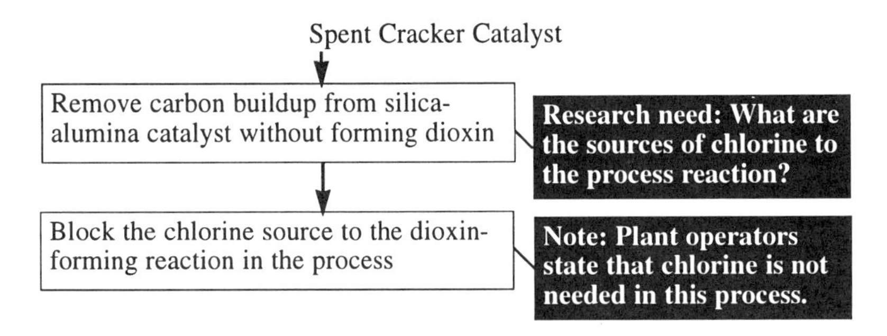

Figure 3. Cracking Unit Pollution Prevention. (Data from Reference 3.)

The chlorine source is verified, and questions about blocking dioxin formation can be defined, in reforming process catalyst regeneration. See Figure 4. Dioxin forms and emits when petroleum refiners add chlorinated solvents and burn aromatic carbon deposits from a platinum/rhenium catalyst under pressure *(35,36,37,38,39)*. Alternatives that protect or repair the catalyst, while removing carbon deposits without incinerating chlorinated solvents, can be assessed by measurements of process chemistry and on-site operations. Given the analytical detection power of high volume sampling and analysis methods *(14,24,30)*, anecdotal suggestions that a dry process might eliminate dioxin from this source can be verified.

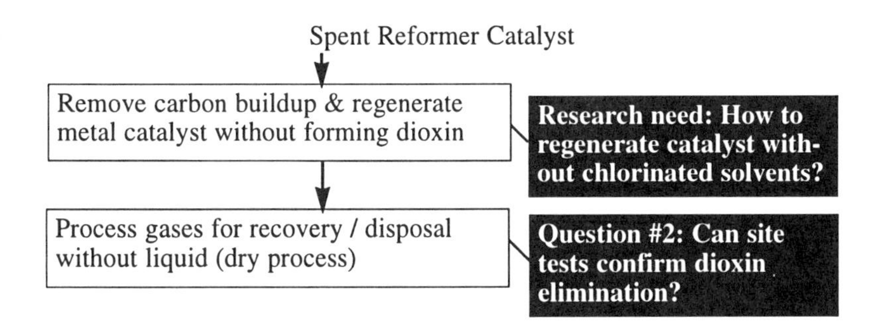

Figure 4. Reforming Unit Pollution Prevention. (Data from Reference 3.)

Alternatives to the introduction of chlorinated compounds into incineration processes can be defined for the health care industry. As shown in Figure 5, a well-defined option for this multi-plant production system would phase out medical use of polyvinyl chloride, segregate waste, recycle, and treat the major portion of infectious waste by autoclave and/or microwave *(25,44)*. Safe alternative methods to disinfect wastes, especially pathological wastes, are the subject of ongoing inquiries.

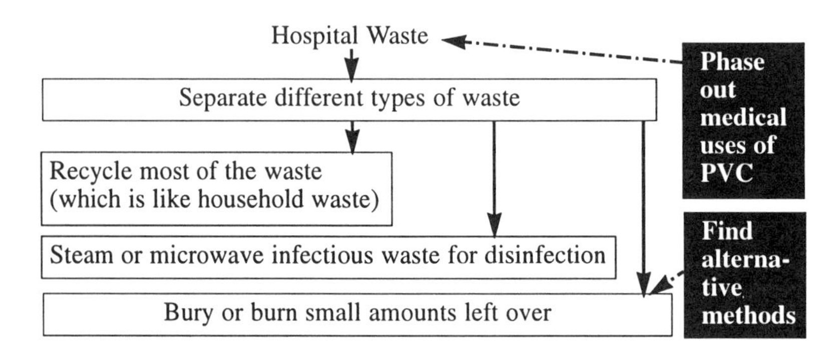

Figure 5. Incinerator Pollution Prevention. (Data from Reference 3.)

Alternatives that block the chlorine source to dioxin formation can be verified in pulp paper plants. Figure 6 outlines a "totally chlorine free" (TCF) process scheme: extended cooking of the pulp with oxygen; peroxide instead of chlorine bleaching; and recycling the water freed of chlorinated organic wastes *(13,25)*. The effectiveness of this TCF option was distinguished from another option that replaces chlorine gas with chlorine dioxide, forms free chlorine, and causes a measurable increase in dioxin as compared with TCF *(25)*. This example shows an advanced stage in the inquiry, where the identified chlorine source is blocked, and zero dioxin is verified.

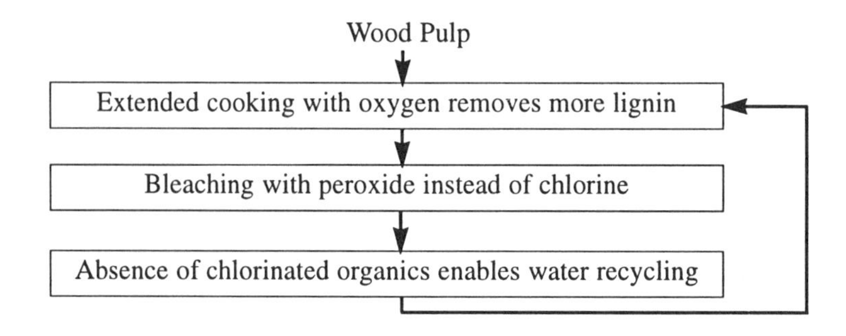

Figure 6. Paper Plant Pollution Prevention. (Data from Reference 3.)

Pollution Prevention Action Priorities

Despite their many limitations, the data support measurement of increased cumulative exposure in humans from dioxin emission caused by identified industrial processes involving chlorine in the region. These processes can be measured to define and verify alternatives for zero dioxin. Present measurements show that, at some individual facilities, known dioxin production from a known root cause adds dioxin to this cumulative exposure. At other potential sources, dioxin formation, its root causes, and/or releases to the bay require confirming investigation.

Known and potential sources are ranked in Table II according to whether dioxin production by the source, a root cause creating dioxin, and its addition (even if unquantified) to the cumulative exposures, is known. At a petroleum refinery in Richmond, dioxin production is confirmed by direct measurements of PCDD/F compounds in on-site cracking and reforming processes, root causes of formation are identified (e.g., chlorine addition to reformer catalyst regeneration), and releases to the bay are confirmed in effluent and at relatively high levels in on-site runoff *(3,28, 32,35,37)*. For this facility and others at the top of Table II, steps to block dioxin formation, usually by blocking its chlorine source, can be defined now to eliminate confirmed additions to S.F. Bay food chain exposure.

Near the bottom of the table, wood burning is an example of priorities for different actions. Measurements elsewhere detect dioxin in post-industrial wood fires, but preindustrial data *(13,21)*, and global forest contamination with industrial chlorine compounds today *(43)*, suggest wood burning is not in itself a root source. Specific, preventable Bay Area chlorine sources which cause dioxin formation in wood burning remain to be identified, and no measurements confirm dioxin release from this source to the bay. For this potential source and others nearer the bottom of Table II, source confirmation investigation may be needed to find preventable root causes of dioxin exposure before pollution prevention options can be defined.

Safety Precaution

Safety precaution is fundamental to pollution prevention, suggesting some next steps for investigation. Independently verifiable results are crucial to the scientific method, and necessary to ensure that safer alternatives are identified given a significant conflict in time scales for investment. This conflict arises from societal interests in safety from trans-generation exposures to dioxin, and business interests in a much shorter-term return on investment. Including workers in the investigations can provide more plant operating data to identify and address job exposure and safety. Since dioxin has no industrial use, eliminating it should be jobs-neutral, and training in the new methods can avoid job displacement as well as job safety problems. Potential zero dioxin alternatives that involve changes in chemical use or release should be assessed for safety in acute, chronic and trans-generation exposure.

The new discovery of dioxin in catalytic cracking raises the possibility of other unidentified sources of exposure. This can be addressed by fish tissue and angler

Table II. Dioxin Pollution Prevention Action Inventory

Source, location	*Dioxin confirmation*	*Cause of formation*	*Release to Bay*
Petroleum refinery, Richmond	Measurement[a]	Identified	Confirmed[c]
Medical incinerator, Oakland	Measurement[a]	Identified	Confirmed[c]
Petroleum refinery, Avon	Measurement[a]	Identified	Confirmed[c]
Petroleum refinery, Rodeo	Measurement[a]	Identified	Confirmed[c]
Petroleum refinery, Benicia	Measurement[a]	Identified	Confirmed[c]
Gray iron foundry, Oakland	Measurement[a]	Identified	Confirmed[d]
Sludge incinerator, Martinez	Measurement[a]	Identified	Confirmed[d]
Oil re-refinery, Newark	Measurement[a]	Identified	Confirmed[d]
Portland cement kiln, Cupertino	Measurement[a]	Identified	Confirmed[d]
Sludge incinerator, Palo Alto	Measurement[a]	Identified	Confirmed[d]
Sludge incinerator, San Mateo	Measurement[a]	Identified	Confirmed[d]
Gray iron foundry, Union City	Measurement[a]	Identified	Confirmed[d]
Drum furnace, San Francisco	Measurement[b]	Identified	Confirmed[c]
Petroleum refinery, Martinez	Measurement[b]	Identified	Confirmed[c]
Old/in use PCBs, Bay Area-wide	Measurement[b]	Identified	Confirmed[c]
CO boiler/incinerator, Martinez	Measurement[b]	Identified	Confirmed[c]
Scrap PWB furnace, Gilroy	Measurement[b]	Identified	Unconfirmed
Drum furnace, Hayward	Measurement[b]	Identified	Unconfirmed
Aluminum scrap furnace, Oakland	Measurement[b]	Identified	Unconfirmed
Scrap PWB furnace, Santa Clara	Measurement[b]	Identified	Unconfirmed
Scrap wire furnace, Oakland	Measurement[b]	Identified	Unconfirmed
Coke-fired utility, C.C. County	Measurement[b]	Identified	Unconfirmed
Scrap PWB furnace, San Jose	Measurement[b]	Identified	Unconfirmed
Drum furnace, Richmond	Measurement[b]	Identified	Unconfirmed
Scrap wire furnace, Oakland	Measurement[b]	Identified	Unconfirmed
Ind. wood incinerator, Windsor	Measurement[b]	Identified	Unconfirmed
Diesel fuel, Bay Area-wide	Measurement[b]	Unconfirmed	Confirmed[d]
Residential wood fires, Bay Area	Measurement[b]	Unconfirmed	Unconfirmed
Haz. waste facility, E. Palo Alto	Process chem.	Unconfirmed	Confirmed[d]
Organic chemicals plant, Pittsburg	Process chem.	Unconfirmed	Unconfirmed

NOTES: Dioxin confirmation measurements are either *(a)* from tests at the specific source or *(b)* from tests at similar sources elsewhere; confirmed releases to S.F. Bay are either *(c)* confirmed with evidence of relatively high concentration or amount or *(d)* confirmed with unknown potential for relatively high concentration or amount.
SOURCE: Reprinted with permission from reference 3. Copyright 1998.

breast milk monitoring before and after dioxin is zeroed out at known sources, and by source analysis for dioxin compounds. The latter should be a short-term priority for dioxin-like PCBs. PCBs contribute most of the total dioxin-like TEQ in S.F. Bay fish, and the mass of tetra- through hepta-chlorinated PCBs is much larger than that of PCDD/F measured in stack tests at several current sources *(3,38)*.

Dioxin sources in other regions, and exposure to other toxic chemicals should be addressed by assessing the applicability of the inventory method to other regions and other persistent, bioaccumulative chemicals. Emerging evidence of increasing exposure to polybrominated diphenyl ethers provides a compelling example of this need *(7)*. Such assessments should encompass a broad safety concern identified by this analysis: They should address the safety inherent in a method for preventive actions based on verifiable criteria, relative to methods which delay such actions while attempting to measure fully the risk of future harm from present toxic chemical releases – an impossible task using present tools.

Literature Cited

1. Nichols, F. H.; Cloern, J. E.; Luoma, S. N.,; Peterson, D. H. *Science (Washington, D.C.).* **1986,** *volume 231,* pp 525-648.
2. Smith, S.E.; Kato, S. *In San Francisco Bay: the Urbanized Estuary;* Conomos, T. J.; Leviton, A. E.; Eds.; Pacific Division/American Association for the Advancement of Science: San Francisco, CA, 1979; Vol. 1, pp 445-468.
3. Karras, G. *On the Hook for Zero Dioxin;* Report No. 98-2; Communities for a Better Environment: San Francisco, CA, 1998.
4. Cohen, A. N. *Fishing for Food in San Francisco Bay;* Part I; Save San Francisco Bay Association: Oakland, CA, 1995.
5. Wong, K. A.; Nakatani, K. *Fishing for Food in San Francisco Bay;* Part II; Save San Francisco Bay Association: Oakland, CA, 1997.
6. Chiang, A. *A Seafood Consumption Survey of the Laotian Community of West Contra Costa County, CA;* Asian Pacific Env. Network: Oakland, CA, 1998.
7. Noren, K.; Meironyte, D. *Organohalogen Compounds.* **1998,** *volume 38,* pp 1-4.
8. San Francisco Bay Regional Water Quality Control Board; California Department of Fish and Game; State Water Resources Control Board. *Contaminant Levels in Fish Tissue from San Francisco Bay;* Final Report; San Francisco Bay Regional Water Quality Control Board: Oakland, CA, 1995.
9. Petreas, M. *Final Report According to Contract PHI#642 (SFEI-Fish Study);* Final Report to the San Francisco Estuary Institute; State Department of Toxic Substances Control Hazardous Materials Laboratory: Berkeley, CA, 1998.
10. Chadwick, H. K.; Urquhart, K. A. F.; Regalado, K. *Selenium Verification Study 1988-1990;* Report 91-2-WQWR Prepared Under Interagency Agreement #9-114-300-0; State Water Resources Control Board: Sacramento, CA, 1991.
11. *Establishing a Regional Task Force and Policy on Dioxin, Public Health and the Environment;* Resolution; City of Oakland, CA: 1999.
12. *Establishing Dioxin Pollution as a High Priority for Immediate Action for the City and County of San Francisco in Order to Restore Water, Air and Total Environment Quality;* Resolution; City and County of San Francisco, CA: 1999.
13. Zook, D. R.; Rappe, C. *In Dioxins and Health;* Schecter, A.; Ed.; Plenum Press: New York, 1994; Vol. 1, pp 79-113.

14. Afghan, B. K.; Carron, J.; Goulden, P. D.; Lawrence, J.; Leger, D.; Onsuka, F.; Sherry, J.; Wilkinson, R. *Can. J. Chem.* **1987,** *volume 65,* pp 1086-1097.
15. Telliard, W. A.; McCarty, H. B.; Riddick, L. S. *Chemosphere.* **1993,** *volume 27 (1-3),* pp 41-46.
16. Startin, J. R.; *In Dioxins and Health;* Schecter, A.; Ed.; Plenum Press: New York, 1994; Vol. 1, pp 115-136.
17. Schecter, A.; *In Dioxins and Health;* Schecter, A.; Ed.; Plenum Press: New York, 1994; Vol. 1, pp 449-485.
18. Papke, O.; Herrmann, T; *Influence of Various Food Consumption Habits on the Human PCDD/PCDF Body Burden;* Unpublished paper presented before the Div. of Env. Chem., ACS, at Anaheim, CA, March 21-25, 1999.
19. Silbergeld, E. K.; deFur, P. L.; *In Dioxins and Health;* Schecter, A.; Ed.; Plenum Press: New York, 1994; Vol. 1, pp 51-78.
20. Birnbaum, L. S. *TCDD: Human Health Toxicity Assessment and WHO TEFs;* Prepared by Linda S. Birnbaum, Ph.D., D.A.B.T., Associate Director for Health, National Health and Environmental Effects Research Laboratory; U.S. Environmental Protection Agency: Research Triangle Park, NC, 1998.
21. Webster, T.; Commoner, B. *In Dioxins and Health;* Schecter, A.; Ed.; Plenum Press: New York, 1994; Vol. 1, pp 1-50.
22. Brzuzy, L. P.; Hites, R. A. *Environ. Sci. Technol.* **1996,** *vol. 30 (6),* pp 1797-1804.
23. Broman, D.; Naf, C.; Rolff, C.; Zebuhr, Y. *Environ. Sci. Technol.* **1991,** *volume 25 (11),* pp 1850-1863.
24. Lee, R. G. M.; Jones, K. C. *Environ. Sci. Technol.* **1999,** *vol. 33(5),* pp 705-712.
25. Commoner, B.; Cohen, M.; Bartlett, P. W.; Dickar, A.; Eisl, H.; Hill, C.; Rosenthal, J. *Dioxin Fallout in the Great Lakes and Zeroing Out Dioxin in the Great Lakes: Within Our Reach;* Submitted to the International Joint Commission of the U.S. and Canada; Center for the Biology of Natural Systems, Queens College: Flushing, NY, 1996.
26. Cleverly, D.; Monetti, M.; Phillips, L.; Cramer, P.; Heit, M.; McCarthy, S.; O'Rourke, K.; Stanley, J.; Winters, D. *Presentation at Dioxin '96, 16th International Symposium on Chlorinated Dioxins and Related Compounds, August 12-16 in Amsterdam, The Netherlands;* Short paper in *Organohalogen Compounds.* **1996,** *volume 28,* pp 77-82.
27. Petreas, M. *Draft Final Report According to Contract #97-T1500;* Report to the Office of Environmental Health Hazard Assessment; State Department of Toxic Substances Control Hazardous Materials Laboratory: Berkeley, CA, 1998.
28. San Francisco Bay Regional Water Quality Control Board. *Survey of Storm Water Runoff for Dioxins in the San Francisco Bay Area;* California Regional Water Quality Control Board, San Francisco Bay Region: Oakland, CA, 1997.
29. Analytical results submitted by Quanterra Inc. to the San Francisco Public Utilities Commission from projects 302541 and 303251 and sample collection data. Provided under the California Public Records Act in February, 1999.
30. Gotz, R.; Enge, P.; Friesel, P.; Roch, K.; Kjeller, L.-O.; Kulp, S. E.; Rappe, C. *Chemosphere.* **1994,** *volume 28 (1),* pp 63-74.

31. Paustenbach, D. J.; Wenning, R. J.; Mathur, D. *Organohalogen Compounds.* **1996,** *volume 28,* pp 111-116.
32. California Air Resources Board. *Summary of Results of PCDD, PCDF, and PCB Emissions Tests at FCCU at Chevron Products Company, Richmond Refinery;* Draft Report; California Air Resources Board: Sacramento, CA, 1997.
33. Block, C.; Dams, R. *J. Radioanalytical Chem.* **1978,** *volume 46,* pp 137-144.
34. Truex, T. J.; Norbeck, J. M.; Smith, M. R.; Arey, J.; Kado, N.; Okamoto, B.; Kiefer, K.; Kuzmicky, P.; Holcomb, I. *Evaluation of Factors that Affect Diesel Exhaust Toxicity;* Final Report to Calif. Air Resources Board, Contract 94-312; Center for Env. Res. and Technology, U.C. Riverside: Riverside, CA, 1998.
35. Thompson, T. S.; Clement, R. E.; Thornton, N.; Luyt, J. *Chemosphere.* **1990,** *volume 20(10-12),* pp 1525-1532.
36. Dioxin test report submitted by T. P. Royer, Exxon Co. USA Benicia Refinery, to S. R. Ritchie, Cal. Regional Water Quality Control Board, S. F. Bay Region. Dated August 19, 1991. Provided under the California Public Records Act.
37. U.S. Environmental Protection Agency, *Preliminary Data Summary for the Petroleum Refining Category;* Report No. EPA-821-R-96-015; U.S. Environmental Protection Agency: Washington, D.C., 1996. Appendix G and pp. 49-52.
38. California Air Resources Board. *Determination of Emissions from the No. 3 Reformer at Tosco Corporation, Avon Refinery, Martinez;* Draft Report; California Air Resources Board: Sacramento, CA, 1998.
39. Radian Corp. *Results of Dioxin Testing on the Catalytic Reformer Unit #1 Exhaust, Texaco Refinery, Bakersfield, California;* Final Report Prepared for Texaco Refining and Marketing, Inc.; Radian Corp.: Sacramento, CA, 1991.
40. Versar, Inc. *Formation and Sources of Dioxin-Like Compounds;* A Background Issue Paper Prepared for the U.S. Environmental Protection Agency, National Center for Environmental Assessment; Versar, Inc.: Springfield, VA, 1996.
41. U.S. Environmental Protection Agency. *The Inventory of Sources of Dioxin in the United States;* Report No. EPA/600/P-98/002Aa, External Review Draft; U.S. Environmental Protection Agency: Washington, D.C., 1998.
42. Kerr, G. T. *Scientific American.* July, 1989, pp 100-105.
43. Simonich, S. L.; Hites, R. A. *Science (Washington, D.C.).* **1995,** *volume 269,* pp 1851-1854.
44. California Air Resources Board. *Proposed Dioxins Control Measure for Medical Waste Incinerators;* Stationary Source Division Staff Report; California Air Resources Board: Sacramento, CA, 1990.
45. California Water Resources Control Board. *Chlorinated Dibenzo-p-dioxin and Dibenzofuran Contamination in California from Chlorophenol Wood Preservative Use;* Report No. 88-5WQ; Water Resources Control Board: Sacramento, CA, 1988.
46. Strosher, M. *Investigation of Flare Gas Emission in Alberta;* Final Report to Environment Canada, Conservation and Protection, the Alberta Utilities Board, and the Canadian Association of Petroleum Producers; Environmental Technologies, Alberta Research Council: Calgary, Alberta, Canada, 1996.

Chapter 12

U.S. Environmental Protection Agency New Chemicals Program PBT Chemical Category

Screening and Risk Management of New PBT Chemical Substances

Kenneth T. Moss, Robert S. Boethling, J. Vincent Nabholz, and Charles M. Auer

Office of Pollution Prevention and Toxics, U.S. Environmental Protection Agency, Washington, DC 20460

PBT chemical substances possess characteristics of persistence (P) in the environment, accumulation in biological organisms (bioaccumulation (B)), and toxicity (T) that make them priority pollutants and potential risks to humans and ecosystems. EPA has developed a category of PBT chemical substances, for the purposes of facilitating the assessment of new chemical substances under Toxic Substances Control Act (TSCA) section 5(e) prior to their entry into the marketplace. The category statement includes the boundary conditions, such as fish bioconcentration/bioaccumulation factors and environmental persistence values, that would determine inclusion in (or exclusion from) the category, and standard hazard and fate tests to address P, B, and T concerns for the chemical substances fitting the category description. Chemicals exceeding the boundary conditions for P, B, and T, based upon data or estimates from predictive tools and structure-activity relationships (SAR), will be identified for control as needed to reduce exposure, require testing to confirm a chemical's PBT status, and guide final regulatory action. Establishment of this category thus provides a vehicle by which the Agency may gauge the flow of PBT chemical substances through the TSCA New Chemicals Program and measure the results of its risk screening and risk management activities.

Background

On November 17, 1998, the EPA published in the Federal Register a notice of availability and solicitation of public comment on a "Multimedia Strategy for Priority Persistent, Bioaccumulative, and Toxic Pollutants" (63 FR 63926). This Strategy formalizes a process for integration and coordination of Agency PBT-related activities under "an enduring cross-office system that will address cross-media issues associated with priority PBT pollutants" (see www.epa.gov/pbt/pbtstrat.htm). It was understood, however, that individual Agency program offices needed to operate within the parameters of their legislative mandates and established regulatory and policy frameworks. For some programs such as the Toxics Release Inventory under section 313 of the Emergency Planning and Community Right-to-Know Act (EPCRA; Public Law 99-499, 42 U.S.C.A. § 11023) and the New Chemicals Program under section 5 of the Toxic Substances Control Act (TSCA; Public Law 94-469; 15 U.S.C.A. § 2604), actions involving PBTs are a historical reality and their experience has, in fact, largely shaped the Strategy.

This chapter explores the development of technical criteria for a category of PBT chemical substances under the TSCA New Chemicals Program. The structure of the TSCA new chemicals review process and the tools used to implement it flow logically from its statutory purpose and suggest that the category approach outlined here is the most appropriate means of addressing potential concerns for substances possessing PBT characteristics.

Overview of the TSCA New Chemicals Process

Under section 5(a) of TSCA, persons must notify EPA at least 90 days before manufacturing or importing a new chemical substance for non-exempt purposes. A new chemical substance, as defined in section 3(9) of TSCA, is any chemical that is not included on the Inventory compiled under section 8(b) of TSCA.

Section 5 of TSCA gives EPA 90 days to review a Premanufacture Notice (PMN). However, the review period can be extended under section 5(c) for "good cause;" it may also be suspended voluntarily by the mutual consent of EPA and the PMN submitter. During the review period, the Agency may conclude that the information available to the Agency is insufficient to permit a reasoned evaluation of the human health and environmental effects of that PMN substance, that the manufacturing, processing, distribution in commerce, use, or disposal of the substance may present an unreasonable risk of injury to human health or the environment, and/or that the PMN substance will be produced in substantial quantities and there may be significant or substantial human exposure to the substance or the PMN substance may reasonably be anticipated to enter the environment in substantial quantities. As a result, EPA may take action under section 5(e) or (f) to prohibit or limit the production, processing, distribution in commerce, use, and disposal of new chemical substances that raise health or environmental concerns, pending the development of testing necessary to address those concerns. If EPA has not taken action under section 5(e) or (f), the PMN submitter may manufacture or import the new chemical substance when the review period expires.

No later than 30 days after the PMN submitter initiates manufacturing or importing, it must provide EPA with a notice of commencement of manufacture or import. Section 8(b) of TSCA provides that, upon receipt of such a notice, EPA must add the substance to the TSCA Inventory. Thereafter, other manufacturers and importers may engage in activities involving the new substance without submitting a PMN, unless the Agency has used its Significant New Use Rule (SNUR) authority under section 5(a)(2) to designate that a use of a chemical substance is a "significant new use." Section 5(a)(1)(B) of TSCA would then require persons to submit a Significant New Use Notice (SNUN) to EPA at least 90 days before they manufacture, import, or process the substance for that use. The required SNUN provides EPA with the opportunity to evaluate the intended use, and if necessary, to prohibit or limit that activity before it occurs.

More information on the TSCA new chemicals program is available through the Internet at www.epa.gov/opptintr/newchms.

History

Since 1979, EPA has reviewed over 30,000 TSCA section 5 submissions for new chemical substances. During the intervening years, EPA has implemented various initiatives which have enabled the Agency to review a greater number of new chemicals more efficiently. In 1987, for example, EPA's Office of Toxic Substances (now the Office of Pollution Prevention and Toxics) first used its accumulated experience to group chemical substances with similar physicochemical, structural, and toxicological properties into working categories (Ref. *1*; see also www.epa.gov/opptintr/newchms/chemcat). These categories, including the subject one for PBT chemical substances, are developed by EPA based on available data and experience reviewing PMNs on similar substances. Such groupings enable both PMN submitters and EPA reviewers to benefit from the accumulated data and decisional precedents and to facilitate the assessment of new chemical substances.

Establishment of this PBT category alerts potential PMN submitters to possible assessment or regulatory issues associated with PBT new chemicals review. It also provides a vehicle by which the Agency may gauge the flow of PBT chemical substances through the TSCA New Chemicals Program and measure the results of its risk screening and risk management activities for this category of chemical substances; as such, it is a major element in the Agency's overall strategy to further reduce risks from PBT pollutants.

Development of this TSCA new PBT chemicals policy has occurred in coordination with US national, US/Canada binational, and international efforts to identify and control the environmental release of persistent organic pollutants (POPs). The PBT category description, in the form of an October 5, 1998 proposed Federal Register policy statement (63 FR 53417), was provided to the United Nations Environment Program (UNEP) Governing Council's Criteria Expert Group (CEG) for POPs, established at the first session of the Intergovernmental Negotiating Committee (INC). The CEG is an open-ended technical working group with a mandate to present to the INC proposals for science-based criteria and a procedure for identifying additional POPs as candidates for future international action. The CEG is to incorporate criteria pertaining to persistence, bioaccumulation, toxicity and exposure in different regions and should take into account the potential for regional and global transport, including dispersion mechanisms for the atmosphere and the hydrosphere, migratory species and the need to reflect possible influences of marine transport and tropical climates. At its first meeting, October 26-30, 1998 in Bangkok, the CEG recommended that the INC consider developing a provision encouraging countries and regions to include in their new chemicals schemes elements relating to development and introduction of new chemical POPs. The US described its proposed TSCA PBT new chemicals category, and the full text of the October 5, 1998 policy statement was distributed to all delegations as a Conference Room Paper. The second meeting of the INC (January 25-29, 1999 in Nairobi) accepted the CEG's recommendation and will consider it further in its deliberations. At the third meeting of the INC (September 6-11, 1999 in Geneva) additional discussion occurred on a proposed provision concerning the review of new chemicals and the consideration of persistence, bioaccumulation and toxicity in that review.

The final policy statement on this PBT category was published in the US Federal Register on November 4, 1999 (64 FR 60194) and represents the first formal statement of US national policy regarding new chemical POPs. Under our domestic program, the policy statement provides guidance criteria for persistence, bioaccumulation and toxicity for new chemicals and advises the industry about our regulatory approach for chemicals meeting the criteria. Internationally, the October 5, 1998 proposed policy statement alerted the parties involved in negotiation of the POPs Convention to the need for inclusion of a new chemicals provision in the Convention. The issuance of the final policy statement reaffirms US leadership on this issue and serves as a model for other countries in taking steps to discourage the introduction of POPs as new chemicals and pesticides.

Evaluation Criteria and Process for PBT Chemical Substances

Generally, persistent bioaccumulators are chemical substances that partition to water, sediment or soil and are not removed at rates adequate to prevent their bioaccumulation in aquatic or terrestrial species. EPA has developed the following specific identification criteria and associated process for use in evaluating new chemical substances. The category description draws upon ongoing international efforts as well as Agency efforts to craft a coordinated and scientifically supportable approach to identifying PBT chemical substances.

	NEW CHEMICALS PROGRAM PBT CATEGORY CRITERIA	
	TSCA Section 5 Control Action Level	
	Moderate	High
Persistence (transformation half-life in aquatic environment)	>2 months	>6 months
Bioaccumulation* (Fish BCF or BAF)	≥1000	≥5000
Toxicity	Develop toxicity data where necessary, based upon various factors, including concerns for Persistence, Bioaccumulation, other physical/chemical factors, and toxicity.	

*Chemicals must also meet criteria for MW (<1000) and cross-sectional diameter (<20Å, or <20 x 10^{-8} cm).
Adapted from the Federal Register notice on November 4, 1999 (64 FR 60194).

In determining the thresholds for this policy, EPA concluded that it would be appropriate to reflect the levels of concern that the various PBT chemicals presented, based on the differing degrees to which the chemicals persist and bioaccumulate. The Agency ultimately chose to adopt a two-tier approach, and to establish two separate thresholds to reflect the chemicals' varying potentials to persist and bioaccumulate, as well as to reflect the Agency's belief that the different levels of regulatory action under TSCA are warranted for the two tiers. As discussed in EPA's proposed and final rules to lower the reporting thresholds for PBT chemicals subject to reporting under section 313 of EPCRA (January 5, 1999; 64 FR 687 and October 29, 1999; 64 FR 58666), EPA found that generally the criteria selected by various US and international regulatory bodies for either persistence or bioaccumulation clustered around two values. For persistence in water, soil, and sediment, the criteria were grouped around half-lives of 1 to 2 months and 6 months, and for persistence in air, either 2 or 5 days. Bioaccumulation criteria were grouped around bioconcentration or bioaccumulation factor (BCF/BAF) values of 1,000 and 5,000. The half-life/persistence criterion for aquatic environments of > 2 months represents a chronic exposure to aquatic organisms, as well as approximating the duration of some standard bioconcentration (28-56 days) and chronic toxicity (14-90 days) tests, and is therefore thought to be adequate for detecting many long-term toxic effects as well as any tendency for a substance to accumulate in fatty tissue of aquatic organisms. The BAF/BCF measures the potential for a chemical to accumulate in living organisms relative to its concentration in the surrounding environment. BCF/BAF is estimated using calculations based on octanol-water partition coefficients (Kow), although data can also be provided from field or laboratory measurements. Chemical substances having a BCF or BAF >1000, equivalent to log Kow of 4.2 (Ref. *2,3,4*), are characterized by a tendency to accumulate in organisms (Ref. *5)*.

The preamble to the above EPCRA proposed rule states "Bearing in mind that one of Congress' articulated purposes for EPCRA section 313 was to provide local communities with relevant information on the release and other waste management activities of chemicals in their community that may present a hazard, EPA determined that the criteria that were most consistent with these purposes were, for persistence, half-lives of 2 months for water, sediment, and soil, and 2 days in air, and for bioaccumulation, BCF/BAF values of 1,000 or greater" (64 FR 692). EPA is making a similar determination for the PBT new chemicals policy under TSCA. The PMN process is one of EPA's premier Pollution Prevention programs and plays a critical role in making sure that new chemical substances do not present unacceptable risk when they are commercialized. Given this, the TSCA new chemicals program is and must be conservative by nature, which suggests that a half-life shorter than 6 months and a BCF criterion lower than 5000--values that were selected solely or primarily to isolate substances *already widely acknowledged* to be POPs–are appropriate for regulatory scrutiny of new chemicals under TSCA. Note that the CEG, at the October 26-30, 1998 Bangkok meeting described above (see "History"), developed indicative numerical values as bracketed criteria text which included persistence of 2 vs. 6 months in water and log Kow of 4 vs. 5 (roughly equivalent to BCF of 1000 vs. 5000, respectively).

Releases to all environmental media, such as air emissions from stacks, wastes disposed of in landfills or on land, and waste discharged into water, are factored into the Agency's determination of potential risk posed by a given PMN chemical substance's total environmental load. In making this determination of potential risk the Agency may employ multimedia fate models in order to account for all potential sources and loadings, environmental transformation processes, and intermedia partitioning, in an integrated fashion. Multimedia fate models like the Environmental Quality Criteria (EQC) model (Ref. *6*) require compartmental half-lives for air, water, soil and sediment, which cannot necessarily be interpreted as half-lives for any specific process such as biodegradation. The EQC model is based on the fugacity approach and subsequently applied to numerous environmental processes. It uses an "evaluative environment" in which environmental parameters such as bulk compartment dimensions and volumes (e.g., total area, volume of soil and sediment, etc.) are standardized, so that overall persistence for chemicals with different properties and rates of transformation may be compared on an equal basis. In general, measured values of toxicity, chemical properties, compartmental transformation half-lives, etc., provided the data are of acceptable quality, are preferred over those that are predicted or estimated via a model or computer program. Data on air half-lives for input to models would be either measured or derived from the Atmospheric Oxidation

Program (AOP; Ref. *7*) or similar methodology. Studies by Boethling et al. (Ref. *8*) and Federle et al. (Ref. *9*) suggest that half-lives in bulk soil may be assumed for screening purposes to be about the same as for surface water, and that sediment half-lives may be assumed to be 3-4 times longer. EPA's current *suggested* approach to finding water half-life is to use the Ultimate Survey Model (USM) in the EPI BIOWIN program (Ref. *10*). Estimation of bulk compartment half-lives from USM model data requires several assumptions, including that (1) biodegradation is the only significant fate process in water, soil, sediment; (2) water and soil half-lives are the same; and (3) sediment is dominated by anaerobic conditions and therefore sediment half-life is four times longer than water half-life.

The toxicity rating for a PBT chemical applies to repeated exposures which result in human or environmental toxicity, including, for example, systemic toxicity, mutagenic damage, reproductive toxicity, or developmental toxicity. An example of this is chronic toxicity towards aquatic organisms of organotins from contaminated marine environments, which ultimately resulted in the use of tributyl tin in marine anti-fouling paints being regulated. Repeated exposures from drinking water or contaminated food result from a chemical after it has been released into the environment. The classic PBT problems (i.e., PCBs and DDT) have been associated with food chain contamination.

New Chemicals Program Control Strategy for PBT Chemical Substances

Where EPA is unable to adequately determine the potential for bioaccumulation, persistence in the environment, and toxicity which may result from exposure of humans and environmental organisms to a possible PBT chemical substance, the Agency may find it appropriate to control the manufacturing, importing, processing, distributing in commerce, using, or disposing of the PMN substance in the US pending the development of information necessary for a reasoned evaluation of these effects.

Control action under TSCA section 5(e) may be needed in varying degrees, based upon level of risk concern. Agency control actions taken under TSCA section 5(e) for chemical substances meeting these criteria would be based upon the level of certainty for the PBT properties of a PMN substance (e.g., measured vs. estimated values), the magnitude of Agency concerns, and conditions of expected use and release of the chemical. For example, new chemical substances meeting the PBT criteria listed under "Moderate" could be addressed via a negotiated consent agreement under which necessary testing is triggered by specific production limits. While the PMN submitter would be allowed to commercialize the substance, certain controls could be stipulated, including annual Toxics Release Inventory-type reporting on environmental releases of the PMN substance and specific limits on exposures, releases or uses. The "High" criteria are equivalent to those that have been used internationally to identify POPs. For these PMN substances, the concern level is higher and the Agency would look carefully at any and all environmental releases. Because of the increased concern, more stringent control action would be a likely outcome, up to a ban on commercial production until data are submitted which allow the Agency to determine that the level of risk can be appropriately addressed by less restrictive measures. The above described control actions represent just one body of possible decisions and should not be considered as exclusive of other risk management options.

The Agency will consider P and B and T, individually and together, and exposure in making risk-based judgments. Risk, specific to the PMN substance as well as its risk relative to substitutes currently on the market, is predicted as a function of the potential hazard of the substance and the expected exposure. For exposure-based determinations, EPA will use a case-by-case approach for making findings by applying other considerations (i.e., toxicity or physical/chemical properties) and consider P and B aspects as factors which might argue for regulatory action under TSCA section 5(e) at lower levels of production or exposure/release than are described in the guidelines for the new chemicals program's exposure-based policy (referenced and discussed in the October 5, 1998 proposed policy statement). Overall, companies are not being prevented from developing and using new substances that are judged to be potential PBT chemicals, but EPA may require certain controls (e.g., stipulating that there will be no release of the PMN chemical to the environment) or testing as a result of its assessments.

In order to be so identified as a PBT new chemical, all three parameters must be satisfied. The Agency has adopted a 1 to 3 rating system for each of P, B, and T although, because the Agency's health hazard evaluation of chronic toxicity is a qualitative one, there are no quantitative parameters given for toxicity (T) within the PBT category, as there are for P and B. Chemicals exceeding the boundary conditions for P, B, and T will be identified for control as needed to reduce exposure and require testing to confirm a chemical's PBT status. Once P and B levels have been confirmed, the T level will be confirmed. Each of P and B and T are weighed in the Agency's assessment. The testing strategy outlined below is intended to build the case, starting with testing to establish persistence and bioaccumulation, and then determining toxicity and confirming a chemical's status as a PBT chemical in tier 3. Once a chemical becomes distributed in the environment at low concentrations, the combination of persistence and bioconcentration in organisms can result in residues high enough to approach a toxic dose. The first two tiers focus on P and B because of the critical role these aspects play in PBT determinations and because of their relatively lower cost. Thus, chronic toxicity testing, which is expected to be the most expensive testing, is reserved until tier 3 where it serves to establish PBT status. Although the early tier P and B testing may either obviate the need for toxicity testing or result in more directed and cost-effective toxicity testing, the need for toxicity testing is considered in each testing tier and will be obtained in lower tiers where needed on a case-specific basis. As with all new chemicals reviewed by the Agency under TSCA, the potential toxicity of the chemical is determined from test data, if any, or by analogy to structurally similar chemicals.

If chemical has a low or moderate Kow (i.e., "B1," with fish BCF estimated as less than 1000), a company should not expect to conduct the ready biodegradation study, because the B1 rating results in it not being classified as a "PBT chemical." For example, some surfactants could be P3B1T3, highly persistent in the environment and chronically toxic to organisms, but with low bioaccumulation potential. However, Agency action may still be taken for less than complete PBT chemicals such as these, either for traditional, "non-PBT" risks, or to obtain more information on high production volume and release chemicals. Similarly, calcium would also not be considered a PBT chemical, as it would be ranked P3B3T1; it is persistent in the environment, it bioaccumulates, but it is not considered toxic. Although the Agency does not promote the environmental discharge of more persistent materials, the environmental "desirability" of a given chemical often depends on a balance of various factors, including toxicity and ability of the chemical to bioaccumulate. Like the previous surfactant example, the Agency may nonetheless take action on a P3B3T1 chemical (not calcium per se).

Testing Strategy for PBT Chemical Substances

Where EPA is unable to adequately determine the potential for bioaccumulation, persistence in the environment, and toxicity which may result from exposure of humans and environmental organisms to a possible PBT chemical substance, the Agency may find it appropriate to prohibit a company from manufacturing, importing, processing, distributing in commerce, using, or disposing of the PMN substance in the United States pending the development of information necessary for a reasoned evaluation of these effects. The following testing strategy describes test data which, if not otherwise available, EPA believes are needed to evaluate the potential persistence, bioaccumulation, and toxicity of a PBT chemical substance. The tests are tiered; depending upon the circumstances, such as magnitude of environmental releases, results of testing, or structure-activity relationships (SAR), testing could begin above Tier 1 or additional, higher levels of testing may be required. EPA's Office of Prevention, Pesticides and Toxic Substances (OPPTS) harmonized test guidelines referenced below are found on the Internet at http://www.epa.gov/OPPTS_Harmonized/ and, where available, the comparable Organization for Economic Cooperation and Development (OECD) test guideline (see http://www.oecd.org/ehs/test/testlist.htm) is also provided.

EPA realizes that often there are a number of different but acceptable means to providing testing information. However, EPA's acceptance of a guideline not specified and/or use of data generated under such guidelines depends on multiple factors including the specifics of the test substance, purpose of the testing, familiarity with specific procedures and equipment, validation of the method, etc. EPA requires that testing be conducted according to TSCA Good Laboratory Practice Standards at 40 CFR part 792 and using methodologies generally accepted at the time the study is initiated. Before starting to conduct any such study, the PMN submitter must obtain approval of test protocols from EPA by submitting written protocols. Published test guidelines specified below provide general guidance for development of test protocols, but are not themselves acceptable protocols.

***Tier 1.* Based upon SAR and professional judgment, the Agency identifies a new chemical substance as a possible PBT chemical substance.**

Log Kow	Liquid chromatography (OPPTS 830.7570/OECD 117) or generator column (OPPTS 830.7560) method
Ready biodegradability	Ready biodegradability (OPPTS 835.3110/OECD 301) 6 methods (choose one, or an equivalent test): DOC Die-Away, CO_2 Evolution, Modified MITI (I), Closed Bottle, Modified OECD Screening, Manometric Respirometry *or* Sealed-vessel CO_2 production test (OPPTS 835.3120)
Hydrolysis in water (if, based upon SAR, susceptibility to hydrolysis is suspected)	OPPTS 835.2110 (OECD 111)
Results	If the measured log Kow is <4.2 or if the test chemical passes the ready biodegradability test (i.e., not persistent in the environment), no further PBT-related testing is required. If the measured log Kow is greater than or equal to 4.2 and the chemical does not pass the ready biodegradability test, the chemical would require tier 2 testing. If hydrolysis testing is conducted and results in a half-life of <60 days, further testing may not be needed, but the need for testing must be determined after consideration of factors specific to the case, such as physical/chemical properties, persistence and bioaccumulative qualities of hydrolysis products, and the nature of the expected releases.

Tier 2. **Biodegradability and Bioaccumulation**

Biodegradability	Shake-flask die-away test (OPPTS 835.3170) or an equivalent test.
Bioaccumulation	Fish bioconcentration test (OPPTS 850.1730/OECD 305), or an equivalent test). Measured BCF should be based on 100 percent active ingredient and measured concentration(s).
Results	If the measured biodegradation half-life is >60 days and measured BCF is >1000, tier 3 testing will be required. If only one condition is met, releases and exposure are further considered to determine if additional testing is required.

Tier 3. **Toxicity/advanced environmental fate testing.**

Human health toxicity	Combined repeated dose oral toxicity with the reproductive/developmental toxicity screening test (OECD No. 422) in rats. Other health testing will be considered where appropriate.
Environmental fate	Sediment/water microcosm biodegradation test (OPPTS 835.3180).
Chronic environmental toxicity	Fish (rainbow trout) and daphnids should be tested according to 40 CFR 797.1600 (same as OPPTS 850.1400/OECD 210) and 40 CFR 797.1330 (same as OPPTS 850.1300/OECD 202), respectively. Additional testing to evaluate other biota (e.g., avian, sediment dwelling organisms) or other effects (e.g., endocrine disrupting potential) will be considered where appropriate.

Adapted from the Federal Register notice on November 4, 1999 (64 FR 60194).

Conclusions

EPA's strategic goal is to identify and reduce risks to human health and the environment from current and future exposures to priority PBT pollutants. What has been described here is the U.S. EPA's approach to evaluation, control, and information collection for new PBT chemicals entering commerce, using criteria consistent with those used in other programs within EPA and internationally. Although discrete criteria values are described, the TSCA PMN review program operates more on the principle of "weight of evidence" than "check the box." EPA applies professional judgment, Structure Activity Relationships (SAR), computer modeling and assessment methods to identify PBTs in much the same way as for any other chemical substance in the PMN review process. Using predictive tools (in the absence of test data) and professional judgment, EPA leans towards a "reasonable worst case" risk assessment when there is lack of chemical-specific data. Industry always has the option of assisting and enhancing the Agency's determinations by submitting scientifically valid test data.

References

1. USEPA. 1988. Letter from Charles L. Elkins to Geraldine V. Cox (Chemical Manufacturers Association). Office of Toxic Substances, USEPA (September 22, 1988).
2. Veith GD and Kosian P. 1982. Estimating bioconcentration potential from octanol/water partition coefficients, in Physical Behavior of PCB's in the Great Lakes (MacKay, Paterson, Eisenreich, and Simmons, eds.), Ann Arbor Science, Ann Arbor, MI.
3. Bintein, S., Devillers, J. and Karcher, W. 1993. Nonlinear dependence of fish bioconcentration on n-octanol/water partition coefficient. *SAR and QSAR in Environ. Research* 1:29-39.
4. Meylan WM, Howard PH, Boethling RS, Aronson D, Printup H, and Gouchie S. 1999. Improved method of estimating bioconcentration/bioaccumulation factor from octanol/water partition coefficient. *Environ. Toxicol. Chem.* 18: 664-672.
5. Smrchek, J.C. and M.G. Zeeman. 1998. Assessing Risks to Ecological Systems from Chemicals, pp. 24-90. In, P. Calow (ed.), Handbook of Environmental Risk Assessment and Management, Blackwell Science Ltd., Oxford, UK.
6. Mackay, D., Di Guardo, A., Paterson, S., Cowan, C.E. 1996. Evaluating the environmental fate of a variety of types of chemicals using the EQC model. *Environ. Toxicol. Chem.* 15:1627-1637.
7. Meylan WM, Howard PH. 1993. Computer estimation of the atmospheric gas-phase reaction rate of organic compounds with hydroxyl radicals and ozone. *Chemosphere* 26: 2293-2299.
8. Boethling RS, PH Howard, JA Beauman and ME Larosche. 1995. Factors for intermedia extrapolation in biodegradability assessment. *Chemosphere* **30:** 741-752).
9. Federle TW, SD Gasior and BA Nuck. 1997. Extrapolating mineralization rates from the ready CO2 screening test to activated sludge, river water, and soil. *Environ. Toxicol. Chem.* 16: 127-134.
10. Boethling RS, Howard PH, Meylan W, Stiteler W, Beauman J, Tirado N. 1994. Group contribution method for predicting probability and rate of aerobic biodegradation. *Environ. Sci. Technol.* 28: 459-465.

The above references are included in the official record of a policy statement on

"Category for Persistent, Bioaccumulative, and Toxic Chemical Substances" published in the US Federal Register, under the public docket control numbers OPPTS-53171 and OPPTS-53171A. The proposed policy statement was published in the US Federal Register on October 5, 1998 (63 FR 53417), and the final policy statement on November 4, 1999 (64 FR 60194). The public version of the official record, which includes printed, paper versions of any electronic comments submitted during an applicable comment period, and references cited in the notices, is available from the TSCA Nonconfidential Information Center, Rm. NEB-607, Waterside Mall, 401 M St., SW., Washington, DC 20460. The Center is open from 12 noon to 4 p.m., Monday through Friday, excluding legal holidays. The telephone number of the Center is 202-260-7099 and the email address is oppt.ncic@epa.gov.

In addition, the following web sites are helpful sources for the various legal and legislative texts cited in this chapter (all citations begin with http://):

USEPA Federal Register notices: www.epa.gov/fedrgstr/
UNEP POPs activities: irptc.unep.ch/pops/
EPCRA (42 U.S.C.A. § 11023) and TSCA (15 U.S.C.A. § 2604) US laws: uscode.house.gov/usc.htm
Title 40 of the US Code of Federal Regulations (40 CFR), Chapter I (EPA), subchapter R (TSCA): www.epa.gov/docs/epacfr40/chapt-I.info/subch-R/

Chapter 13

The U.S. Environmental Protection Agency Waste Minimization Prioritization Tool: Computerized System for Prioritizing Chemicals Based on PBT Characteristics

Mark D. Ralston, Daniel L. Fort, Jay H. Jon, and James K. Kwiat

U.S. Environmental Protection Agency, Ariel Rios Building, 1200 Pennsylvania Avenue, NW, Washington, DC 20460

Introduction

One of the most difficult tasks facing decision-makers in industry, government, or environmental groups is the prioritization of chemicals for further action. Prioritization of chemicals based on their tendency to be persistent, bioaccumulative, and toxic (PBT) to humans or ecosystems is increasingly a guiding principle in environmental management. The Waste Minimization Prioritization Tool (WMPT) is a computerized system developed by the U.S. EPA to help decision-makers prioritize chemicals for pollution prevention efforts based on PBT concerns. Unlike current international PBT initiatives that focus on perhaps a dozen chemicals of international concern or other screening systems that provide detailed information on only a few hundred chemicals, the WMPT performs screening-level prioritization among several thousand chemicals.

Background on The WMPT

EPA's Office of Solid Waste (OSW) and EPA's Office of Pollution Prevention and Toxics (OPPT) developed the WMPT to help meet the objectives of EPA's Waste Minimization National Plan (U.S. EPA, 1994c) and OSW's waste minimization objectives under the Government Performance and Results Act. A beta-test version of the tool (and accompanying User's Guide) was made available for review in June 1997. (U.S. EPA, 1997 c,d) In September 1997, EPA organized a focus group meeting of technical experts from industry, environmental groups, and states to provide peer review of the WMPT. In November 1997, EPA convened an agency workgroup to seek consensus on addressing core scientific and technical issues raised in peer and focus group review of the WMPT. The workgroup recommended a variety of specific revisions to the beta-test version of the WMPT, resulting in a revised system. The Revised WMPT was then used as the starting point for developing EPA's Draft RCRA PBT Chemical List. (US EPA, 1998d) This list, when final, will serve as the focus of EPA's voluntary national hazardous waste minimization program.

Overview of WMPT Chemical Scoring

The purpose of the WMPT scoring algorithm is to develop chemical-specific scores that can be used for a screening-level ranking of chemicals. As a screening-level tool, the WMPT is designed to generally error on the side of inclusiveness (i.e., avoiding false negatives) where there is uncertainty in the underlying data. It is critical to note that, given the uncertainties associated with this type of screening-level tool, the results from the WMPT should not, under any circumstances, be viewed as a substitute for a detailed risk assessment.

As illustrated in Figure 1, the scoring algorithm is designed to generate an overall chemical score that reflects a chemical's potential to pose chronic risk to either human health or ecological systems. A measure of human health concern is derived, consistent with the risk assessment paradigm, by jointly assessing the chemical's human toxicity and potential for exposure. Similarly, a measure of the ecological concern is derived by jointly assessing the chemical's ecological toxicity and potential for exposure. The higher of the human health and ecological concern scores is used as the overall chemical score.

The Human Health Concern score is derived by adding two scores, one reflecting the chemical's toxicity to humans and the other the chemical's potential for exposure. While designed as a screening-level tool, WMPT's structure is generally consistent with the risk assessment paradigm that guides current risk assessment practices. Thus, WMPT's scoring algorithm is modeled after the general risk calculation equation used by U.S. EPA and others, where a chemical risk is derived by combining estimates of the chemical's toxicity with estimates of the chemical's actual or potential exposure. WMPT uses a few relatively simple measures to represent a chemical's toxicity and exposure potential, consistent with a screening-level approach.

The Human Toxicity score is derived by taking the higher of two scores: (1) Cancer Effects, and (2) Noncancer Effects. The Human Exposure Potential score is derived as the sum of two scores: (1) Persistence and (2) Bioaccumulation Potential. Similar to the Human Health Concern score, the Ecological Concern score is derived by adding two scores, one reflecting the chemical's toxicity to aquatic ecosystems and the other the chemical's potential for exposure. Ecological Toxicity is scored currently based only on Aquatic Toxicity. The Ecological Exposure Potential score is derived in the same way as (and is equal to) the Human Exposure Potential score.

Persistence, bioaccumulation potential, and toxicity scores are generated and then aggregated to obtain an overall chemical score. These scores are generated by evaluating certain "data elements." For example, for human toxicity noncancer effects, the scoring algorithm uses various data elements indicative of the chemical's capacity to cause chronic adverse effects in human receptors and the magnitude and severity of those effects (e.g., Reference Dose and Reference Concentration). See Table V for additional examples.

The WMPT is designed to take advantage of the best available data for chemical scoring. Data elements are grouped into categories of data preference (e.g., highest, high, medium, low, lowest). Data preference hierarchies reflect both the type of data element as well as specific data sources. For example, measured BCF values

Figure 1
Overview of the WMPT Scoring Algorithm

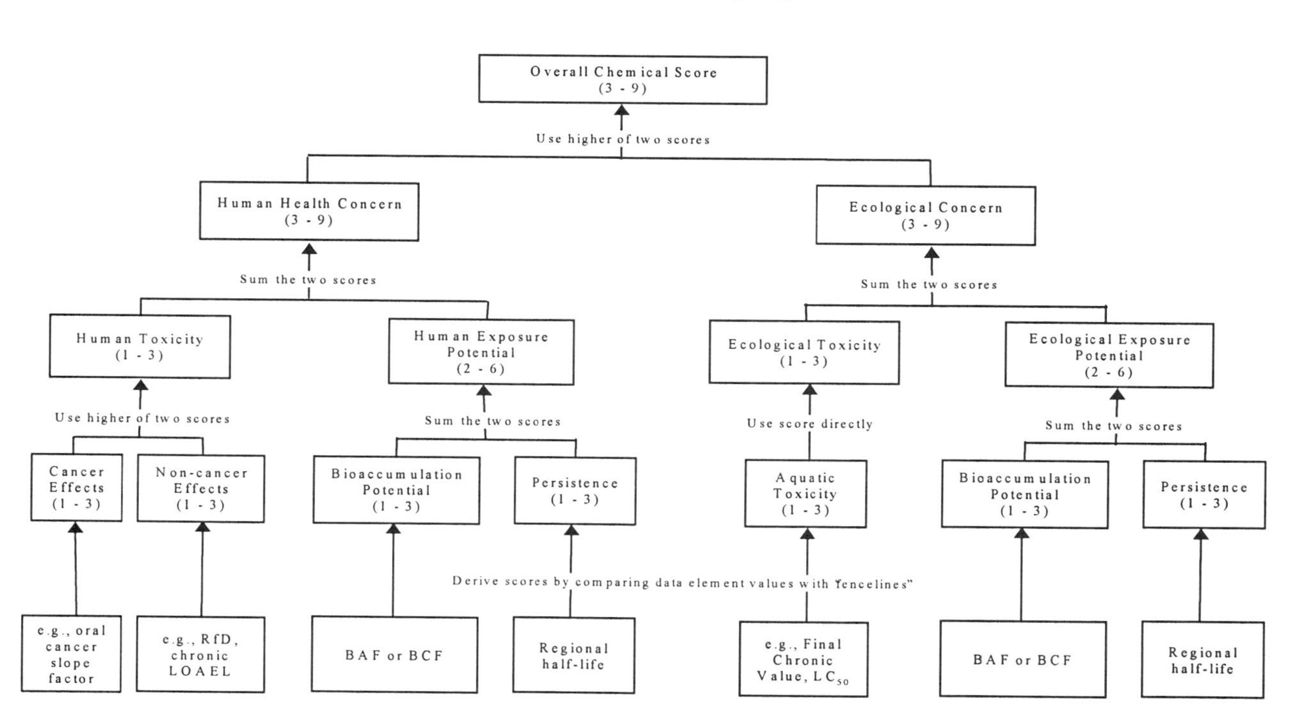

from specific data sources are preferred over estimated BCF values. Each data element and data source in the data preference hierarchy was positioned based on several attributes of the underlying data. In general, the highest preference data elements and data sources were those that: relied on measured (rather than predicted) data, reflected Agency consensus or significant external peer-review, were relatively recent, and were well documented. In cases where a chemical had values for several data elements within a given data preference level, the value providing the highest score was selected. Thus, the WMPT is programmed so that higher preference data is used before lower preference data. The use of data hierarchies in theWMPT has allowed a large number of chemicals to be scored for PBT characteristics while using the highest preference data available. For the sake of brevity, the specific rationale for the placement of data elements and data sources in each data preference hierarchy is not included here; it can be found in the the *Waste Minimization Prioritization Tool Spreadsheet Document*. (U.S. EPA, 1998h)

In selecting the types of data elements used in the WMPT, U.S. EPA has worked to maintain consistency with approaches used in other chemical screening methods and systems, particularly those developed and used within the Agency. For example, the highest preference data element used to score Aquatic Toxicity is the Final Chronic Value (FCV), a measure of chronic aquatic toxicity. The use of an FCV as the highest preference data element is consistent with methods used in other U.S. EPA initiatives such as the Great Lakes Water Quality Initiative. (U.S. EPA, 1995a) Similarly, most of the measures used in scoring human toxicity are standard data elements used by U.S. EPA and other agencies in risk assessment and screening-level procedures.

Persistence, bioaccumulation potential, and toxicity are generally scored using a "fenceline" approach. The fenceline scoring approach involves comparing the value for a given chemical data element against predefined threshold values (or fencelines) for that data element to obtain scores of 1 (low concern), 2 (medium concern), or 3 (high concern). For example, based on the based on the human toxicity noncancer effects fencelines shown in Table VI, a chemical with a Reference Dose of 0.005 mg/kg/day would score 2 for noncancer effects.

Where possible, the fenceline values used in WMPT are based on Agency program expertise (i.e., based on accepted thresholds indicating low, medium, and high levels of concern) without consideration of the distribution of the underlying data. These fencelines are consistent with approaches in similar chemical screening methods and systems used in the Agency and are based on the expert judgement of scientists within OPPT who have reviewed thousands of chemicals under the Toxic Substance Control Act. Where fencelines based on program expertise were not available, fenceline values were established based on the distribution of the underlying data, using a 1:2:1 binning approach. For example, persistence scores were assigned as follows: the 25 percent of the chemicals with the highest regional half-life values were assigned a "high" score of 3, the 25 percent of the chemicals with the lowest regional half-life values were assigned a "low" score of 1, and the remaining 50 percent of the chemicals were assigned a "medium" score of 2.

Details on most of data elements, data preference hierarchies, and fencelines used for scoring persistence, bioaccumulation potential, human toxicity, and ecological toxicity are provided below. For the sake of brevity, some data element and preference hierarchies were not provided (for example, those for physical/ chemical property data used as input to the multimedia partitioning model). For further information, see the *Waste Minimization Prioritization Tool Spreadsheet Document*. (U.S. EPA, 1998h)

Persistence Scoring

To score persistence, the WMPT scoring algorithm uses a steady-state, non-equilibrium, multimedia partitioning model based on an evaluative multimedia model described by Mackay et al. (Mackay, 1992, and Mackay, 1996) to estimate the regional half life of each chemical (i.e., overall environmental persistence). This regional half life estimate is based on the half lives in four compartments (air, water, soil, and sediment) and other chemical property data that influence partitioning among these compartments. As shown in Table I, measured half life data are used in preference to predicted data. Where measured data are unavailable, half life data are estimated using several models. The approach used by WMPT was adopted to reflect internationally-recognized methods for screening-level estimation of environmental persistence. Elemental metals are assigned a persistence score of 3 (high).

Persistence Data Elements and Data Preference Hierarchy

Data preference hierarchies were established for many of the data elements used as inputs to the multimedia partitioning model. The data hierarchy for half life data elements is shown in Table I for illustration.

Persistence Fencelines.

Due to a lack of empirical thresholds representing levels of moderate or high concern for persistence, a distributional approach was selected to setting fencelines for regional half life. The distribution selected was 1:2:1 (i.e., the upper 25% in the high score bin, the middle 50% in the medium score bin, and the lower 25% in the low score bin). Chemicals with complete data sets were run in the spreadsheet partitioning model and the output was divided according to a 1:2:1 distribution. The persistence fencelines derived using this distribution-based approach are presented in Table II.

Bioaccumulation Scoring

Bioaccumulation Data Elements and Data Preference Hierarchy.

Several chemical-specific measures can be used to evaluate a chemical's potential to bioaccumulate. These measures or data elements include measured and estimated bioaccumulation factors (BAFs) and measured and estimated bioconcentration factors (BCFs). The WMPT data preference hierarchy is shown in

Table I. Data Preference Hierarchy for Half Life Data Elements

Data Source	*Data Element*	*Preference Level*
Howard et al, *Handbook of Environmental Degradation Rates* (Howard et al, 1990)	• Measured Half-life Data (air, water, soil, sediment (anaerobic))	Highest
MacKay et al *Illustrated Handbook of Physical- Chemical Properties and Environmental Fate for Organic Chemicals* (Mackay et al, 1991-97)	• Measured Half-life Data (air, water, soil, sediment)	High
The Agriculture Research Service Pesticide Properties Data Base (USDA, 1998)	• Measured Half-life Data (soil (aerobic), sediment (anaerobic))	Medium
Atmospheric Oxidation Program (AOPWIN) (Meylan and Howard, 1993)	• Estimated Half-life Data (air)	Low
Ultimate Survey Model (USM) (Howard and Meylan, 1995)	• Estimate of Biodegradation Time (water; surrogate for soil and sediment)	Low
HYDROWIN (Syracuse Research Corporation, undated b)	• Estimated Hydrolysis Rate (water; surrogate for soil and sediment)	Low

Table II. Persistence Fencelines

	Scoring Fencelines (Score)			*Fenceline Setting Method*
Data Element	*High (3)*	*Medium (2)*	*Low (1)*	
Regional Half Life (hr)	> 580	580 - 140	< 140	1:2:1 Distribution

Table III. Measured data were used in preference to predicted data, and BAFs were used in preference to BCFs.

Bioaccumulation Fencelines.

The fencelines used in the Revised WMPT for scoring chemicals based on BAFs and BCFs are presented in Table IV. These fencelines are derived from the thresholds used originally by OPPT in its evaluation of new and existing chemicals (U.S. EPA, 1992).

Human Toxicity Scoring

Human Toxicity represents adverse effects to human health from chronic chemical exposures. Human Toxicity is scored based on Cancer Effects and Noncancer Effects. The higher of these two scores becomes the Human Toxicity score.

Human Toxicity-Noncancer Data Elements and Data Preference Hierarchy.

The human noncancer toxicity score is based on data elements representing human noncancer toxicity, such as Reference Doses (RfDs), Reference Concentrations (RfCs), No Observed Adverse Effects Levels (NOAELs), Lowest Observed Adverse Effects Levels (LOAELs), minimal risk levels (MRLs), and reportable quantities (RQs). The WMPT data preference hierarchy is shown in Table V. Most of the data were measured data, and higher preference was assigned to those data elements and sources where Agency consensus was achieved or significant Agency or external peer review took place.

Human Toxicity-Noncancer Fencelines.

Due to a lack of empirical thresholds representing levels of moderate or high concern for human noncancer effects, the RfD, RfC, MRL, REL, and subchronic NOAEL and LOAEL fencelines from all sources in the Revised WMPT are based on a 1:2:1 distribution. To develop the distribution, the lowest (representing the greatest unit risk) oral value and the lowest inhalation value for each chemical were used. Thus, when both an oral and inhalation value were available for a given chemical, both values were used because not all chemicals that exhibit toxicity via inhalation exposure routes exhibit toxicity via oral exposure routes, and vice-versa. Inhalation

Table III. Bioaccumulation Data Preference Hierarchy for the WMPT

Data Source	*Data Element*	*Preference Level*
DRAFT HWIR (U.S. EPA, 1995e)	• Measured BAF	Highest
Mercury Report to Congress (U.S. EPA, 1997b)	• Measured BAF	Highest
DRAFT HWIR (U.S. EPA, 1995e)	• Measured BCF	High
Ambient Water Quality Criteria documents (U.S. EPA, 1980, 1984 a-d, 1987a)	• Measured BCF	High
SRC ISIS BCF File (SRC, undated c)	• Measured BCF	High
DRAFT HWIR (U.S. EPA, 1995e)	• Predicted BAF	Medium
DRAFT HWIR (U.S. EPA, 1995e)	• Predicted BCF	Low
BCFWIN (Part of EPI Suite) (SRC, 1997)	• Predicted BCF	Low

Table IV. Bioaccumulation Potential Scoring Fencelines

	Scoring Fencelines (Score)			
Data Element	*High (3)*	*Medium (2)*	*Low (1)*	*Fenceline Setting Method*
Measured and predicted BAFs and BCFs	≥ 1,000	≥ 250 to < 1,000	< 250	Program expertise

Table V. Human Noncancer Toxicity Data Preference Hierarchy

Data Source	*Data Element*	*Preference Level*
IRIS (U.S. EPA, 1998c)	• Reference Dose (RfD) • Reference Concentration (RfC)	Highest
ATSDR (ATSDR, 1998)	• Oral Minimal Risk Level (MRL) • Inhalation MRL	High
HEAST (U.S. EPA, 1997a)	• RfD • RfC	High
OERR RQ Database (U.S. EPA, 1994b)	• Chronic Toxicity Reportable Quantity (RQ)	Medium
Toxic Substances Control Act (TSCA) Section 4 Data (U.S. EPA, 1998f)	• Subchronic NOAEL • Subchronic LOAEL • Developmental NOAEL • Developmental LOAEL	Medium
DRAFT Cal/EPA Noncancer Chronic RELs (Cal EPA, 1997b)	• Inhalation Reference Exposure Levels (RELs, similar to RfCs)	Medium
TSCA Section 8(e) (U.S. EPA, 1998g)	• Triage screening results	Low
CESARS (Ontario MOEE and MI DNR, 1998)	• Oral Mammalian Sublethality Score (based solely on NOAEL)	Low
OPPT database (U.S. EPA, 1998e)	• Human Health Structure Activity Team (SAT) Rankings	Lowest

values were then converted to oral equivalents using the following formula provided in the HEAST (U.S. EPA, 1997) documentation:

RfD (mg/kg/day) = RfC (mg/m3) * 20 m3/day (inhalation rate) * 1/70 kg (body mass)
= RfC * 20/70

The 25th and 75th percentiles of the distribution of these values were calculated. The inhalation fencelines were calculated using the above equation in reverse. The subchronic NOAEL and LOAEL fencelines were calculated by multiplying the oral fencelines by 1,000 and 10,000, respectively.

The CESARS oral mammalian sublethality score is derived from a subchronic oral NOAEL threshold using a six-bin scoring system. The six-bin CESARS scoring system was converted to a three-bin WMPT score based on the underlying cut-off values for subchronic NOAELs in the WMPT and in CESARS. The fencelines used in the Revised WMPT for scoring chemicals are presented in Table VI.

Human Toxicity-Cancer Data Elements and Data Preference Hierarchy.

Human Cancer Effects is scored based on cancer slope factors or potency factors and weight-of-evidence associated with the cancer potency or slope factors. The first step in scoring a cancer effect is to determine if a cancer weight-of-evidence (WOE) exists for the chemical. If a WOE is available for the chemical and cancer slope factor (CSF), unit risk, or RQ potency factor data are available, the chemical is scored using the appropriate CSF, unit risk, or RQ potency factor fencelines for the associated WOE. If no WOE is available for the chemical, but CSF, unit risk, or RQ potency factor data are available, the chemical is assigned a WOE of B and is assigned a score using the WOE = B fencelines. If the chemical is associated with a WOE but not with CSF, unit risk, or RQ potency factor data, the chemical is scored based solely on the WOE. A chemical with a WOE of A is assigned a high concern score (3), and a chemical with a WOE of B or C is assigned a medium concern score (2). Chemicals with WOE of D or E are not scored for cancer effects in the Revised WMPT.

The Revised WMPT data preference hierarchy is shown in Table VII. All of the data were measured data, and higher preference was assigned to those data elements and sources where Agency consensus was achieved or significant Agency or external peer review took place.

Human Toxicity-Cancer Fencelines.

WOE classifications were converted to a common WMPT WOE classification for use in scoring. Due ot a lack of empirical thresholds representing levels of moderate or high concern for human cancer effects, oral and inhalation CSF fencelines and unit risk fencelines are based on a 1:2:1 distribution. The distribution was developed using only chemicals that had both a cancer risk estimate and a WOE classification of A or B. Where more than one oral CSF was available for a specific chemical, the highest (representing the greatest unit risk) value was selected. Where more than one inhalation value was available for a specific chemical, the highest (representing the greatest unit risk) value was selected. Thus, where both an inhalation and an oral value were available for a given chemical, both values were used because not all chemicals that exhibit toxicity via inhalation exposure routes exhibit toxicity via oral exposure routes, and vice-versa. The inhalation unit risk values were converted to CSF equivalents using the following formula provided in the HEAST (July 1997) documentation:

$$\text{slope factor (mg/kg/day)}^{-1} = \text{unit risk } (\mu g/m^3)^{-1} / [20 \text{ m}^3/\text{day (inhalation rate)} * 1/70 \text{ kg (body mass)} * 10^{-3} \text{ mg}/\mu g] = \text{unit risk} / [(20*0.001)/70]$$

The 25th and 75th percentiles were calculated. The inhalation fencelines were calculated using the above equation in reverse. The fenceline values applied to cancer risk values for carcinogens with WOE classifications of A or B were multiplied by 10 to set the fencelines for the same values for carcinogens with WOE classifications of C.

The fencelines for RQ potency factors are based on a 1:2:1 distribution. All RQ potency factors associated with chemicals with a WOE of A or B were used in the distribution. The 25th and 75th percentiles were calculated. The fenceline values applied to RQ potency factor value for carcinogens with WOE classifications of A or B were multiplied by 10 to set the fencelines for the same values for carcinogens with WOE classifications of C. The fencelines used in the Revised WMPT for scoring chemicals are presented in Table VIII.

Ecological Toxicity Scoring

Ecological Toxicity Data Elements and Data Preference Hierarchy.

A variety of ecological toxicity data elements were included in the WMPT, as shown in Table IX. Higher preference was assigned to those data elements and sources that reflected the use of measured data, the use of chronic data, and Agency consensus or significant Agency or external peer review.

Ecological Toxicity Fencelines.

The fencelines used to score Aquatic Toxicity in the Revised WMPT are all based on EPA OPPT program expertise. Table X presents the Revised WMPT Aquatic Toxicity fencelines. These fencelines can be categorized into four sets: (1) chronic toxicity, (2) acute toxicity, (3) aquatic toxicity RQ, and (4) TSCA 8(e) Triage concern level. The fencelines used to score the chronic aquatic toxicity data elements

Table VI. Human Toxicity - Noncancer Effects Fencelines

Data Element	*Scoring Fencelines*			*Fenceline Setting Method*
	High (3)	*Medium (2)*	*Low (1)*	
IRIS Reference Dose (RfD)	< 0.0006 mg/kg/day	0.0006 - 0.06 mg/kg/day	> 0.06 mg/kg/day	1:2:1 Distribution
IRIS Reference Concentration (RfC)	< 0.002 mg/m3	0.002 - 0.2 mg/m3	> 0.2 mg/m3	1:2:1 Distribution
Minimal Risk Level (MRL) -- Oral	< 0.0006 mg/kg/day	0.0006 - 0.06 mg/kg/day	> 0.06 mg/kg/day	1:2:1 Distribution
Minimal Risk Level (MRL) – Inhalation	< 0.002 mg/m3	0.002 - 0.2 mg/m3	> 0.2 mg/m3	1:2:1 Distribution
HEAST Reference Dose (RfD)	< 0.0006 mg/kg/day	0.0006 - 0.06 mg/kg/day	> 0.06 mg/kg/day	1:2:1 Distribution
HEAST Reference Concentration (RfC)	< 0.002 mg/m3	0.002 - 0.2 mg/m3	> 0.2 mg/m3	1:2:1 Distribution
Reportable Quantity (RQ)	≤ 10 lb	100 lb, 1,000 lb	≥ 5,000 lb	Program expertise
TSCA 4 Subchronic NOAEL	< 0.6 mg/kg/day	0.6 - 60 mg/kg/day	> 60 mg/kg/day	Based on new RfD fencelines (1,000x)
TSCA 4 Subchronic LOAEL	< 6 mg/kg/day	6 - 600 mg/kg/day	> 600 mg/kg/day	Based on new RfD fencelines (10,000x)

Table VI. Human Toxicity - Noncancer Effects Fencelines

Data Element	*Scoring Fencelines*			*Fenceline Setting Method*
	High (3)	*Medium (2)*	*Low (1)*	
TSCA 4 Developmental NOAEL	< 50 mg/kg/day	50 - 250 mg/kg/day	> 250 mg/kg/day	Program expertise
TSCA 4 Developmental LOAEL	< 500 mg/kg/day	500 - 2,500 mg/kg/day	> 2,500	Program expertise
Reference Exposure Level (REL)	< 2 μg/m3	2 - 200 μg/m3	> 200 μg/m3	1:2:1 Distribution
TSCA 8(e) Submission	3	2	1	Program expertise
CESARS Oral Mammalian Sublethality Score	≥ 8	6, 4	≤ 2	CESARS scores converted based on subchronic oral NOAEL fencelines
Human Health Structure Activity Team Rank	High	Medium	Low	Program expertise

Table VII. Human Cancer Toxicity Data Preference Hierarchy

Data Source	*Data Element*	*Preference Level*
	WEIGHT-OF-EVIDENCE	
IRIS (U.S. EPA, 1998c)	• Weight-of- Evidence	Medium
HEAST (U.S. EPA, 1997a)	• Weight-of- Evidence	Medium
IARC (IARC, 1998)	• Weight-of- Evidence	Medium
National Toxicology Program (NIEHS, 1998)	• Weight-of- Evidence	Medium
	CANCER SLOPE FACTOR (i.e., CANCER POTENCY FACTOR)	
IRIS (U.S. EPA, 1998c)	• Oral Slope Factor • Inhalation Unit Risk	Highest
HEAST (U.S. EPA, 1997a)	• Oral Slope Factor • Inhalation Slope Factor	High
EPA Cancer Data Documents (U.S. EPA, 1988-93)	• Oral Slope Factor	High
CERCLA Section 102 background document (U.S. EPA, 1989b)	• RQ Potency Factor	Medium
DRAFT Cal/EPA Standards and Criteria Work Group List of Cancer Potency Factors (Cal EPA, 1997a)	• Oral Slope Factor • Inhalation Slope Factor	Medium

Table VIII. Fencelines and Rationale for Human Toxicity - Cancer Effects Data Elements

Data Element	*Scoring Fencelines*			*Fenceline Setting Method*
	High (3)	*Medium (2)*	*Low (1)*	
EPA WOE Score	A	B or C	-NA-	Program expertise
IARC WOE Score	1	2A or 2B	-NA-	Converted to EPA WOE scores; based on EPA WOE score fencelines
NTP WOE Score	-NA-	CE or SE or EE or P or E	-NA-	Converted to EPA WOE scores; based on EPA WOE score fencelines
IRIS Oral Cancer Slope Factor (WOE A or B)	> 4.6 $(mg/kg/day)^{-1}$	4.6 - 0.046 $(mg/kg/day)^{-1}$	< 0.046 $(mg/kg/day)^{-1}$	1:2:1 Distribution
IRIS Oral Cancer Slope Factor (WOE C)	> 46 $(mg/kg/day)^{-1}$	46 - 0.46 $(mg/kg/day)^{-1}$	< 0.46 $(mg/kg/day)^{-1}$	Based on new oral cancer slope factor fencelines for WOE A and B (10x)
IRIS Inhalation Unit Risk (WOE A or B)	> 0.0013 $(\mu g/m^3)^{-1}$	0.0013 - 0.000013 $(\mu g/m^3)^{-1}$	< 0.000013 $(\mu g/m^3)^{-1}$	1:2:1 Distribution

Continued on next page.

Table VIII. Fencelines and Rationale for Human Toxicity - Cancer Effects Data Elements

Data Element	*Scoring Fencelines*			*Fenceline Setting Method*
	High (3)	*Medium (2)*	*Low (1)*	
IRIS Inhalation Unit Risk (WOE C)	> 0.013 $(\mu g/m^3)^{-1}$	0.013 - 0.00013 $(\mu g/m^3)^{-1}$	< 0.00013 $(\mu g/m^3)^{-1}$	Based on inhalation unit risk fencelines for WOE A and B (10x)
HEAST Oral Cancer Slope Factor (WOE A or B)	> 4.6 $(mg/kg/day)^{-1}$	4.6 - 0.046 $(mg/kg/day)^{-1}$	< 0.046 $(mg/kg/day)^{-1}$	1:2:1 Distribution
HEAST Oral Cancer Slope Factor (WOE C)	> 46 $(mg/kg/day)^{-1}$	46 - 0.46 $(mg/kg/day)^{-1}$	< 0.46 $(mg/kg/day)^{-1}$	Based on new oral cancer slope factor fencelines for WOE A and B (10x)
HEAST Inhalation Cancer Slope Factor (WOE A or B)	> 4.6 $(mg/kg/day)^{-1}$	4.6 - 0.046 $(mg/kg/day)^{-1}$	< 0.046 $(mg/kg/day)^{-1}$	1:2:1 Distribution
HEAST Inhalation Cancer Slope Factor (WOE C)	> 46 $(mg/kg/day)^{-1}$	46 - 0.46 $(mg/kg/day)^{-1}$	< 0.46 $(mg/kg/day)^{-1}$	Based on new oral cancer slope factor fencelines for WOE A and B (10x)
EPA Cancer Data Oral Cancer Slope Factor (WOE A or B)	> 4.6 $(mg/kg/day)^{-1}$	4.6 - 0.046 $(mg/kg/day)^{-1}$	< 0.046 $(mg/kg/day)^{-1}$	1:2:1 Distribution

Table VIII. Fencelines and Rationale for Human Toxicity - Cancer Effects Data Elements

Data Element	*Scoring Fencelines*			*Fenceline Setting Method*
	High (3)	*Medium (2)*	*Low (1)*	
EPA Cancer Data Oral Cancer Slope Factor (WOE C)	> 46 $(mg/kg/day)^{-1}$	46 - 0.46 $(mg/kg/day)^{-1}$	< 0.46 $(mg/kg/day)^{-1}$	Based on new oral cancer slope factor fencelines for WOE A and B (10x)
RQ Potency Factor (WOE A or B)	> 100 $(mg/kg/day)^{-1}$	100 1.3 $(mg/kg/day)^{-1}$	< 1.3 $(mg/kg/day)^{-}$	1:2:1 Distribution
RQ Potency Factor (WOE C)	> 1,000 $(mg/kg/day)^{-1}$	1,000 - 13 $(mg/kg/day)^{-1}$	< 13 $(mg/kg/day)^{-1}$	Based on RQ potency factor fencelines for WOE A and B (10x)
Cal/EPA Inhalation Cancer Slope Factor (WOE A or B)	> 4.6 $(mg/kg/day)^{-1}$	4.6 - 0.046 $(mg/kg/day)^{-1}$	< 0.046 $(mg/kg/day)^{-1}$	1:2:1 Distribution
Cal/EPA Inhalation Cancer Slope Factor (WOE C)	> 46 $(mg/kg/day)^{-1}$	46 - 0.46 $(mg/kg/day)^{-1}$	< 0.46 $(mg/kg/day)^{-1}$	Based on new oral cancer slope factor fencelines for WOE A and B (10x)

Continued on next page.

Table VIII. Fencelines and Rationale for Human Toxicity - Cancer Effects Data Elements

Data Element	*Scoring Fencelines*			*Fenceline Setting Method*
	High (3)	*Medium (2)*	*Low (1)*	
Cal/EPA Oral Cancer Slope Factor (WOE A or B)	> 4.6 $(mg/kg/day)^{-1}$	4.6 - 0.046 $(mg/kg/day)^{-1}$	< 0.046 $(mg/kg/day)^{-1}$	1:2:1 Distribution
Cal/EPA Oral Cancer Slope Factor (WOE C)	> 46 $(mg/kg/day)^{-1}$	46 - 0.46 $(mg/kg/day)^{-1}$	< 0.46 $(mg/kg/day)^{-1}$	Based on new oral cancer slope factor fencelines for WOE A and B (10x)

(i.e., FCVs, SCVs, measured chronic values, and estimated chronic values) are identical. The fencelines used to score the acute toxicity data elements (i.e., FAVs, CMCs, measured LC50s and EC50s, and predicted LC50s and EC50s) are also identical, with the exception of the aquatic toxicity RQ. The TSCA 8(e) Triage concern level fencelines are qualitative in nature.

WMPT Scoring Results

Distribution of Scores

Distributions of WMPT scoring results are shown in Figures 2-6. Figure 2 shows the distribution of overall chemical concern scores (i.e., the higher of human and ecological concern scores). Out of 4,157 chemicals included in the WMPT database, 2,895 had persistence, bioaccumulation, and toxicity data and received overall concern scores. The remaining 1,262 chemicals had only partial scoring information (e.g., a toxicity score was missing) and therefore did not receive overall concern scores. About 9 percent of the 2,895 chemicals with overall concern scores received scores of 8 or 9, indicating a relatively high level of concern. About 26 percent of chemicals received scores of 3 or 4, indicating relatively low levels of concern. The remainder of the chemicals, about 65 percent, received concern scores in the middle. The roughly bell-shaped distribution is largely a result of the use of 1:2:1 distributions to set fencelines for the underlying persistence and toxicity data, as discussed above.

Please note that this division of results into high, medium, and low overall concern is provided primarily to give a general sense of the distribution of scores. Given the screening nature of this tool, there is significant uncertainty in the underlying data values and chemical scores. Consequently the scores are only rough approximations of levels of concern. While it can be reasonably argued that chemicals at the high end of the scale present greater concerns than chemicals in the middle or low end, it is more difficult to argue that a chemical with an overall score of 8 is a significantly greater concern than a chemical with a score of 7.

Figure 3 shows the distribution of WMPT persistence scores. Of the 4,157 chemicals in the database, 2,918 received persistence scores. Since persistence scores of 3 (high), 2 (medium), and 1 (low) were derived using a 1:2:1 distribution, roughly 25 percent of the chemicals that were scored received scores of 3, 25 percent received scores of 1, and 50 percent received scores of 2. It was possible to score a relatively large number of chemicals for persistence given the use of predictive methodologies.

The distribution of bioaccumulation scores is shown in Figure 4. A total of 3,350 chemicals had bioaccumulation scores. The scores in this case were skewed toward the low end due to the use of expert judgment, rather than a 1:2:1 distribution, to establish scoring fencelines. Of the chemicals that were scored, about 8 percent scored 3, 8 percent scored 2, and 84 percent scored 1. As with persistence, it was possible to score a fairly large number of chemicals by using predictive methodologies.

Table IX. Ecological Toxicity Data Preference Hierarchy

Data Source	*Data Element*	*Preference Level*
SQC Documents (U.S. EPA, 1996a)	• Sediment Quality Criteria (SQC) Tier I Final Chronic Value (FCV)	Highest
GLWQI Criteria Documents (U.S. EPA, 1995a)	• Great Lakes Water Quality Initiative (GLWQI) Tier I FCV	Highest
Draft Quality Criteria for Water (AWQC Documents) (U.S. EPA, 1987b, 1995b) and Ecotox Thresholds ECO Update (U.S. EPA, 1996a)	• Ambient Water Quality Criteria (AWQC) FCV	Highest
Ecotox Thresholds ECO Update (U.S. EPA, 1996a) and Hazardous Waste Identification Rule (HWIR) documents (U.S. EPA, 1995 c,d)	• GLWQI Tier II methodology Secondary Chronic Value (SCV)	High
OPPT (U.S. EPA, 1994a, 1996b)	• Measured Chronic Data (EC50, EC10, LC50, or GMATC)	High
OPPT (U.S. EPA, 1996b)	• Estimated Chronic Data (EC10 or GMATC)	High
GLWQI Criteria Documents (U.S. EPA, 1995a)	• GLWQI Tier I Final Acute Value (FAV)	Medium
Draft Quality Criteria for Water (AWQC Documents) (U.S. EPA, 1995b)	• AWQC Acute CMC	Medium

Table IX. Ecological Toxicity Data Preference Hierarchy

Data Source	*Data Element*	*Preference Level*
OERR RQ database (U.S. EPA, 1989a) and OPPT EPCRA 313 TRI database (U.S. EPA, 1998b)	• Aquatic Toxicity Reportable Quantity (RQ)	Low
AQUIRE (U.S. EPA, 1998a)	• Measured Chronic Data (EC50, EC10, LC50, or GMATC)	Low
AQUIRE (U.S. EPA, 1998a)	• Measured Acute Data (LC50 or EC50)	Low
TSCA 8(e) U.S. EPA, 1998g)	• Triage screening results	Low
ECOSAR (SRC, undated a)	• Estimated Chronic GMATC • Estimated Acute Data (LC50 or EC50)	Lowest

Table X. Ecological Toxicity Fencelines

Data Element	*Scoring Fencelines*		
	High (3)	*Medium (2)*	*Low (1)*
Sediment Quality Tier I FCV	< 0.1 mg/L	0.1 – 10 mg/L	> 10 mg/L
GLWQI Tier I FCV I	< 0.1 mg/L	0.1 – 10 mg/L	> 10 mg/L
AWQC FCV	< 0.1 mg/L	0.1 – 10 mg/L	> 10 mg/L
GLWQI Tier II SCV	< 0.1 mg/L	0.1 – 10 mg/L	> 10 mg/L
OPPT=s Measured Chronic Value	< 0.1 mg/L	0.1 – 10 mg/L	> 10 mg/L
OPPT=s Predicted Chronic Value	< 0.1 mg/L	0.1 – 10 mg/L	> 10 mg/L
GLWQI Tier I FAV	< 1 mg/L	1 – 100 mg/L	> 100 mg/L
AWQC CMC	< 1 mg/L	1 – 100 mg/L	> 100 mg/L
Measured Chronic Value from AQUIRE	< 0.1 mg/L	0.1 – 10 mg/L	> 10 mg/L
Aquatic Toxicity RQ	1, 10 pounds	100, 1000 pounds	5000 pounds
Measured Acute Value (LC50 or EC50)	< 1 mg/L	1 – 100 mg/L	> 100 mg/L
TSCA Section 8(e) Triage Screening Result for Aquatic Toxicity Study Types	High	Medium	Low

Table X. Ecological Toxicity Fencelines

Data Element	*Scoring Fencelines*		
	High (3)	*Medium (2)*	*Low (1)*
ECOSAR Predicted Chronic Value	< 0.1 mg/L	0.1 – 10 mg/L	> 10 mg/L
ECOSAR Prediction of No Toxic Effects at Saturation (NTS)	Not applicable	Not applicable	1
ECOSAR Predicted Acute Value (LC50 or EC50)	< 1 mg/L	1 – 100 mg/L	> 100 mg/L

Figure 2
Distribution of WMPT Human/Ecological Concern Scores

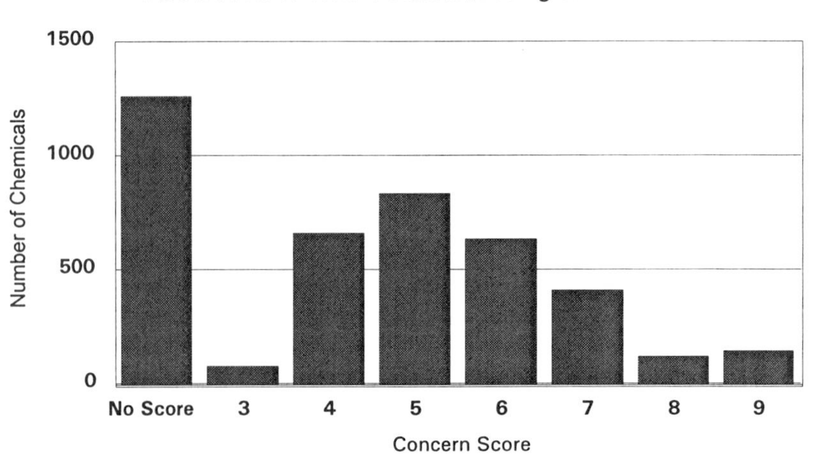

Figure 3
Distribution of WMPT Persistence Scores

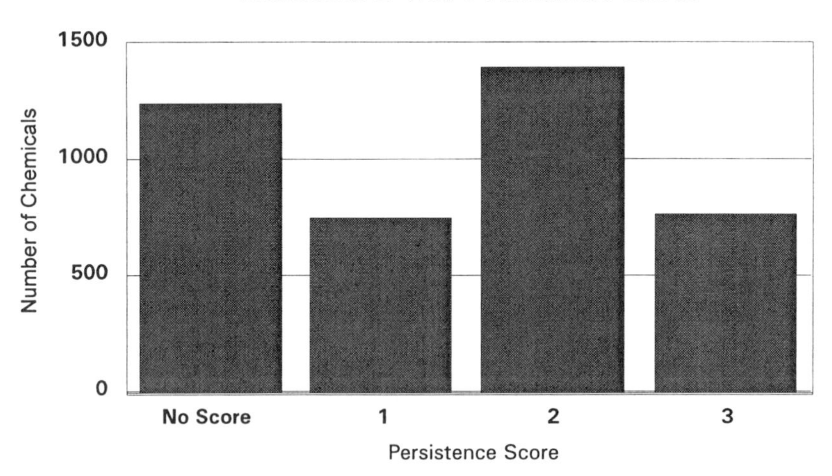

Figure 4
Distribution of WMPT Bioaccumulation Potential Scores

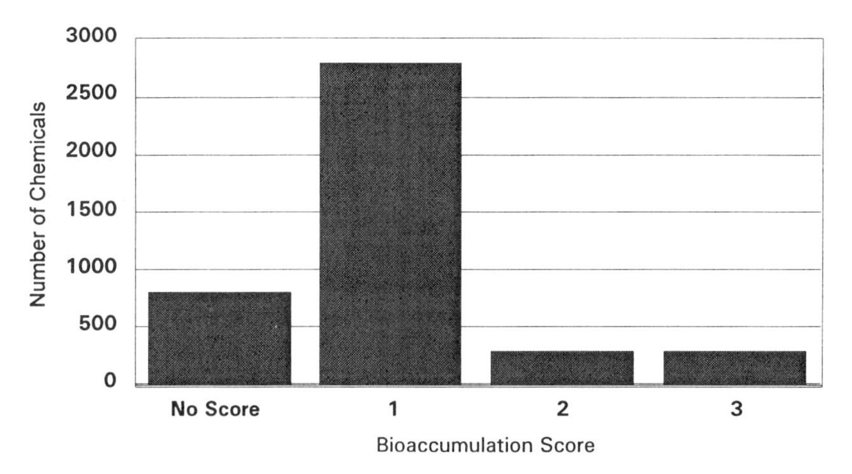

The distribution of human toxicity scores is shown in Figure 5. Given that predictive methodologies were not employed in scoring human toxicity, far fewer chemicals received scores. Of the 791 chemicals that had measured data and received scores, about 27 percent scored 3, 58 percent scored 2, and 16 percent scored 1.

Finally, Figure 6 shows the distribution of scores for ecological toxicity. A total of 3,466 chemicals received scores, and of these 41 percent received a score of 3, 37 percent received a score of 2, and 21 percent received a score of 1. This distribution is skewed toward the high end, perhaps in part due to the selection of measured and predicted values for the most sensitive endpoints (i.e., the lowest toxicity values) when scoring chemicals.

Strengths and Limitations of WMPT Scoring Approach

One of the strengths of the WMPT scoring approach is that it allows for screening of thousands of chemicals based on the best-available measured and predicted data. By evaluating different types of data from a variety of data sources using a common three-bin scoring system, it allows much broader comparison of chemicals than would be possible by comparing them using only a single type of data (.e.g., RfDs for human toxicity). Few other screening tools currently have this breadth of coverage. Another strength is that it focuses on several of the key parameters for estimating human and ecological risk: persistence, bioaccumulation, and toxicity. The WMPT also considers ecological toxicity in addition to human toxicity when evaluating chemical concerns, which few other screening systems currently do; an implicit assumption here is that ecological endpoints should receive equal priority to human endpoints. Finally, the WMPT has benefitted from numerous comments provided during extensive public and expert review processes.

One of the limitations of the WMPT is the limited discrimination provided by a three-bin scoring scheme; in some cases there may be more than an order of magnitude of difference between values within a given bin (e.g., the high bin may contain both high and very high values). The use of bins may also result in "border effects" where two values that are relatively close together get separated into adjacent bins. Another limitation is that the WMPT scoring approach arguably provides better discrimination in scoring at the high and low ends of the scoring spectrum than in the middle; given the use of 1:2:1 distributions for setting fencelines for many of the WMPT data elements, approximately half of the chemicals end up in the middle range of the scoring distribution. Finally, given that good measured fate and toxicity data are generally available for only a few hundred of the best-characterized chemicals, the WMPT relies heavily on predictive methodologies to score chemicals. While use of predicted data may arguably provide greater consistency in comparing chemicals (avoiding the site-specific nature of a lot of measured data), measured data from good studies may provide a closer reading of actual chemical concerns (e.g., better reflecting metabolism of chemicals in the case of bioaccumulation data). The

Figure 5
Distribution of WMPT Human Toxicity Scores

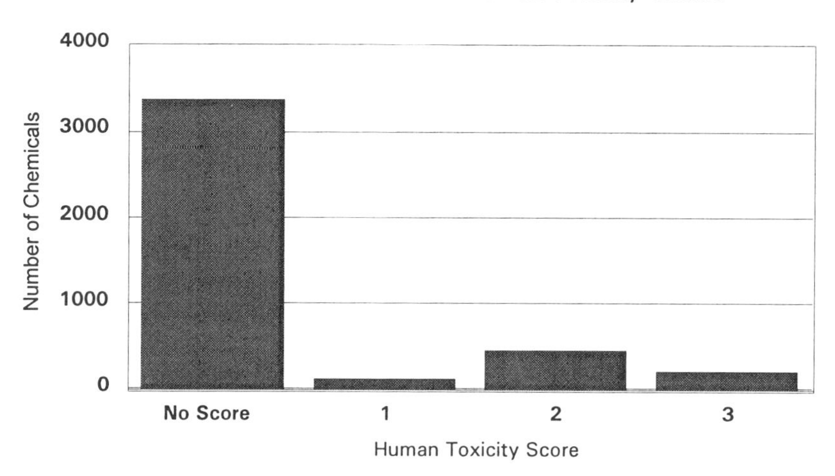

Figure 6
Distribution of Ecological Toxicity Scores

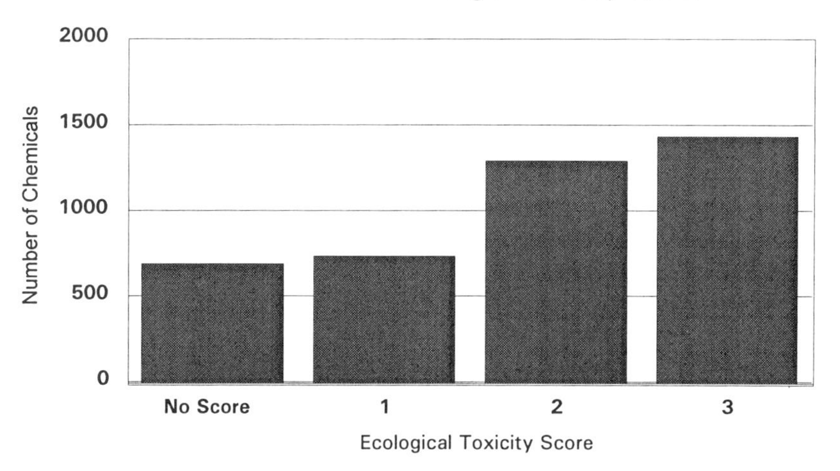

collection of measured fate and toxicity data for a larger number of chemicals will help to improve predictions from screening tools like the WMPT.

Current Status of the WMPT

The WMPT is currently being revised in response to public comments on the Agency's Draft RCRA PBT List. Scoring information for the subset of WMPT chemicals that are candidates for the PBT list will be released along with the revised list. The Agency is currently debating whether and how to re-release the full WMPT. Among the challenges in providing this type of screening tool for public use are clearly spelling out appropriate uses of the tool (e.g., it is intended for screening and not risk assessment), providing appropriate stewardship of the underlying data and algorithm as the state of the science advances (e.g., doing periodic data retrievals to update values), and deciding how best to make the information available (e.g., through files that can be downloaded or via direct access on the Internet). Further information on the status of the WMPT can be found at **www.epa.gov/wastemin.**

GLOSSARY:

AQUIRE: Aquatic Toxicity Informaiton Retrieval
BAF:Bio accumulation factor
BCF:bio conenctration factor
CESARS: Chemical Evaluation Search and Retrieval System
ECOSAR: Ecological Structrue Activity Relations
HEAST: Health Efects Assessment Summary Tables
HWIR: Hazardous Waste Identification Rule
IARC: Inter Agency
IRIS: Integrated Risk Information System
ISIS: Integrated Scientific Information System
LOAEL: Chronic Lowest observed adverse effects level
MRL: Minimum Risk Level
NOAEL: chronic no obersrved adverse effects level
OERR: Office of Emergency and Remedial Response, US EPA
OSW: Office of Solid Waste, US EPA
OPPT: Office of Pollution Prevention and Toxics, US EPA
PBT: Persistant Bioaccumulative and Toxic
RCRA: Resource, Conservation, and Recovery Act
REL: Recommended exposure limit
RFC: Reference concentration
RQ: Reportable Quantity
WOE: Weight-of-Evidence

References

1. Agency for Toxic Substances and Disease Registry (ATSDR). 1998. MRL, Minimal Risk Level. URL http://atsdr1.atsdr.cdc:8080/mrls.html.
2. International Agency for Research on Cancer (IARC). 1998. IARC Monographs on the Evaluation of Carcinogenic Risks to Humans. URL http://193.51.164.11/monoeval/grlist.html.
3. National Institute of Environmental Health Sciences (NIEHS). 1998. National Toxicology Program Weight of Evidence data files.
4. Ontario Ministry of Environmental and Energy (Ontario MOEE) and Michigan Department of Natural Resources (MI DNR). 1998. CESARS, Chemical Evaluation Search and Retrieval System.
5. U.S. Department of Agriculture (USDA). 1998. Agricultural Research Service Pesticide Properties Data Base. May 1995 update. URL http://www.arsusda.gov/rsml/ppdb2.html
6. U.S. EPA. 1998a. Aquatic Information Retrieval Database, AQUIRE. Office of Research and Development National Health and Environmental Effects Research Laboratory.
7. U.S. EPA. 1998b. Emergency Planning and Community Right-to-Know Act Section 313 Toxics Release Inventory Chemicals Screening Database. Washington, DC: Office of Pollution Prevention and Toxics.
8. U.S. EPA. 1998c. IRIS, Integrated Risk Information System. URL http://www.epa.gov/iris.
9. U.S. EPA. 1998d. Notice of Availability of Draft RCRA Waste Minimization PBT Chemical List. Federal Register, vol. 63, No. 216, p. 60332, November 9, 1998.
10. U.S. EPA. 1998e. Office of Pollution Prevention and Toxics Human Health Structure Activity Team Ranking.
11. U.S. EPA. 1998f. Toxic Substances Control Act Section 4 NOAEL and LOAEL data table. Washington, DC: Office of Pollution Prevention and Toxics.
12. U.S. EPA. 1998g. Toxic Substances Control Act Section 8(e) Triage data. Washington, DC: Office of Pollution Prevention and Toxics.
13. U.S. EPA. 1998h. Waste Minimization Prioritization Tool Spreadsheet Document for the RCRA Waste Minimization PBT Chemical List Docket (#F-98-MMLP-FFFFF). Washington, DC: Office of Solid Waste. September. URL http://www.epa.gov/wastemin.
14. California Environmental Protection Agency (Cal EPA). 1997a. Air Toxics Hot Spots Program Risk Assessment Guidelines: Technical Support Document for Describing Available Cancer Potency Factors, October 1997, Public and Scientific Review Panel Draft.
15. California Environmental Protection Agency (Cal EPA). 1997b. Draft Cal/EPA Office of Environmental Health Hazard Assessment Technical Support Document for the Determination of Noncancer Chronic Reference Exposure Levels. Table 8. October.

16 Syracuse Research Corporation (SRC). 1997. BCFWIN, bioconcentration factor estimation program, version 2.01. Part of Estimation Programs Interface, version 2.30. September. URL http://esc.syrres.com.
17 U.S. EPA. 1997a. HEAST, Health Effects Assessment Summary Tables.
18 U.S. EPA. 1997b. Mercury Study Report to Congress. December.
19 U.S. EPA. 1997c. Waste Minimization Prioritization Tool (Beta Version 1.0) Software. EPA530-C-97-003.
20 U.S. EPA. 1997d. Waste Minimization Prioritization Tool (Beta Version 1.0): User's Guide and System Documentation. EPA530-R-97-019.
21 Mackay, D., A. Di Guardo, S. Paterson, and C. E. Cowan. 1996. Evaluating the environmental fate of a variety of types of chemicals using the EQC model. Environmental Toxicology and Chemistry 15:1627-1637.
22 U.S. EPA. 1996a. Ecotox Thresholds. ECO Update. Washington, DC: Office of Emergency and Remedial Response. EPA540/F-95/038.
23 U.S. EPA. 1996b. Environmental Toxicity Profiles. Washington, DC: Office of Pollution Prevention and Toxics.
24 Howard, P. and W. Meylan. 1995. User's Guide for the Biodegradation Probability Program. Version 3, March 1995. Syracuse, NY: Syracuse Research Corporation.
25 U.S. EPA. 1995a. Great Lakes Water Quality Initiative Criteria Documents for the Protection of Aquatic Life in Ambient Water. EPA 820/B-95/004. Washington, DC: Office of Water.
26 U.S. EPA. 1995b. Quality Criteria for Water. Draft. Washington, DC: Office of Water, Health and Ecological Criteria Division.
27 U.S. EPA. 1995c. Supplemental Technical Support Document for HWIR: Risk Assessment for Human and Ecological Receptors. Volume 1. Washington, DC: Office of Solid Waste. November.
28 U.S. EPA. 1995d. Technical Support Document for HWIR: Risk Assessment for Human and Ecological Receptors. Volume 1. Washington, DC: Office of Solid Waste. August.
29 U.S. EPA. 1995e. Technical Support Document for the Hazardous Waste Identification Rule: Risk Assessment for Human and Ecological Receptors. From Table 5-18, Bioaccumulation Factors and Bioconcentration Factors for the Limnetic Ecosystem by Trophic Level, and Table 5-22, Biota-Sediment Accumulation Factors and Bioaccumulation Factors for the Littoral Ecosystem by Trophic Level. August.
30 U.S. EPA. 1994a. Addition of Certain Chemicals; Toxic Chemical Release Reporting; Community Right-to-Know. Federal Register. January 12, 1994. pp 1788-1859.
31 U.S. EPA. 1994b. RQ, Reportable Quantities Database.
32 U.S. EPA. 1994c. Waste Minimization National Plan. EPA530-R-94-045.
33 Meylan, W. and P. Howard. 1993. Atmospheric Oxidation Program. Version 1.85. URL http://esc.syrres.com.
34 Mackay, D., S. Paterson, and W.Y. Shiu. 1992. Generic models for evaluating the regional fate of chemicals. Chemosphere 24:695-717.

35 U.S. EPA. 1992. Classification Criteria for Environmental Toxicity and Fate of Industrial Chemicals. Washington, DC: Office of Pollution Prevention and Toxics, Chemical Control Division.

36 Mackay, D. et al. 1991-97. Illustrated Handbook of Physical-Chemical Properties and Environmental Fate for Organic Chemicals. Lewis Publishers Inc, Boca Raton, Florida. 5 volumes.

37 Howard, P. et al. 1990. Handbook of Environmental Degradation Rates. Heather Taub Printup editor, Lewis Publishers, Michigan.

38 U.S. EPA. 1989a. Reportable Quantities Database. Washington, DC: Office of Emergency and Remedial Response.

39 U.S. EPA. 1989b. Technical Background Document to Support Rulemaking Pursuant to CERCLA Section 102. Volumes 1, 2, and 3. Office of Solid Waste and Emergency Response.

40 U.S. EPA. 1988-93. Various reports prepared by Office of Health and Environmental Assessment and Environmental Criteria and Assessment Office.

41 U.S. EPA. 1987a. Ambient Water Quality Criteria for Zinc - 1987. PB87-153581. February.

42 U.S. EPA. 1987b. Update #2 to Quality Criteria for Water. Washington, DC: Office of Water.

43 U.S. EPA. 1984a. Ambient Water Quality Criteria for Cadmium - 1984. PB85-227031. January 1985.

44 U.S. EPA. 1984b. Ambient Water Quality Criteria for Chromium - 1984. PB85-227478. January 1985.

45 U.S. EPA. 1984c. Ambient Water Quality Criteria for Copper - 1984. PB85-227023. January 1985.

46 U.S. EPA. 1984d. Ambient Water Quality Criteria for Mercury - 1984. PB85-227452. January 1985.

47 U.S. EPA. 1980. Ambient Water Quality Criteria for Lead. PB81-117681. October.

48 Syracuse Research Corporation (SRC). undated. ECOSAR Class Program. Version 0.99.

49 Syracuse Research Corporation (SRC). undated. HYDROWIN, the Aqueous Hydrolysis Rate Program. Version 1.63.

50 Syracuse Research Corporation (SRC). undated. ISISbase database of fish BCF values.

New Chemicals

Chapter 14

Short Chain Chlorinated Paraffins: Are They Persistent and Bioaccumulative?

Derek Muir[1], Don Bennie[1], Camilla Teixeira[1], Aaron Fisk[2], Gregg Tomy[3], Gary Stern[3], and Mike Whittle[4]

[1]National Water Research Institute, Environment Canada, Burlington, Ontario L7R 4A6, Canada
[2]Canadian Wildlife Service, Environment Canada, Ottawa, Ontario K1A 0H3, Canada
[3]Freshwater Institute, Fisheries and Oceans Canada, Winnipeg, Manitoba R3T 2N6, Canada
[4]Fisheries and Oceans Canada, Burlington, Ontario L7R 4A6, Canada

Short-chain chlorinated paraffins (SCCPs) are polychlorinated-[C_{10} to C_{13}]-n-alkanes which are used as additives in metal working fluids and flame retarding applications. They have physical properties similar to many persistent organic pollutants (POPs). In this study levels of SCCPs were measured in effluents, sediments, water, air, and fish from Lake Ontario and in beluga whale (*Delphinapterus leucas*) blubber from the St. Lawrence River estuary and the Canadian arctic. SCCPs were detected in all samples but generally at levels much lower than PCBs. There was also evidence for biotransformation of SCCPs. The results suggest SCCPs are not as persistent or bioaccumulative as many POPs.

Introduction

Short-chain chlorinated paraffins (SCCPs) are polychlorinated-[C_{10} to C_{13}]-n-alkanes with chlorine content from 50 to 70% by mass. SCCPs are used mainly as extreme temperature additives in metal working fluids for a variety of engineering and metal working operations such as drilling, machining/cutting, drawing and stamping (*1,2*). They also are used in paints and sealants and have a minor but environmentally significant use as fat liquors in the leather working industry. SCCPs are part of a larger group of polychlorinated-n-alkanes (PCAs) with carbon chain lengths up to C_{30} and chlorine content of 30-70%. The higher molecular weight products are used as flame retardant plasticizers, especially in PVC, as well as additives to improve water resistence and flame retardancy in adhesives, paints, rubber and sealants. Global production capacity for all PCAs is estimated to be 300 kT (*3,4*), however, quantities produced may be significantly lower because capacity in Europe and the USA exceeded demand during the 1990s (*5*). Published information on historical production of SCCPs is difficult to find. Howard et al. (*6*) reported quantities used in the USA from 1945 to 1972. In the mid-1990's, SCCP production represented about 1/5 of total PCA, i.e. about 18 kT per year in the USA (*7*). From these various sources (*4-8*) it is possible to construct a tentative global historical use profile (Figure 1). Due to concerns about the toxicity, bioaccumulation and persistence of SCCPs, use of these chemicals is declining as users switch to alternative products (*9*). For e.g. use of SCCPs in Germany as metal working fluids has declined since 1985 and substitution was virtually complete by 1999 (*10*). Production in the US, Europe and Japan, the major consumers of PCAs globally, remained relatively constant throughout the 1990's (*5,7*).

In the United States, SCCPs have been placed on the US Environmental Protection Agency (EPA) Toxic Release Inventory (*7,9*), and in Canada they are under consideration for classification as "Track 1", Priority Toxic Substances under the Canadian Environmental Protection Act (*9*). In Europe, the European Commission has completed a risk assessment of SCCPs and a risk reduction strategy consisting of restrictions on marketing and use of SCCPs in metal and leatherworking is being put into place for all members of the European Union (*9*). Sweden has implemented a phase out of SCCPs which is to be completed in 2000 (*9*). The Oslo-Paris Commission (OSPAR) has recommended a phase out of SCCPs in paint and coatings, sealants, metalworking fluids and plastics (*9*).

SCCPs did not replace PCBs used in electrical equipment in the 1970's, despite having some similarities in physical properties and flame retardant characteristics (*6*). SCCPs were not good PCB replacements for uses requiring high heat stability (e.g. capacitors, transformers) because of much lower thermal stability.

The SCCPs are of particular interest because they have the greatest potential for environmental release, and the highest toxicity of PCA products (*1-3, 11*). Assessments of SCCPs by Environment Canada (*3*), Swedish Chemicals Inspectorate (*12*), International Program on Chemical Safety (*13*), and US Environmental Protection Agency (*14*) have cited toxicity to aquatic invertebrates and carcinogenicity in rats and mice as the major concerns with regard to continued manufacture and use of these compounds. These assessments also identified lack of environmental measurements and information on degradation rates as major knowledge gaps. This chapter reviews the sources, physical properties and recent environmental measurements of SCCPs in North America and assesses their potential for persistence and bioaccumulation.

Industrial Synthesis, Other Possible Sources and Emissions

SCCPs are produced by chlorination of C_{10}-C_{13} n-alkanes, using molecular chlorine, either of the liquid paraffin or in a solvent, typically carbon tetrachloride (*4,6,8,15*) Depending upon the n-alkane feedstock, the reaction takes place at temperatures between 50 and 150°C, at elevated pressures, and/or in the presence of UV light (*4,15,16*). After chlorination the product is stripped of solvent, residual chlorine and reaction products (e.g. HCl) by gas sparging. Final products are mixtures that are viscous, colorless or yellowish dense oils (*6*).

It is possible that SCCPs could be formed as byproducts of other industrial syntheses or processes involving chlorine (*6*). While the presence of SCCPs in chlorinated solvents (e.g. CCl_4, perchloroethylene, methylene chloride) is unlikely because most are distilled or fractionated, SCCPs, or compounds resembling them in molecular weight and physical properties, could be byproducts of incomplete polymerization of chlorinated ethylene monomers (*6*). The possible inadvertant production of compounds resembling commercial SCCPs (i.e. >50% Cl by weight) during aqueous chlorination e.g. from hydrocarbons in waste water or drinking water treatment is highly unlikely because of dilute conditions (*6*).

The release of SCCPs into the environment could occur during production, storage, transportation, industrial use and carryoff on manufactured products, release from plastics, paints and sealants in which they are incorporated, leaching, runoff or volatilization from landfill, sewage sludge amended soils, or other waste disposal sites. Of these, however, the major releases are thought to be from production and from industrial usage (*3, 12*). The use of SCCPs in metal working fluids and in leather fat liquors may lead to significant emissions to the environment from equipment washoff and leaching from treated materials. Disposal and burning of waste containing SCCPs may be another potential source of entry of these compounds into the environment. Land filling of products such as plastics, textiles, painted materials, paint cans and oils containing SCCPs may result in slow leaching and/or volatilization from these matrices (*3, 11,17*).

Complexity of SCCPs

SCCPs are multi-congeneric mixtures, one or two orders of magnitude more complex than PCB mixtures. When analysed by high resolution gas chromatography or liquid chromatography, SCCPs generally elute over a wide retention time range, and individual components are not resolved (*18,19*) (Figure 2). Assuming a maximum of one chlorine per carbon, since a second chlorine atom does not readily substitute for hydrogen at a carbon already bound to chlorine and taking into account that isomers with 4 Cl or less are not present, the theoretical number of positional isomers possible for C_{10}-C_{13} alkanes ranges from 327 for penta to decachloro-n-decane to 4159 (*18,19*) for penta- to dodecachlorotridecane (7). GC-MS analysis of commercial SCCPs shows that hexa- and heptachloro- decanes, undecanes and dodecanes are the predominant isomers. The actual substitution pattern on the carbon chain is unknown. Less chlorinated alkanes are likely to have 1,3,5-type substitution due to steric considerations, but SCCPs with high chlorine contents (e.g. 70%) must have numerous congeners with vicinal chlorines (*6*). The lack of CCl_2 groups in commercial PCA mixtures with less than 60% Cl has been confirmed by NMR (*20*).

Analytical methods

The complexity of industrial mixtures of SCCPs creates major problems for their quantitative analysis in environmental samples. Only a very limited number of individual congeners are available (*21*) and their representativeness is unknown. Thus most methods of quantitation rely on the commercial mixtures. Gas chromatography-low resolution quadrupole mass spectrometry (GC-LRMS), the most commonly employed technique lacks specificity. Procedures based on monitoring the ubiquitous m/z 70-73 ions, i.e., $Cl_2^{-\cdot}$ and HCl_2^- (22) (using electron capture negative ion (ECNI) mass spectra), present the problem that many other

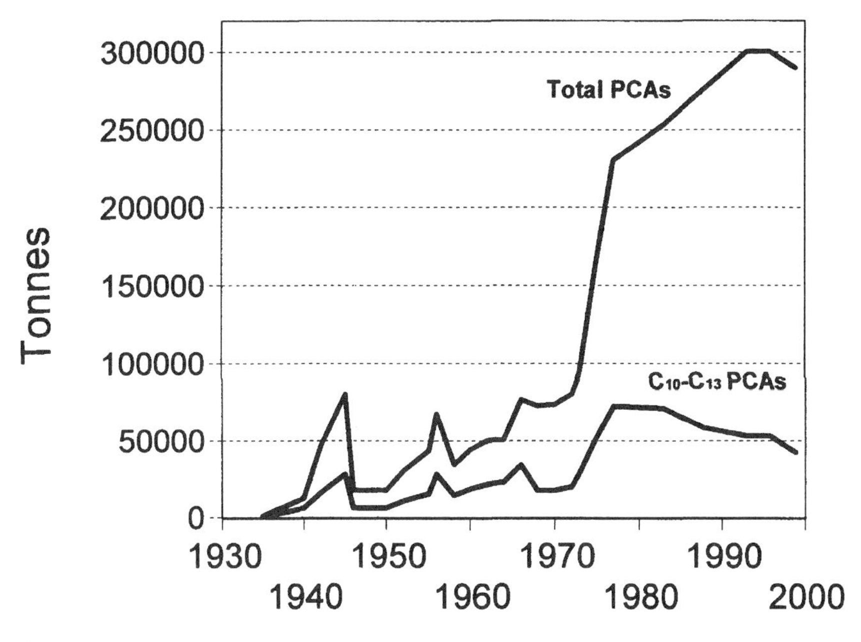

Figure 1. Estimated global production of all chlorinated n-alkanes (PCAs) and of SCCPs (C_{10}-C_{13}-PCAs.) Based on data from references *4-8*.

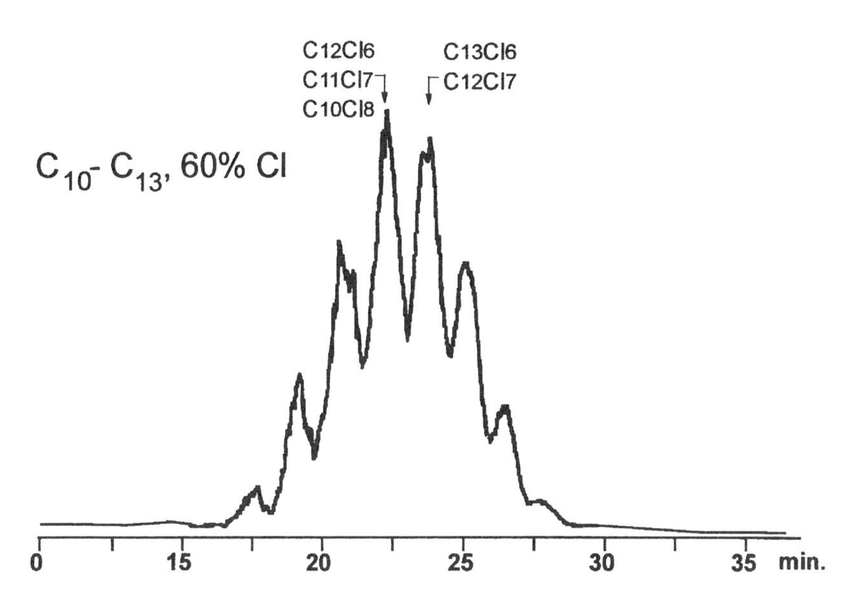

Figure 2. High resolution GC-Electron Capture negative ion high resolution MS total ion chromatogram of a short chain chlorinated paraffin mixture. Column: 30m x 0.25mm i.d. DB-5ms Temp. Program: 80 °C, ramp 7 °C/min to 260 °C, hold 8 min., ramp 10 °C/min to 280 °C, hold 20 min.

persistent aliphatic OCs fragment to yield such ions, e.g., p,p'-DDT, p,p'-DDE, lindane, dieldrin, aldrin and endrin. Thus, if these contaminants are not selectively removed from the sample matrix during the extraction or clean-up procedures, they would ultimately contribute to the response of the quantitation ion, $Cl_2^{-\cdot}$ (m/z 70), and lead to an overestimation in the level of SCCPs in samples. Many of the GC-quadrupole MS methods recently employed also lack the sensitivity to measure trace amounts of SCCPs because they have used full scan rather than selected ion techniques (*23-25*).

Tomy et al. (*18*) reported on the use of GC/ECNI-high resolution (HR) MS to overcome some of the problems associated with earlier methods. Using an ion source temperature of 120°C the authors were able to selectively increase the abundance of the [M-Cl]- ion relative to the ubiquitous and structurally noncharacteristic $Cl_2^{-\cdot}$ and HCl_2^- ions. The two most intense ions of the [M-Cl]- species for C_{10} (Cl_5 to Cl_{10}), C_{11} (Cl_5 to Cl_{10}), C_{12} (Cl_6 to Cl_{10}) and C_{13} (Cl_7 to Cl_9) were monitored at a resolving power 12 000 in retention time windows. The ECNI-HRMS method was compared in an interlab study in which seven participating laboratories were asked to quantify a solution of a commercially available PCA product containing 70% chlorine by mass (PCA-70), a synthetic SCCP mixture consisting of purified products derived from the chlorination of 1,5,9-decatriene (PCA-1) and two fish sample extracts (*26*). Participants used different instrumental types and conditions including three quadrupole systems and four magnetic sector instruments. Two laboratories performed the analysis at a high resolving power (RP>10000) while others used the mass spectrometer operating at nominal resolution. All laboratories performed the analysis under ECNI conditions. The results from the study showed that there was a high degree of variability associated with the quantitative methods used for the determination of SCCPs. Results from various laboratories are likely to be comparable only within a factor of 2 at best. Additional variability is also introduced when different commercial formulations are used as external standards. For the environmental samples, the reported measurements for the first fish sample were more consistent than those made on the second. The authors suggest this could have resulted from the fact that the former measurements were all made by laboratories that corrected for co-eluting interferences, while the latter measurements included laboratories that chose not to do so. As a result, it was recommended that procedures involving use of low resolution mass spectrometry should try to take into account, and correct for, the possible presence of other organochlorine contaminant interferences (*27*).

Extraction and isolation of SCCPs poses fewer problems for analysts. SCCPs are readily extracted under the same conditions as other organochlorines of similar chlorine content such as PCBs. SCCPs can generally be separated from lipids using size exclusion chromatography and also from PCBs on silica gel or Florisil columns (*18, 27, 28*).

Physical properties

SCCPs are hydrophobic substances. Drouillard *et al.* (*29,30*) reported water solubilities of individual PCA congeners, measured by the generator column technique, ranging from 37 to 994 μg/L (Table 1). Octanol-water partition coefficients (log K_{OW}'s) for SCCPs, determined by slow-stirring are in the 5.9 to 6.2 range, classifying them as lipophilic, similar to DDT but less than PCBs of similar molecular weight. Henry's Law Constants (HLC) for C_{10}-C_{12} PCAs range from 0.7 to 18 Pa m^3/mol which is similar to HLCs for chlorinated pesticides (HCH, toxaphene, p,p'-DDT) and implies partitioning from water to air, or from soils to air, depending on environmental conditions and prevailing concentrations in each compartment.

Table 1. Physical properties of individual SCCPs

Congener[1]	$S_W \pm SE$[2] μg/L	HLC ± SE [3] Pa m^3 /mol	log K_{OW} [4]
$C_{10}H_{18}Cl_4$	668±75	18±4.6	5.93
$C_{10}H_{17}Cl_5$ (a,b)	678±80	4.9±0.71	6.04-6.20
$C_{10}H_{17}Cl_5$ (c,d)	995±119	2.6±0.64	6.04-6.20
$C_{11}H_{20}Cl_4$	575±93	6.3±1.8	5.93
$C_{11}H_{19}Cl_5$ (a,b)	546±106	1.46±0.33	6.20-6.40
$C_{11}H_{19}Cl_5$ (c,d,e)	962±300	0.68±0.2	-
^{14}C-$C_{12}Cl_{5-7}$	36.7±6.0[3]	1.37±0.07	6.8

[1] Letters after formulas refer to isomers which could be separated by GC.
[2] Determined using the generator column technique (*30*).
[3] Henry's Law constants of chlorinated n-alkanes determined by gas-sparging at 23°C.
[4] Determined by slow stirring (*31*) except for ^{14}C-$C_{12}Cl_{5-7}$ which was by C18-reverse phase HPLC (32).

Bioaccumulation

The hydrophobic nature of SCCPs (log Kow's > 5.0, see Table 1), and results from a limited number of studies, suggest that these compounds bioconcentrate in aquatic organisms and have the potential to biomagnify through aquatic food webs. Bioconcentration (uptake only from the dissolved phase in water) of SCCPs has been found to vary with species, carbon chain length and chlorine content (*7,33*). Bengtsson *et al.* (*33*) found that the rate of uptake of commercial PCA products from water by fish increased with decreasing carbon chain length and chlorine content. The greater uptake rates of the SCCPs were probably due to their greater water solubility. Highly chlorinated SCCPs are predicted to have the greatest bioconcentration factor (BCFs) because they are more hydrophobic and resistant to biotransformation than lower chlorinated PCAs, and their accumulation is not hindered by a large molecular size or extremely high K_{OW} as observed for intermediate and long carbon chain PCAs (*34*). BCFs also vary between species. Madeley *et al.* (*35*) found a BCF (wet weight concentrations) of a $^{14}C_{11}$ SCCP (58% Cl) of about 40,000 in mussels which was approximately 7 times greater than in rainbow trout (*36*), a result probably due to the greater biotransformation in the trout.

As a result of their hydrophobicity, SCCPs should be bioaccumulated predominantly through food chain transfer (*37*). PCAs are readily accumulated from food by fish in laboratory experiments, although, as with bioconcentration, dietary accumulation is influenced by carbon chain length and chlorine content (*34,38*). SCCPs with greater than 60% Cl were found to have equilibrium biomagnification factors (BMFs) > 1 and half-lives ≥ 50 days in juvenile rainbow trout, which implies a potential to biomagnify in aquatic food chains (*34,38*). When compared to the half-lives in rainbow trout for a series of non-metabolizable PCB congeners (i.e. 2,4-substituted) also studied a series of synthetic C_{10}-C_{16}-PCAs of known Cl content and chain length showed intermediate behavior (*38,39*). Most SCCP congeners had lower half-lives and BMFs than PCB congeners with the same log K_{OW} but a few highly chlorinated components had similar half-lives (Figure 3).

Degradation Pathways

Although SCCPs do not degrade under direct photolysis, they may be subject to attack, *via* indirect photolysis, by oxidizing radicals in the troposphere (*3*). Based on Atkinson's OH radical reaction model, theoretical half-lives of PCAs in the atmosphere range from 1.2 to 1.8 days for C_{10}-C_{13} (no information given on Cl content), 0.85 to 1.1 days for C_{14}-C_{17} and 0.5 to 0.8 days for C_{18}-C_{30} (*40*). Indirect photolysis reactions in the aquatic environment that involve attack by free radicals might also occur, although this has not studied.

Rates of hydrolysis of PCAs in natural waters are considered negligible at ambient temperatures (*1-3*). However, reactions involving catalysts, known to be present in the aquatic environment, might induce hydrolysis or oxidation reactions, although no studies have been done to demonstrate this. Reiger and Ballschmiter (*41*) have noted, for example, that the interaction of PCAs with activated alumina results in dehydrochlorination

Results of microbial biodegradation studies suggest that degradation is inhibited by greater chlorine content and longer carbon chains. Zitko and Arsenault (*42*) examined the aerobic and anaerobic biodegradation of two long carbon chain (C_{20}-C_{30}) PCAs (42 and 70% Cl) in a suspension of sea water and decomposing organic matter at room temperature (19-22°C) and found that the rate of biodegradation was higher under anaerobic than aerobic conditions, and that a higher chlorinated PCA (70% Cl) was degraded to a greater extent than a less chlorinated PCA (42% Cl). Madeley and Birtley (*43*) used BOD tests to examine the biodegradation of a range of PCAs with different carbon chain lengths and chlorine contents. They concluded that short carbon chain PCAs (< 60% Cl) appeared to be rapidly and completely degraded, and medium and long carbon chain PCAs with up to 45% Cl degraded more slowly than SCCPs. Omori *et al.* (44) studied the PCA dechlorination potential of a series of soil bacterial strains. Although they could not isolate a bacterial strain which could use PCAs as a sole carbon source, they did find that different strains pretreated with *n*-hexadecane had different dechlorination abilities. A mixed culture (4 bacterial strains) released 15 to 57% of the Cl of five PCA products with the amount of Cl released decreasing with increasing carbon chain length and chlorine content. The

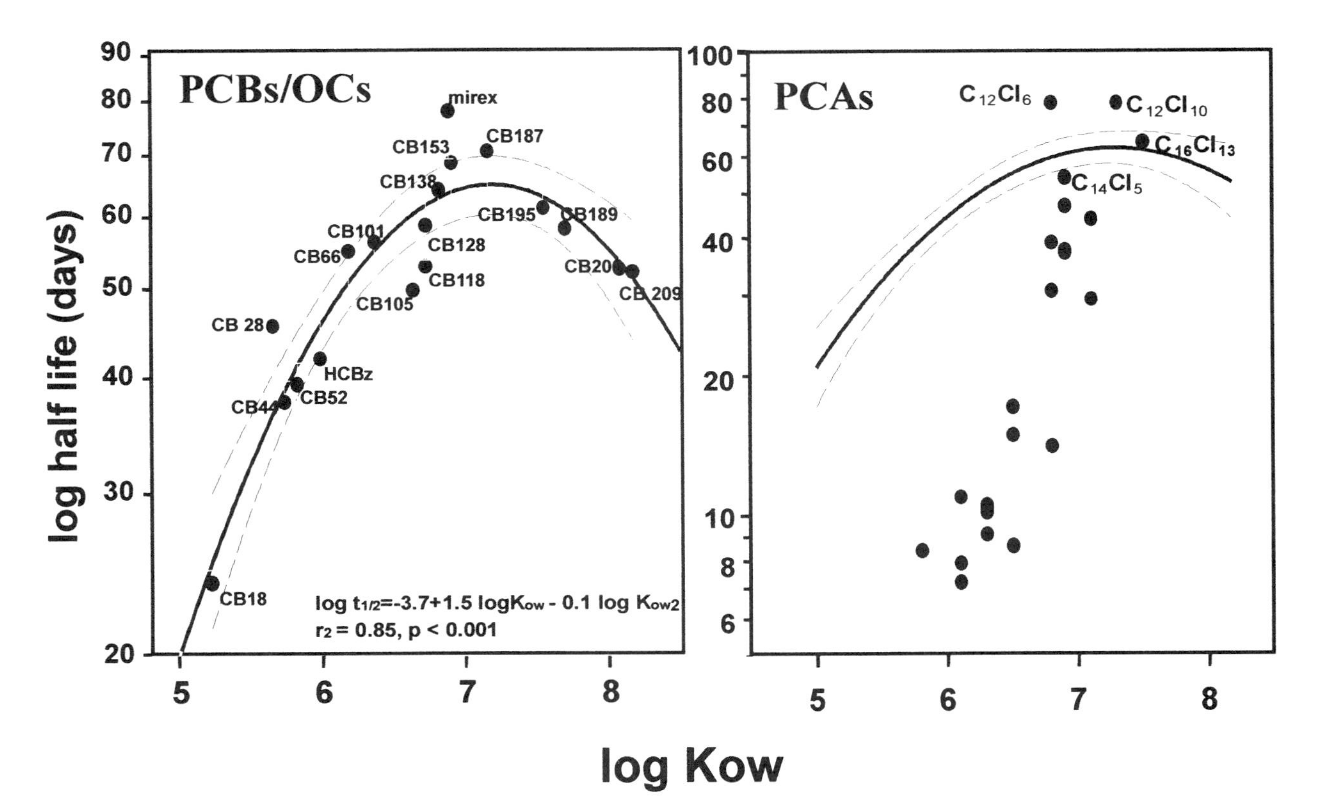

Figure 3 Half-life - log Kow relationship for C_{10}-C_{16} PCAs in rainbow trout following 30 day dietary exposures and comparison with half-lives of recalcitrant PCBs and OC pesticides. The curve in the PCA plot is the same quadratic expression as that in the PCB/OC plot.

authors concluded that 2, 3 or more Cl reduce, as well as the presence of adjacent Cl, reduce the dechlorination of PCA by bacterial strains. Fisk et al. (*32*) examined the biodegradation of two ^{14}C-labelled SCCPs (C_{12}, 56 and 69% Cl) in aerobic sediments by comparing the amount of toluene extractable and non-extractable ^{14}C after 18 and 32 days incubation. Half lives of the C_{12}-PCAs in sediment were 13 and 30 days for the 56 and 69% Cl products, respectively.

Environmental Levels of SCCPs

Concentrations of SCCPs have been measured in water, sediment, air and biota over the past 20 years and were recently reviewed (*11,45*). Early work used thin layer chromatography for semi-quantitative determination of C_{10}-C_{30} PCAs but did not specifically measure the C_{10}-C_{13} PCAs (*17*). Low ug/L levels of PCAs were detected in surface waters of the UK. C_{10}-C_{30} PCAs were generally present in ng/g levels sediments in the UK but a few sites in industrialized areas (e.g. harbours) had levels >1 ug/g dw (*17*). In general much more sampling and analysis for PCAs in waters and sediments has been done in western Europe than in North America. However, in Europe measurements of SCCPs in biota are confined to results of studies in Sweden (*46*). There is a growing body of data on SCCPs in biota in Canada using the HRMS method of Tomy et al. (*18*). Very few results for SCCPs in biotic or abiotic media have been reported in Japan, the other major use region.

In North America, information on levels of SCCPs in surface waters and sediments are quite limited. A survey for the US EPA assessment of SCCPs focussed on surface waters in industrial areas of Illinois, Ohio, Michigan and Indiana because of the presence of metal working industries (*13,25*). Water concentrations were predominately less than 30 ng/L, the lowest concern concentration for aquatic biota, while 25% of 248 samples had levels >30 ng/L. Water collected near a manufacturing facility had SCCP levels ranging from 200-300 ng/L (*25*).

We have recently conducted studies on levels of SCCPs in water, sediments and biota in Canada in order to investigate the current levels and possible long range transport of SCCPs (*18, 28*). In this report we focus on results for SCCPs in waste water treatment effluents, water, air, sediments and fishes in Lake Ontario and in beluga whale blubber from the St. Lawrence estuary and the Canadian Arctic. The St. Lawrence drainage area, and the Great Lakes in particular, have a well documented history of contamination with persistent organic pollutants, therefore it is of interest to see how SCCP levels compare with compounds such as PCBs. Studies of SCCPs in this region have been limited to surveys by the US EPA (*13, 25*) and Environment Canada (*23,24*). These studies used low resolution ECNIMS for analysis of SCCPs. In this study all samples have been analysed by ECNIMS to determine levels of various C_{10}-C_{13} PCA groups and to compare them with other published information.

Material and Methods

Sampling and Extraction

Waste water treatment effluents were obtained in 1996 from 6 municipal treatment plants located in the west end of Lake Ontario (*27*). Final effluents (4L) were collected in glass bottles and then extracted with dichloromethane (DCM). Extracts were exchanged into hexane for cleanup on an alumina column.

Surficial sediments were obtained with an Ekman grab (~0-5 cm depth) from three harbour areas, Toronto (3 locations), Port Credit (1), and Hamilton harbour (3 sites) in 1996 (*27*). Sediment was centrifuged to remove excess water, mixed with sodium sulfate and Soxhlet extracted with DCM.

Lake water (1 m depth) was obtained from the CCGS Limnos in July 1999 from the western end of Lake Ontario (*47*). 100L were pumped through a glass fiber filter (GFF) and extracted using an XAD-2 column. The XAD column was eluted with methanol and then DCM. The extracts were evaporated to near dryness and exchanged into hexane. Air was collected on the same cruise using a hi-vol sampler (June 1999) equipped with polyurethane foam plugs (PUFs) and GFF. PUFs were extracted with DCM and the extract was exchanged into hexane.

Carp (*Cyprinus carpio*) from Hamilton Harbour and lake trout (*Salvelinus namaycush*) from two locations in Western Lake Ontario (Port Credit (northwest) and Niagara-on-the-Lake (southwest)), were collected in 1996 (*27*). Beluga (*Delphinapterus leucas*) blubber samples from carcasses of dead animals from the St. Lawrence River estuary were provided by Dr. R. Bailey (Institute Maurice Lamontagne, Mont Joli PQ). The blubber samples were from animals collected in over the period 1988-91. Blubber from Arctic beluga in Cumberland Sound (southeast Baffin Island) was obtained from hunted animals in 1995. Whole tissue homogenates of fish and thawed blubber were mixed with sodium sulfate and Soxhlet extracted with DCM. Extracts were passed through a gel permeation column (BioBeads S-X3) to remove lipids.

Final cleanup and fractionation of sediments, STP effluents and biota samples was accomplished on an alumina column (1 g activated) (*27*) The SCCPs were separated from PCBs by first eluting the column with 1% diethyl ether hexane followed by 50% diethyl ether hexane. Lake water and PUF extracts were cleaned up and fractionated on silica gel columns (*47*). PCBs were eluted with hexane while SCCPs eluted with 50% DCM/hexane. A known amount of $^{13}C_8$-mirex was added at this stage for use as an internal standard during the GC-MS analysis of PCAs.

Analysis by GC-MS

SCCPs were quantified by HRGC-ECNI-HRMS as described by Tomy et al. (*18*). In brief, analyses were performed on a 5890 Series II gas chromatograph (Hewlett-Packard Instruments), fitted with a DB-5ms fused silica column (30 m x 0.25 mm i.d., 0.25 um film thickness), connected to a Kratos Concept MS (Kratos Instruments Manchester UK) controlled by a Mach 3 data system. SIM was performed at a resolving power of ~12 000 (sufficient to exclude potential interferences from other organochlorines), with a cycle time of 1 sec for each window, and equal dwell times for each ion monitored.

PCBs were determined in separate samples of beluga blubber and lake trout (whole fish) using GC-electron capture detection. Results from beluga blubber and lake trout samples were previously reported by Muir et al. (*48*) and D.M. Whittle (DFO Burlington, unpublished data), respectively.

Results and Discussion

Waste water treatment effluents

SCCPs were detected in all sewage treatment plant final effluents at ng/L concentrations. Higher concentrations were found in samples from treatment plants in industrialised areas, e.g. Hamilton and St. Catherines, compared to non-industrial towns, e.g. Niagara-on-the-Lake (Table 2). These concentrations appear to be typical of municipal waste waters. Reiger and Ballschmiter (*41*) reported C_{10}-C_{13} (62% Cl) SCCP concentrations of 70 to 120 ng/L in water upstream and downstream of a sewage treatment plant in Germany. In the United States, Murray et al. (*25*) reported C_{10}-C_{13} (60% Cl) SCCP concentrations of <150 to 3300 ng/L in water from an impoundment drainage ditch that received effluent from a SCCP manufacturing production plant in Dover OH.

Table 2. SCCPs in final effluent of sewage treatment plants on Lake Ontario (ng/L)

Location	C10	C11	C12	C13	ΣSCCPs
Woodward Ave. STP, Hamilton, ON	128	155	153	11.5	448
Halton Skyway STP, Burlington, ON	38	19	12	<1	68
Stanford WPCP, Niagara Falls, ON	11	34	36	1	82
Port Dalhousie WPCP, St. Catherines, ON	19	39	47	5	110
Port Weller WPCP, St. Catherines, ON	22	27	28	4	80
Niagara-on-the-Lake, WPCP	13	18	27	1	60

[1]C10 to C13 headings refer to concentrations of all chlorodecanes to chlorotridecanes, respectively. ΣSCCPs are the sum of each formula group.

Lake Ontario Water and Overlying Air

SCCPs were present in Lake Ontario surface water (1 m depth) samples from the west basin of the lake at low ng/L concentrations (Table 3). Overlying air had pg/m^3 concentrations of SCCPs with the chlorodecanes and -undecanes predominating. Fugacity ratios suggest that the lake is in disequilibrium with respect to the air and

Table 3. Levels of SCCPs in Lake Ontario surface water and air

Carbon chain group	Water (ng/m^3)[1]	Air (pg/m^3)[1]	fw/fa[2]
C10	168	100	3.9
C11	490	78	2.0
C12	1000	67	3.1
C13	94	4.1	2.3
ΣSCCP	1750	249	-
ΣPCBs	620	660	-

[1]Average of duplicate samples of water in west basin of Lake Ontario. Air = single large sample ($75m^3$) collected in west basin while ship was underway (June 1999).
[2]Fugacity ratio: fw= water conc'n *HLC; fa = Air Conc'n/$4.1x10^{-4}$. Lake wide average.
[3] Sum of 100 congeners determined by GC-ECD (*47*).

C_{10}-C_{12} SCCPs are volatilising from water to air. Concentrations of SCCPs in lake water were about 3x higher than PCBs determined in the same samples while PCBs were about 3x higher in air than SCCPs.

Surface sediments
SCCPs were detectable in all surface sediment samples from harbour areas in western Lake Ontario (Table 4). Highest concentrations were found at the most industrialised site (Windemere basin) which has well documented heavy metal, PAH and PCB contamination. Profiles of SCCPs in sediment consisted mainly of C_{12} and C_{13} chlorinated alkanes with high degree of chlorination. These components are more likely to be particle bound than the chlorodecanes. Ballschmiter (*49*) found SCCPs in river sediments in Germany ranging in concentration from <5 to 83 ng/g dry wt. By contrast, sediments of an impoundment ditch near an industrial facility producing SCCPs had 600 –10000 ng/g dw (*25*). Concentrations of SCCPs in surface films (formed by deposition of organic matter) from industrial and a mixed industrial/residential waste water sewer pipes of a city in southern Germany ranged from 0.5 to 30 ug/g and 1 to 17 ug/g (dry weight), respectively (*41*). Surface

Table 4. SCCPs (ng/g dry wt) in surface sediment from Lake Ontario harbours

Location	**C10**	**C11**	**C12**	**C13**	**ΣSCCP**
Toronto Inner Harbour	0.7	4.7	10	12	27
Toronto Inner Harbour Duplicate	0.6	4.2	9.1	10	24
Toronto, Humber River mouth	0.9	2.5	2.2	0.3	5.9
Port Credit Harbour	1.4	2.2	3.3	0.4	7.3
Hamilton - site 1: west harbour	3.3	14	17	7.0	41
Hamilton - site 1: west harbour duplicate	2.2	9.1	11	4.6	27
Hamilton - site 2: Windemere Basin	11	56	127	90	290
Hamilton - site 3: northeast	2.9	16	41	21	81

sediments from cores in Lakes Winnipeg, Nipigon, Fox and Hazen Lake had concentrations ranging from 10 to 250 ng/g dw (*28*). A sediment core from the west basin of Lake Ontario had surface concentrations (average over 0-5 cm) of 65 ng/g indicating that SCCPs are present in the depositional areas of the lake at similar or higher concentrations than some harbours (*47*).

Lake Ontario fishes
SCCPs were detectable in all samples of carp and lake trout (Table 5). Concentrations in carp from Hamilton harbour were higher than observed in catfish

and yellow perch from western Lake Erie (*18*). Lower concentrations in lake trout than carp were surprising given than they would be expected to feed at a higher trophic level. It is possible that trout have greater capacity to degrade SCCPs than carp. Comparison of the carbon chain groups (C_{10}-C_{13}) suggests that carp have higher proportions of dodecanes and tridecanes than lake trout (Table 5). The pattern of individual homologue groups also varies considerably between lake trout and carp (Figure 4). The pattern in carp resembles that in sediment from the area they were collected in. The pattern in lake trout bears some resemblance to lake water although higher chlorinated members of each carbon chain group, e.g. $C_{11}Cl_9$, $C_{12}Cl_9$, are present at higher proportions in trout than in water.

Concentrations of ΣSCCPs in lake trout were also much lower than ΣPCBs. This is consistent with the observation that individual SCCP congeners had shorter half-lives and BMFs in juvenile rainbow trout than most 2,4-substituted PCB congeners (*38,39*). Tomy et al. (*18*) found SCCP concentrations of 1,100, 300 and 1,200 ng/g ww in yellow perch, catfish and zebra mussels, respectively, from the mouth of Detroit River. SCCPs and PCB levels were similar in perch and zebra mussels but PCBs were 10x higher than SCCPs in catfish. Jansson et al. (*46*) found that PCAs in Baltic Sea herring were similar to levels of PCBs (sum of 7 congeners). On the other hand, arctic char from Lake Vättern in southern Sweden had about 6-fold lower levels of PCAs (570 ng/g lipid) compared to PCBs (3600 ng/g lipid). The PCB/SCCP ratio in Lake Ontario lake trout is nevertheless surprisingly high and may be a reflection of greater induction of cytochrome P450 oxidizing activity in these salmonids, compared to those in Sweden, due to high PCB and OC pesticide exposure.

Table 5. SCCP concentrations (ng/g ww) in whole fish samples from Lake Ontario

Location	N		C10	C11	C12	C13	ΣSCCP	ΣPCB
Hamilton Harbour 96 Carp	3	Mean	14	360	1090	1170	2630	na[1]
		SD	14	350	1070	1140	2560	-
Rainbow Trout, NOTL[2]	5	Mean	3.1	16	30	9.3	59	3180
		SD	2.6	14	26	8.0	51	1360
Rainbow Trout, Port Credit	5	Mean	3.8	20	37	11	73	5150
		SD	2.4	13	24	7.4	47	1660

[1]Not analysed

[2]Niagara-on-the-Lake - southwest Lake Ontario

Beluga whales

SCCPs were detected at ng/g levels in beluga whales from the St. Lawrence estuary and in those from southeast Baffin Island in the Canadian Arctic (Table 6). There were no significant differences in levels of ΣSCCPs or individual carbon chain groups between males or females. In this respect SCCPs show similar behavior to other less recalcitrant OCs such as hexachlorocyclohexanes in cetaceans (*48*). PCBs show much lower levels in males than females due to excretion via lactation (Table 6). ΣSCCP concentrations were 35 and 180x lower than ΣPCB in blubber of female

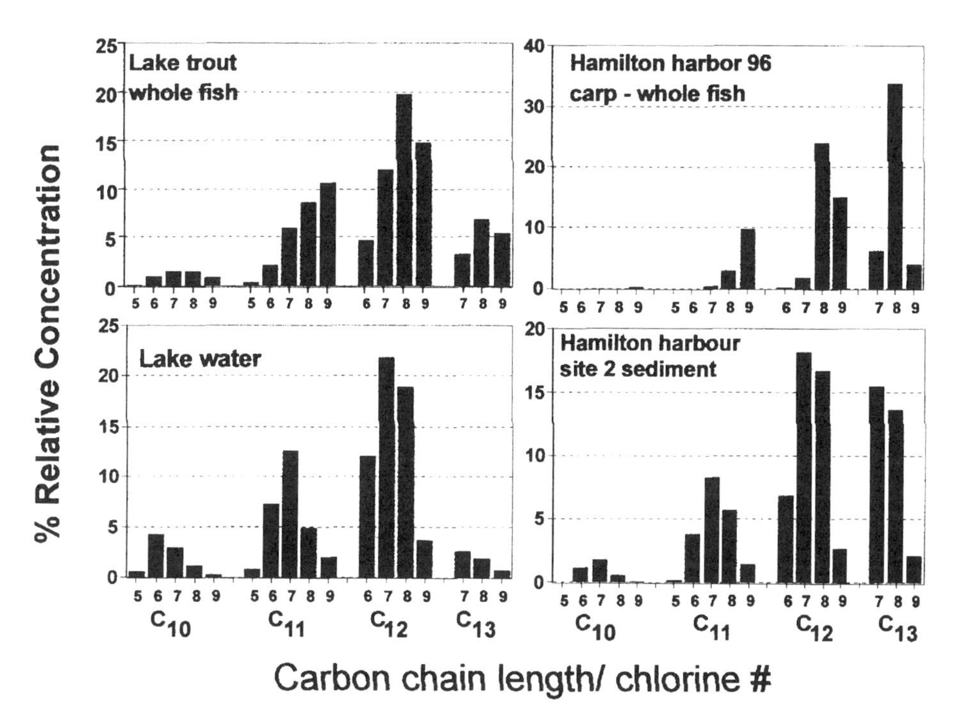

Figure 4. Proportions of carbon chain groups (C_{10}-C_{13}) and individual homologue groups ($C_{10}Cl_5$ etc) in Lake Ontario lake trout, carp, Hamilton harbour sediment and water. Note that the pattern in trout and carp generally resembles that in the water and sediment, respectively, except that higher chlorinated components of each carbon chain group are present in higher proportions.

and male St. Lawrence beluga, respectively, and about 40x lower in Arctic beluga. By contrast, two other chlorinated aliphatics, toxaphene and ΣDDT, were present at similar levels to ΣPCB (*48, 50*). There was no correlation between levels of PCBs or toxaphene and SCCPs. This was unexpected because levels of most recalcitrant OCs in beluga blubber are correlated. It implies that SCCPs undergo a greater degree of biotransformation or have quite different sources, than PCBs.

The St. Lawrence beluga had 6 to 8x higher levels of SCCPs than Arctic beluga. Jansson et al. (*46*) detected PCAs in ringed seals from Svalbard in the Norwegian Arctic (130 ng/g lipid) and found only slightly higher levels of PCAs in grey seals from the Baltic (280 ng/g lipid). These authors also noted that SCCPs did not biomagnify from fish (herring) to seals, indeed, levels were about 5-fold lower on a lipid basis. It is interesting to note that levels in carp and lake trout are much higher than in St. Lawrence beluga when compared on a lipid basis. Unfortunately there are no samples of the prey of beluga for direct comparison.

Table 6. Concentrations (ng/g ww) C_{10}-C_{13} SCCPs and PCBs in beluga whale blubber from the St. Lawrence River estuary and from the Canadian Arctic

Species/collection year		N		C10	C11	C12	C13	ΣSCCPs	ΣPCB
Beluga (1988)	F	9	Mean	190	270	340	140	940	33,630
St. Lawrence estuary			SD	100	140	170	70	480	22,070
	M	3	Mean	170	240	310	130	850	151,000
			SD	120	160	200	80	560	79,900
Beluga (1995)	F	3	Mean	39	35	37	4.0	116	4,390
South East Baffin Is			SD	18	16	17	1.8	52	3,120
	M	3	Mean	47	56	58	6.5	168	7,650
			SD	9.6	12	12	1.3	35	2,870

The carbon chain and homologue profiles of SCCPs in beluga from the Arctic and the St. Lawrence River estuary, as well as a 60% Cl SCCP commercial product are compared in Figure 5. The St. Lawrence animals have a pattern which bears some resemblence to the commercial product although $C_{13}Cl_7$ and $C_{13}Cl_8$ homologues are present in higher proportions and chlorodecanes are lower. The Arctic beluga have much higher proportions of the chlorodecanes and undecanes than the St. Lawrence beluga. The pattern of SCCPs in the Arctic beluga is consistent with long range transport sources i.e. more volatile components are present in higher proportions.

Conclusions

The data for SCCPs in fishes and marine mammals are among the first measurements of SCCPs in biota in North America. Similar to results from Sweden, SCCP contamination appears to be widespread but levels are relatively low in biota compared to PCBs and toxaphene. Laboratory bioaccumulation data (*34,38,39*) point to greater metabolism and shorter half-lives of SCCPs in fishes compared to PCBs

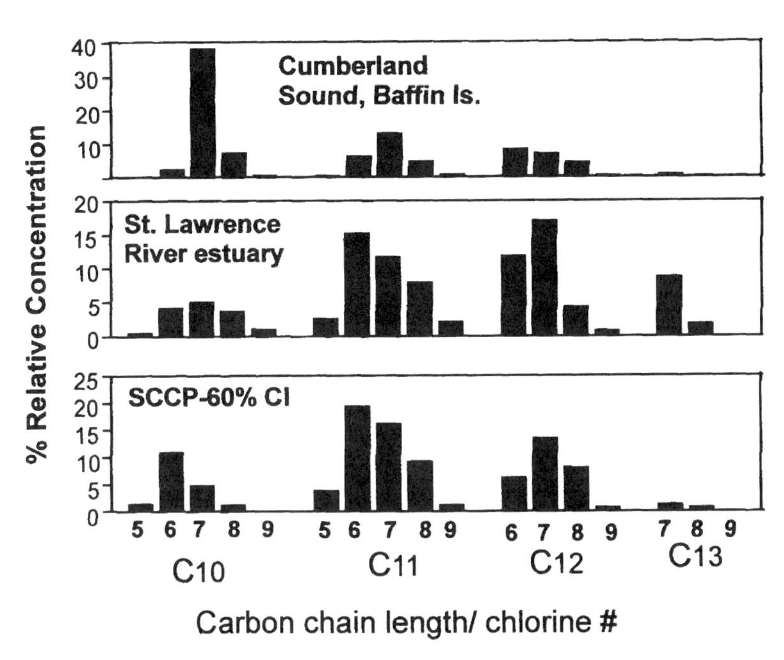

Figure 5. Proportions of carbon chain length and homologue groups in beluga blubber from the Arctic, the St. Lawrence River estuary and the commercial SCCP mixture with 60% chlorine content. Note that the more volatile C_{10} and C_{11} groups predominate in the Arctic beluga while the St. Lawrence beluga resemble the commercial mixture.

and toxaphene congeners. The field and laboratory results suggest that SCCPs are less persistent and bioaccumulative than PCBs but many gaps in the information remain. The volatilization of SCCPs from lake water to air and their presence in Arctic biota suggest SCCPs are relatively mobile and subjected to long range transport.

When compared with other chlorinated organics such as PCBs and chlorinated cyclic aliphatic pesticides, SCCPs appear to illicit fewer acute and chronic toxic effects in mammals, fish and invertebrates (*11*). However, knowledge of chronic toxicity of SCCPs is very limited compared other persistent OCs. Some risk assessments of SCCPs have concluded that there is low ecological risk (*2,14*); others have determined that these products pose a risk to aquatic life and should be phased out (*3,12*). More exposure information for the aquatic and terrestrial environments would be very useful for a complete risk assessment of SCCPs. There is clearly a need for more data for aquatic and terrestrial food chains as well as for half-lives of SCCPs in sediment, surface waters and the atmosphere.

Literature cited

1. Mukherjee, A.B. National board of Waters and the Environment, Helsinki, Finland, **1990**, Series A 66, 53pp.
2. Willis, B.; Crookes, M.J.; Diment, J.; Dobson, S.D. Environmental hazard assessment: chlorinated paraffins, **1994**, Toxic Substances Division. Dept. of the Environment. London UK
3. Environment Canada & Health Canada. Priority substances list Assessment Report: Chlorinated Paraffins. Government of Canada, **1993**, Catalogue No. En 40 215/17E 32pp.
4. Kirk-Othmer Encyclopedia of Chemical Technology. Chlorinated paraffins, **1999,** 4th ed. John Wiley and Sons, Inc.
5. Chemexpo. **1999**. www.chemexpo.com/news/profile990426.cfm
6. Howard, P.H.; Santodonato, J.; Saxena, J. EPA-560/2-75-007, **1975**, NTIS, Springfield Va 22151,109 pp
7. Independent Lubricant Manufacturers Association. www.ilma.org /0897/ paraffin.htm. **1999**, 3 pp.
8. Hardie, D.W.F. Chlorinated paraffins, Kirk-Othmer Encyclopedia of Chemical Technology, 2nd Edition, Vol.5, **1964.**
9. Organization for Economic Co-operation and Development. Short chained chlorinated paraffins and nonyl phenol ethoxylates. ENV/JM(2000)7. **2000**. 10pp.
10. Stolzenberg, H-C. Short chained chlorinated paraffins. UmweltBundesAmt, Berlin, **1999**, presented at the OECD Expert Meeting, Geneva.
11. Tomy, G.T.; Fisk, A.; Westmore, J.B.; Muir, D.C.G. Rev. Environ. Contam. Toxicol., **1998**, 158, 53-128.
12. Swedish National Chemicals Inspectorate, KEMI Report No. 1., **1991**.
13. International Program on Chemical Safety. Environmental Health Criteria 181. **1996**. World Health Organization, Geneva. (www.who.int/pcs/pubs/pub_list.htm)

14. US Environmental Protection Agency. RM2 Exit Briefing on Chlorinated Paraffins and Olefins, **1993**, Washington, DC. 41pp.
15. Windrath, O.M.; Stevenson, D.R. Plastics Compounding, **1985**, pp. 38-52.
16. Zitko, V. In: Handbook of Environmental Chemistry, **1980**, Vol 3A, 149-156, Springer Verlag.
17. Campbell, I.; McConnell, G. Environ. Sci. Technol. **1980,** 14, 1209-1214.
18. Tomy, G.T.; Stern, G.A.; Muir, D.C.G.; Fisk, A.T.; Cymbalisty, C.D.; Westmore, J.B. Anal. Chem., **1997**, 69, 2762-2771.
19. Tomy, G.T. Ph.D. Thesis, **1997**, Dept of Chemistry, University of Manitoba, Winnipeg MB.
20. Gusev, M.N.; Urman, Y.G.; Mochalova, O.A.; Kocharyan, L.A.; Slonim, I.Y. Izv. Akad, Nauk. SSR., Ser. Khim. **1968**, 7, 1549.
21. Parlar, H.; Coelhan, M.; Saraçi, M.; Lahaniatis, E.S.; Lachermeir, C.; Koske, G.; Nitz, S.; Leupold, G. Organohalogen Compounds, **1998**, 35, 395-398.
22. Jansson, B.; Andersson, R.; Asplund, L.; Bergman, Å.; Litzén, K.; Nylund, K.; Reutergårdh, L.; Sellström, U.; Uvemo, U-B.; Wahhlerg, C.; Wideqvist, U. Fresenius J Anal Chem., **1991**, 340, 439-445.
23. Metcalfe-Smith, J.L.; Maguire, R.J.; Batchelor, S.P.; Bennie, D.T. NWRI Contribution No. 95-62, **1995**, National Water Research Institute. Environment Canada, Burlington, ON.
24. Junk, S.A.; Meisch, H-U. Fresen. J. Anal. Chem. **1993**, 347, 361-363.
25. Murray, T.M.; Frankenberry, D.H.; Steele, D.H.; Heath, R.G. Technical Report, U.S. Environmental Protection Agency, **1988**, EPA/560/5 87/012. Vol.1, 150pp.
26. Tomy, G.T.; Stern, S.A.; Muir, D.C.G.; Fisk, A.T.; Westmore, J.B. Anal. Chem. **1999**, 71, 446-451.
27. Bennie D.T.; Sullivan, C.A.;Maguire, R.J. Water Qual. Res. J. Can. **2000** Submitted
28. Tomy, G.T.; Stern, G.A.; Lockhart, W.L.; Muir, D.C.G. Environ. Sci. Technol., 1999, **33**, 2858-2863.
29. Drouillard, K.G.; Muir, D.C.G.; Tomy G.T.; Friesen, K.J. Environ. Toxicol. Chem. **1997**, 17,1252-1260.
30. Drouillard, K.G.; Hiebert, T.; Tran, P.; Muir, D.C.G.; Tomy, G.T.; Friesen, K.J. Environ. Toxicol. Chem., **1997**, 17, 1261-1267.
31. Sijm, D.T.H.M.; Sinnige, T.L. Chemosphere, **1995**, 31, 4427-4435.
32. Fisk, A.T.; Wiens, S.C.; Webster, G.R.B.; Bergman, Å.; Muir, D.C.G. Environ. Toxicol. Chem., **1998**, 17, 2019-2026.
33. Bengtsson, B.; Svenberg, O.; Linden, E.; Lunde, G.; Ofstad, E.B. Ambio, **1979**, 8, 121-122.
34. Fisk, A.T.; Cymbalisty, C.D.; Bergman, Å.; Muir, D.C.G. Environ. Toxicol. Chem., **1996**, 15,1775-1782.
35. Madeley, J.R.; Thompson, R.S.; Brown, D. Imperial Chemical Industries PLC, Devon, UK, **1983,** Brixham Report No. BL/B/2351.
36. Madeley, J.R.; Maddock, B.G. Imperial Chemical Industries PLC, Devon, UK, **1983**, Brixham Report No. BL/B/2310.

37. Thomann, R.V. Environ. Sci. Technol., **1989**, 23, 699-707.
38. Fisk, A.T.; Cymbalisty, C.D.; Tomy, G.T.; Muir, D.C.G., Aquat. Toxicol. **1998**, 43, 209-221.
39. Fisk, A.T.; Norstrom, R.J.; Cymbalisty, C.D.; Muir, D.C.G., Environ. Toxicol. Chem. **1998**, 17, 951-961.
40. Atkinson, R. Chem. Rev. **1986**, 86, 69-201.
41. Reiger, R.; Ballschmiter, K. Fres. J. Anal. Chem., **1995**, 352:715-724.
42. Zitko, V.; Arsenault, E. Fisheries and Marine Service Technical Report, **1974**, No. 491, pp 38.
43. Madeley, J.; Birtley, R. Environ. Sci. Technol. **1980**, 14, 1215-1221.
44. Omori, T.; Kimura, T.; Kodama, T. Appl. Microbiol. Biotechnol. **1987**, 25, 553-557.
45. Muir, D.C.G.; Stern G.A.; Tomy, G.T. In: Handbook of Environmental Chemistry. **1999**. Vol 3K. J. Paasivirta (ed). Springer-Verlag, Berlin. pp. 203-236.
46. Jansson, B.; Andersson, R.; Asplund, L.; Litzén, L.; Nylund, K.; Sellström, U.; Uvemo, U-B.; Wahlberg, C.; Wideqvist, U.; Odsjö, T.; M. Olsson. Environ. Toxicol. Chem., **1993**, 12, 1163-1174.
47. Muir, D.; Bennie, D.; Maguire, J.; Teixeira, C.; Whittle, M.; Tomy, G.; Stern. G. Environ. Sci. Technol., **2000**, Submitted.
48. Muir, D.C.G.; Ford, C.A.; Rosenberg, B.; Norstrom, R.J.; Simon, M.; Béland, P. Environ. Pollut., **1996**, 93, 219-234.
49. Ballschmiter, K. Universität Ulm, Abt Anal Chem und Umweltchemie, **1994**, Unpublished data report.
50. Muir, D.; Braune, B.; DeMarch, B.; Norstrom, R.; Wagemann, R.; Lockhart, L.; Hargrave, B.; Bright, D.; Addison , R.; Payne, J.; Reimer, K. Sci. Total Environ. **1999**, 230, 83-144.

Chapter 15

Polycyclic Musk Fragrances in the Aquatic Environment

Hermann Fromme, Thomas Otto, and Konstanze Pilz

Institute of Environmental Analysis and Human Toxicology (ITox), Invalidenstrasse 60, D–10557 Berlin, Germany (email: Itox@bbges.de)

Abstract

The aim of this study was to obtain data about the contamination of different environmental compartments by polycyclic musks [Galaxolide® (HHCB), Tonalide® (AHTN), Celestolide® (ADBI), Phantolide® (AHMI), and Traesolide® (ATII)] within the framework of an exposure monitoring program. Additionally, the situation as regards eel contamination was to be documented and statements made concerning the enrichment of the above mentioned substances in biological materials. To these ends, a total of 102 surface water samples, 59 sediment samples and 165 eel samples were taken from three Berlin water areas known to be contaminated to different extents with sewage. In addition, samples from the outflows of five municipal sewage works collected over a 30 day period were examined

Results for HHCB gave the following mean values in areas strongly polluted with sewage: surface water 1.59 µg/l; sediment 0.92 mg/kg d.w. and eel 1513 µg/kg f.w. (in the edible portion) (6471 µg/kg lipid). The following average concentrations were found in waters hardly contaminated with sewage: surface water 0.07 µg/l, sediment < 0.02 mg/kg and eel 52 µg/kg f.w. (445 µg/kg lipid). Mean concentrations of 6.85 µg/l (maximum: 13.3 µg/l) could be measured at sewage treatment plants' outlets.

It could be shown that these polycyclics are highly suited to use as indicators of the degree of contamination of waters with organic substances originating from sewage. A mean bioconcentration factor (BCF) on wet weight of 862 (HHCB) and 1069 (AHTN) for the transfer from water to eel under natural conditions could be calculated. The corresponding BCF - values based on the lipid content of eel were 3504 (HHCB) and 5017 (AHTN).

Our measured results for those aquatic regions strongly contaminated with sewage are only slightly below the $PNEC_{water}$ (provisional no effect concentration) values. In waters with low or moderate sewage contamination, the average concentrations for HHCB are respectively a factor of 100 or 30 lower.

Introduction

Originally, natural musks were mainly extracted from exocrine gland secretions of musk ox and musk deer (*Moschus moschiferus*) and used as fragrances. Nowadays, synthetic substances mainly from two groups, namely the aromatic nitro musks (e.g. musk ketone and musk xylene) and the polycyclic musks are commercially produced in quantity by industry. Because of their musk- like odour and their ability to fix other fragrant substances, they are widely employed in, for example, the cosmetic and perfume industries, in many sorts of agents for cleaning, polishing, and washing, in other household uses, and also in scented oils.

The first evidence of aquatic system contamination by nitro musks was observed in 1981 in Japan [1,2], and much later also in Germany [e.g. 3]. Reports of the occurrence of substances with a musk- like odour, but completely different structure to nitro- aromatics, in different environmental compartments were first published in 1994 [4]. In the mean time, the application of these polycyclic fragrances (use world wide ca. 5000 t) already exceeded that of the nitro musks. Meanwhile there are indications that the polycyclic musks are not only to be found in various environmental compartments, but also in the aquatic food chain as well as in fatty tissue and mothers' milk [5]. Little information concerning the toxic effects of this substance class has been published to date [6,7].

Table I. Chemical Name, Trade Name and CAS Number of the Measured Substances.

Abbr.	*Chemical Name*	*Trade Name*	*CAS-No.*
HHCB	1,3,4,6,7,8-hexahydro-4,6,6,7,8,8-hexamethylcyclopenta-(g) 2-benzopyrane	Galaxolide®, Abbalide®	1222-05-5
AHTN	7-acetyl-1,1,3,4,4,6-hexamethyl-tetraline	Tonalide®, Fixolide®	1506-02-1
ADBI	4-acetyl-1,1-dimethyl-6-*tert.*-butylindane	Celestolide®, Crysolide®	13171-00-1
ATII	5-acetyl-1,1,2,6-tetramethyl-3-iso-propyldihydroindane	Traesolide®	68140-48-7
AHMI	6-acetyl-1,1,2,3,3,5-hexamethyl-indane	Phantolide®	15323-35-0

The aim of the examination presented here was to measure the polycyclic musks listed in Table I by way of exposure monitoring in different environmental compartments (surface water, sewage, sediment) and in fish samples. In addition, enrichment data on bioconcentration of this substance class in the aquatic system were to be obtained. The study was part of an examination to present contamination levels of persistent substances in the aquatic environment [8,9].

Berlin's waters consist of the rivers Dahme, Spree and Havel with its lake like broadenings stretch out over 58 km^2, 51 km^2 being fished on a commercial basis. Being situated in a densely populated area (population approaching 4 M) and considering the associated burden of contaminant wastes, their intensive use (e.g. as commercial waterways) and the high nutrient levels, they are subject to high ecological pressure.

Materials and Methods

Study Area / Sample Collection

The samples were collected in 1996 and 1997 in three study areas in Berlin (see Figure 1). One area [A] is that of the rivers Spree and Dahme (with Langer See, Großer Müggelsee, Zeuthener See, Seddinsee and Dämmeritzsee); [B] is the area of the river Havel (with Ober- and Unterhavel, Tegeler See and Großer Wannsee); and the last [C] (with Teltowkanal, Griebnitzsee and Kleiner Wannsee). The mean flow varied from 0.002 - 0.1 m/s in sections of the waters in the area examined. As may be drawn from Figure 1, area [A] has a very low proportion of municipal treated sewage effluents relative to the total water content, area [B] a moderate and area [C] a very high proportion. A yearly total of 310 M. m^3 sewage is fed into these waters.

Altogether, 102 unfiltered surface water samples were collected from a depth of 0.5 m using 1 l amber glass pre-cleaned bottles and kept cool until analysis. 30 proportional daily composite samples of treated sewage effluent were collected from 5 municipal treatment facilities. 59 surface (10 cm depth) sediment samples were collected manually using a grab sampler in the central bed of the river. The samples were kept in glass jars, transported at 4°C from the sampling point to the laboratory and freeze-dried immediately.

341 fish (165 eel, *Anguilla anguilla*; 54 roach, *Rutilus rutilus*; 47 common bream, *Abramis brama*; 33 pikeperch, *Stizostedion lucioperca*; 28 perch, *Perca fluviatilis*; 14 pike, *Esox lucius*) were caught within the same areas of Berlin waters and then filleted and skinned. The whole fish filet was put into food quality polyethylene freezer bags (free from detectable amounts of the analytes of interest) and stored frozen at -20 °C until analysis. The sample preparation was

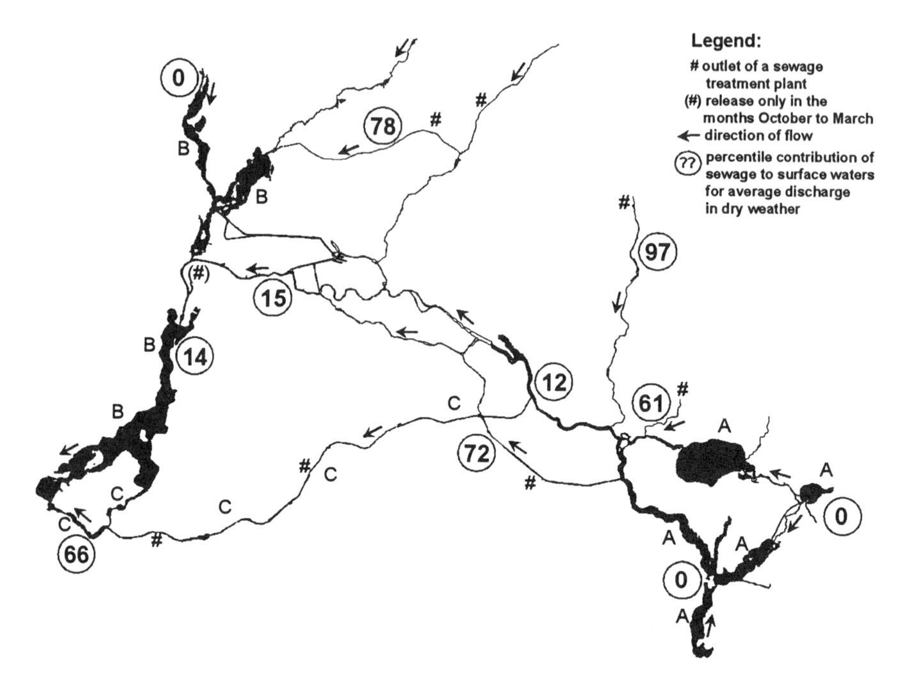

Figure 1. Study Area. A: River Spree and Dahme (with Langer See, Großer Müggelsee, Zeuthener See, Seddinsee and Dämmeritzsee); B: River Havel (with Ober- and Unterhavel and Großer Wannsee); C: Teltowkanal, Griebnitzsee and Kleiner Wannsee.

in accordance with the preparation procedure of the German Food Monitoring Program [10].

Chemicals

All chemicals and solvents used for extraction, cleanup and GC analysis were of pesticide residue analysis grade (Mallinckrodt Nanograde supplied by Baker Chemikalien, Germany). All materials and chemicals used in sample preparation, extraction and cleanup steps were carefully decontaminated by solvent rinse or extraction (e.g. filters, cotton- and glasswool). Glassware, sea sand and sodium sulphate were thermally decontaminated in a furnace at 400 - 600 °C. All reference standards were supplied by Promochem GmbH, Germany. Galaxolide was purchased as technical product consisting of approximately 50 % HHCB (two enantiomers and other isomers in minor concentration).

Extraction, Determination of Lipid Content, Cleanup of the Fish Samples

The analysis of the musk residues was carried out in a similar way to that described for non-polar organochlorine pesticides and PCB [10]. The whole fish filet (muscle tissue) was homogenised and mixed with 15 - 20 g anhydrous sodium sulphate (amount depended on the water content of the homogenate) and 10 g sea sand to form a dry, free flowing product. A Soxhelet extraction with cyclohexane / ethyl acetate 98.5 + 100 (v/v) was then carried out for 10 h. The fat extract was concentrated to 20 ml and the lipid content was determined gravimetrically in an aliquot of 2 ml.

For the cleanup procedure, 5 ml of the extract was cleaned by gel - permeation chromatography (Autoprep 1002A, ABC Laboratories Inc., USA; column: 50 g Bio Beads S-X3, inner diameter 25 mm, height 32 cm; elution solvent: cyclohexane/ethyl acetate 98.5 + 100 v/v [azeotropic at atmospheric pressure]; flow: 5.0 ml/min) followed by fractionation of the collected volume on a miniaturised silica gel column (1 g silica gel 60 particle size 0.063 - 0.2 mm, deactivated with 1.5 percent water; Merk KGaA, Germany). Solvent was removed with a TURBOVAP II ® concentrator (Zymark, USA) to prevent losses of volatile analytes. The whole cleanup procedure was in accordance to the German standard method DFG S 19 [11].

Extraction and Cleanup of the Water Samples

Unfiltered water samples (600 ml), to which 10 g sodium chloride had been added, were extracted using SDE (simultaneous steam - distillation / solvent extraction) [12,13,14]. Cleanup and concentration occur in one single step in SDE. This minimises the solvent and other reagent quantity used and likewise the risk of contamination. Because of their volatility, SDE is a particularly good choice of method for fragrances [15]. The cyclohexane extract (ca. 5 ml) was dried with anhydrous sodium sulfate (Na_2So_4). The extract immediately following removal of solvent with a TURBOVAP II ® concentrator (Zymark, USA) and concentration to a volume of 0.5 ml.

Extraction and Cleanup of the Sediment Samples

10g freeze - dried sediment mixed with 10g NaCl were suspended in 600 ml pure water. Extraction and cleanup were carried out in the same way as for water samples. The cyclohexane extract was concentrated to a volume of 0.5 ml. After adding a small amount of pyrogenic copper powder the extracts were desulfurated in autosampler vials using an ultrasonic bath. The extract could be analysed immediately on completion of this procedure.

Analysis

The polycyclic musk compounds were analysed in the combined second and third fractions by capillary GC/Ion Trap MS using a Finnigan GCQ with Autosampler CTC A200S. GC parameters: splitless injection of 1 µl at 250 °C injector temperature, split opened after 1 min. Capillary column: DB-XLB (J&W Scientific, USA), 30 m length, 0.25 mm inner diameter, 0.25 µm film thickness. Carrier gas: helium with constant velocity 40 cm/sec. Temperature program: start temperature 60 °C held for 1 min., 60 to 160 °C at 30 °C/min. (isothermal for 0.25 min.), 160 to 200 °C at 5 °C/min. (isothermal for 0.25 min.), 200 to 290 °C at 10°C/min. (isothermal for 25 min.). The capillary was coupled directly to the ion source, transfer line temp. was 275 °C. The following MS parameters (see Table II) were used: EI 70 eV, full scan, scan range m/z 60 - 320, scan time 0.4 s (3 microscans). We did not use the MS/MS mode of the GCQ because, in full scan mode, no interfering masses were observable at the retention time of each analyte. The detection limits are given in Table III.

The method's quality was checked by analysis of one blank and one reference material along with each group of 10 fish samples. We used artificially contaminated bovine fat (code no. 999922), used also in the German Food Monitoring Program. No reference materials are commercially available for polycyclic musk compounds. We used spiked butter fat (2.5 g dissolved in 20 ml elution solvent – equivalent to a fish sample with 25 % fat content, 5 ml used for the whole cleanup procedure described above), free from detectable amounts of musk compounds. The mean recovery of the spiked water samples (n = 5) was 84 % (ADBI), 97 % (HHCB), 101 % (AHTN), 82 % (AHMI) and 87 % (ATII) and of the eel samples (n = 5): 80 % (ADBI), 89 % (HHCB), 95 % (AHTN), 78 % (AHMI) and 83 % (ATII).

Table II. Retention Times and Ions used for Quantification of Polycyclic Musk Compounds

Compound	*Retention Time[min.]*	*Mass used for Quantification [m/z]*	*Qualifier Masses*
HHCB	13.30	243	258, 213
AHTN	13.43	243	258, 145, 149, 187, 201
ADBI	11.11	229	244, 173
ATII	13.35	173	258*, 215^{+}
AHMI	11.55	229	244, 187

Note: Molecular ion of each compound is underlined; *: disturbed by molecular ion of HHCB; +: disturbed by HHCB in case of bad chromatographic resolution of HHCB/ATII.

Table III. Detection Limit of the Polycyclic Musk Compounds

Compound	*Detection Limit*		
	Fish (μg/kg)	*Water (μg/l)*	*Sediment (mg/kg)*
ADBI	5	0.005	0.004
HHCB	30	0.020	0.030
AHTN	20	0.010	0.020
AHMI	4	0.005	0.004
ATII	4	0.005	0.004

Results and Discussion

Surface Water

The results of the Berlin surface water examinations are presented in Table IV below. They show a clearly growing contamination tendency with increasing sewage contribution, reaching a maximum mean concentration of 1.59 μg/l HHCB in examination area [C]. The median percentile proportion was 71% for HHCB and 22% for AHTN in samples where all five polycyclics could be measured. The variation in these contributions is quite small. In comparison to levels described in the scientific literature, (Figure 2), which show in parts a considerable variation of single values, our examination resulted in relatively consistent concentration levels. Nevertheless, levels found in the examination area strongly contaminated with sewage [C] have not been observed to date outside of Berlin's waters. Comparison with the results of another group, which examined representative sites in Berlin waters [16], showed good correlation with the contamination data presented, when considering only the results in identical areas of water, despite the different methodology (solid-phase micro extraction).

Sewage

The results of the 24 hour collection samples from sewage plant outlets are presented together in Table V. Very high values were seen, especially for HHCB (maximum: 13.33 μg/l) but also for AHTN (maximum: 4.36 μg/l). Examinations of another group [16] on three Berlin sewage farms in September 1996 gave comparable results with values of 6.3, 10.1 and 10.8 μg/l HHCB (1.95, 5.8 and 5.8 μg/l AHTN). Examinations of sewage treatment plants on the River Ruhr [17] gave median values of 1.2 μg/l HHCB (range: 0.6 - 2.5) indicating a somewhat lower burden, whilst comparable concentrations (1 - 6 μg/l) were found during examinations of three large Swedish sewage plants [19].

Table IV. Concentration of Synthetic Musks in Surface Water Samples (µg/l) from Lakes with High, Moderate or Low Inputs of Sewage Water (Mean = arithmetic mean; SD = standard deviation; 90. P. = 90th percentile; Max = maximum; n = number of samples, values in brackets are percent > detection limit)

		HHCB	*AHTN*	*ADBI*	*ATII*	*AHMI*
Mean	low	0.07	0.02			
	moderate	0.23	0.07			
	high	1.59	0.53	0.02	0.07	0.07
SD	low	0.06	0.01			
	moderate	0.20	0.06			
	high	0.72	0.25	0.01	0.03	0.04
Median	low	0.05	0.02			
	moderate	0.15	0.05			
	high	1.48	0.47	0.02	0.06	0.07
90. P.	low	0.14	0.03			
	moderate	0.49	0.14			
	high	2.73	0.91	0.04	0.09	0.11
Max	low	0.32	0.06	0.01	0.01	
	moderate	0.81	0.27	0.01	0.05	0.03
	high	3.15	1.10	0.06	0.13	0.17
n	low	34 (91)	34 (62)	34 (3)	9 (11)	9 (0)
n	moderate	40 (100)	40 (98)	40 (8)	15 (53)	15 (27)
n	high	28 (100)	28 (100)	28 (89)	8 (100)	8 (100)

Table V. Concentration of Synthetic Musks in Sewage Water Effluents(µg/l) (Mean = arithmetic mean; SD = standard deviation; 90. P. = 90th percentile; Max = maximum; n = number of samples, values in brackets are percent > detection limit)

	HHCB	*AHTN*	*ADBI*	*ATII*	*AHMI*
Mean	6.85	2.24	0.11	0.31	0.27
SD	2.64	0.86	0.04	0.20	0.13
Median	6.65	2.16	0.12	0.26	0.23
90. P.	10.80	3.36	0.17	0.62	0.36
Max	13.33	4.36	0.21	0.70	0.58
n	30 (100)	30 (100)	30 (100)	25 (96)	12 (100)

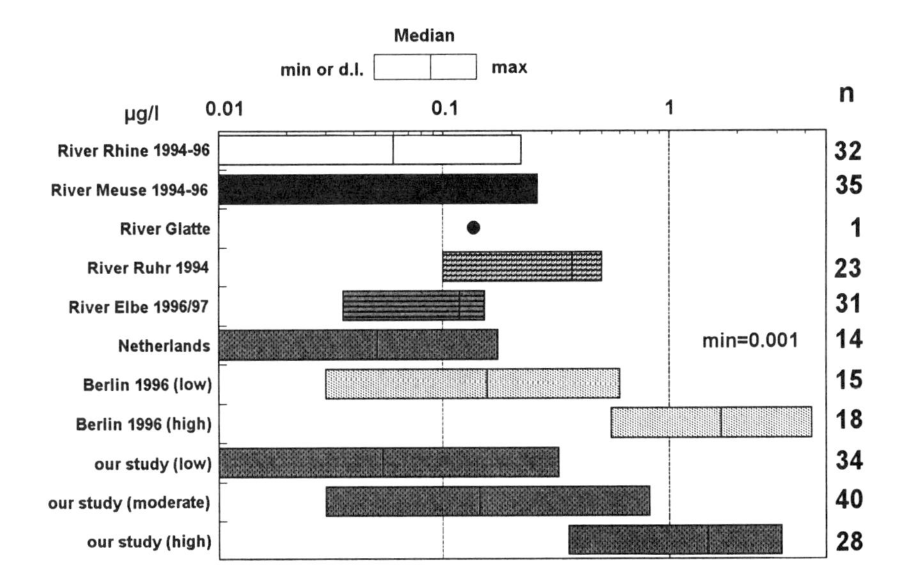

Figure 2. Concentration of HHCB (Galaxolide®) in Surface Water Samples from Lakes and Rivers. Low, moderate and high means the proportion of sewage effluents in the aquatic system (d.l.: detection limit; cited in [6], River Elbe [18] and Berlin [16])

Sediments

Results of sediment sample analyses are collected in Table VI. Contamination in this environmental compartment also appears to be linked to the related burden in the aqueous phase. In particular, samples from regions contaminated with sewage (area C) show a great range in HHCB contents from 0.030 to 2.2 mg/kg d.w. This may be caused by inhomogeneous sample materials, containing varying proportions of organic substances. Otherwise, in a study on the River Ruhr only one single sample has been examined (cited in [6]). The level 0.15 mg/kg d.w. is in accordance with the levels observed by ourselves.

Eel Samples

Eel available for analysis (n = 165) had a median weight of 230 g (maximum: 724 g) and length of 49 cm (maximum: 70 cm). Average fat content was 22% (maximum: 51%). Eel sample results for the three areas examined are collected in Table VII (as µg/kg f.w.) and in Table VIII (as µg/kg lipid). Whilst HHCB could be measured in all samples with values of up to 4800 µg/kg f.w. (18462 µg/kg lipid), other polycyclic compounds (AHMI and ADBI) generally

**Table VI. Concentration of Synthetic Musks in Sediment Samples (mg/kg dry weight)
(Mean = arithmetic mean; SD = standard deviation; 90. P. = 90th percentile; Max = maximum; n = number of samples, values in brackets are percent > detection limit)**

		HHCB	*AHTN*	*ADBI*	*ATII*	*AHMI*
Mean	low		0,02			
	moderate	0.22	0.26	0.010	0.021	0.011
	high	0.92	1.10	0.025	0.101	0.036
SD	low		0.01			
	moderate	0.16	0.21	0.009	0.019	0.010
	high	0.70	0.85	0.031	0.076	0.028
Median	low		0.02			
	moderate	0.23	0.24	0.008	0.021	0.009
	high	0.91	0.93	0.005	0.100	0.031
90. P.	low		0.03			
	moderate	0.38	0.52	0.021	0.044	0.022
	high	1,90	2.21	0.066	0.202	0.068
Max	low	0.03	0.04			
	moderate	0.52	0.61	0.023	0.051	0.026
	high	2.20	2.60	0.068	0.220	0.093
n	low	19 (5)	19 (63)	19 (0)	19 (0)	19 (0)
n	moderate	20 (85)	20 (95)	20 (55)	20 (55)	20 (50)
n	high	20 (90)	20 (90)	20 (80)	20 (85)	20 (80)

could only be found above the detection limit in water areas with a high sewage component. In samples where all five polycyclics were detected, based on the total contents, the following mean distribution pattern emerged: HHCB 61% (49 - 68%), AHTN 31% (23 - 36%) and ADBI, ATII and AHMI together 8%. At the present moment, few examination results have been published for eel samples. Only two studies, with samples from the Ruhr [20] and upper Rhein [21], included the examination of eels under natural conditions. In these cases, HHCB was found at levels of 97 and 126 µg/kg f.w. (Ruhr eel) and with mean levels of 43 µg/kg fw (10 - 120 µg/kg fw.) in Rhine eel. The average concentration levels in Berlin eel samples of 52 and 117 µg/kg fw. HHCB (low or moderate contaminated area) were comparable with the above values. Eel from Berlin waters highly contaminated with sewage showed, in contrast, clearly higher mean values of 1473 µg/kg f.w. HHCB. Such concentrations were previously only known from eel in sewage settlement ponds. Contamination values were 1530 to 19200 µg/kg f.w. [20] and 722 and 859 µg/kg fw. [cited in 22].

Table VII. Concentration (µg/kg fw.) of Synthetic Musks in Eel Samples from Lakes with High, Moderate or Low Inputs of Sewage Water (Mean = arithmetic mean; SD = standard deviation; 90. P. = 90th percentile; Max = maximum; n = number of samples, values in brackets are percent > detection limit)

		HHCB	*AHTN*	*ADBI*	*ATII*	*AHMI*
Mean	low	52				
	moderate	117	53		12	
	high	1513	723	5	123	113
SD	low	38				
	moderate	134	73		11	
	high	990	450	3	52	47
Median	low	50				
	moderate	77	32		8	2
	high	1473	668	4	125	110
90. P.	low	79				
	moderate	210	112		27	24
	high	2812	1380	8	181	156
Max	low	260	60		14	
	moderate	740	420	2	39	45
	high	4800	2300	17	190	210
n	low	54 (83)	54 (2)	54 (0)	26 (15)	26 (0)
n	moderate	53 (85)	53 (64)	53 (2)	14 (78)	14 (36)
n	high	58 (100)	58 (100)	58 (48)	10 (100)	10 (100)

Fish Samples

The results and summary statistics for HHCB and AHTN from the examination of a total of 176 fish samples- excluding eels-are compiled in Table IX (and Table X, relating to fat content). As can clearly be seen from Figure 3, concentrations above the determination limit in fish taken from those Berlin waters with low or medium levels of sewage contamination, could only be found in very few cases for AHTN, and for HHCB at higher values only in common bream (78%), pike (50%) and perch(47%): The median value was 40 µg/kg fw. (3330 µg/kg lipid). These values, even including those for common bream were only just above the limit of determination.

In contrast, HHCB and AHTN could be found in almost all samples from fish with relatively low fat content (perch: 0.4 - 1%; common bream: 0.7 - 3%; roach: 1 - 1.9%; pike: 0.6%; pike perch: 0.3 - 0.7%) which had been taken from waters strongly polluted with sewage. Median values were for perch: 200 µg/kg fw. (33.3 mg/kg lipid), for common bream: 1571 µg/kg fw. (90.1 mg/kg lipid), for roach: 168 µg/kg fw. (13.0 mg/kg lipid), for pike: 366 und 370 µg/kg fw. (66.5 und 61.7 mg/kg lipid) and for pike perch: 190 µg/kg fw. (47.3 mg/kg lipid).

Table VIII. Concentration (µg/kg lipid) of Synthetic Musks in Eel Samples from Lakes with High, Moderate or Low Input of Sewage Water (Mean = arithmetic mean; SD = standard deviation; 90. P. = 90th percentile; Max = maximum; number of samples see Table VII

		HHCB	*AHTN*	*ADBI*	*ATII*	*AHMI*
Mean	low	445				
	moderate	658	287		77	
	high	6470	3060	20	469	431
SD	low	830				
	moderate	649	379		125	
	high	4115	1807	14	204	194
Median	low	198				
	moderate	426	186		36	20
	high	5830	2833	21	474	407
90. P.	low	810				
	moderate	1405	545		107	118
	high	11483	5170	32	673	641
Max	low	5714	1429			
	moderate	3750	2500		500	500
	high	18462	8846	68	731	808

The, in part, very high values in the edible parts of common bream with a maximal value of 3224 µg/kg fw. (143.3 mg/kg lipid) clearly stand out. The unusually large range of the results from areas contaminated with sewage can be observed in Figure 3.

The range of results of HHCB and AHTN in edible parts of freshwater and coastal fishes (excluding eels) in the scientific literature is presented in Table XI.

The maximum contents in fish from areas of low to moderate pollution are, for comparison purposes, 15.3 mg/kg lipid (Median: 3.3) for common bream and 11.8 mg/kg lipid for roach. Comparison with data already published in the literature shows that even fish from these areas show, in part, clearly higher levels of polycyclic musk than previously described. It may also be seen from Table XI that the pollution situation in North America differs from that in Europe. Because of their lower usage in America, the levels of polycyclics in biological materials are much lower.

Table IX. Concentration (µg/kg fw.) of HHCB and AHTN in Fish Samples from Lakes with an High or Low Input of Sewage Water (Min = Minimum; Max = Maximum; n = Number of Samples, Values in Brackets are Percent > Detection Limit)

	Perch	*Common Bream*	*Roach*	*Pike*	*Pikeperch*
		HHCB (low)			
Min	< d.l.	< d.l.	< d.l.	< d.l.	< d.l.
Median		40			
Max	122	260	260	98	113
n	19 (47)	37 (78)	48 (48)	12 (50)	25 (20)
		HHCB (high)			
Min	114	261	106		83
Median	200	1571	168		190
Max	1215	3426	1018		1574
n	9 (100)	10 (100)	6 (100)	2*	8 (100)
		AHTN (low)			
Min	< d.l.	< d.l.	< d.l.	< d.l.	< d.l.
Max		42	50		
n	19 (0)	37 (16)	48 (27)	12+	25 (0)
		AHTN (high)			
Min	36	62	41		< d.l.
Median	47	324	64		37
Max	332	851	339		362
n	9 (100)	10 (100)	6 (100)	2*	8 (75)

NOTE: * only two values (366 and 370 µg/kg fw. HHCB; 44 and 60 µg/kg fw. AHTN; +single value, 21 µg/kg fw. AHTN

General, non- compartmental observations

The average contributions of each polycyclic to the total content, using only samples where all 5 substances could be determined, is shown graphically in Figure 4. For water and eel samples, the contribution of HHCB was 71% and 61% respectively, and of AHTN 22% and 31% respectively. In the sediment samples, however, both HHCB und AHTN are present at almost the same

Table X. Concentration (mg/kg lipid) of HHCB and AHTN in Fish Samples from Lakes with an High or Low Input of Sewage Water (Min = minimum; Max = maximum; d.l. = detection limit)

	Perch	*Common Bream*	*Roach*	*Pike*	*Pikeperch*
		HHCB (low)			
Min	< d.l.	< d.l.	< d.l.	< d.l.	< d.l.
Median		3.3			
Max	13.0	15.3	11.8	14.2	113.0
		HHCB (high)			
Min	13.6	10.2	9.1	*	18.0
Median	33.3	90.1	13.0		47.3
Max	159.9	143.3	55.3		383.9
		AHTN (low)			
Min	< d.l.	< d.l.	< d.l.	< d.l.	< d.l.
Max		2.7	2.0	+	
		AHTN (high)			
Min	4.8	2.4.	3.5	*	< d.l.
Median	7.1	18.4	4.5		10.0
Max	43.7	35.3	18.4		88.3

NOTE: * only two values, 66.5 / 61.7 mg/kg lipid HHCB and 8.0 / 10.0 mg/kg lipid AHTN; + = single value, 2.2 mg/kg lipid

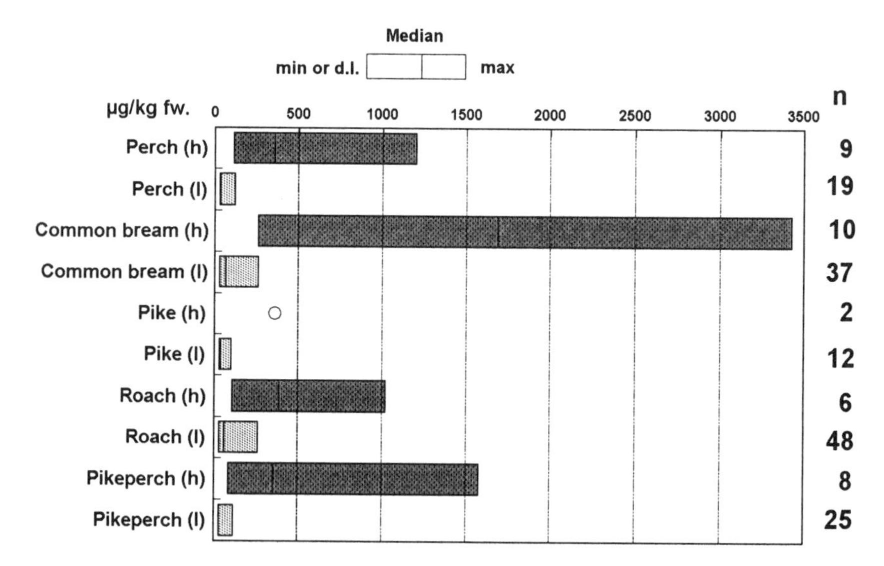

Figure 3. Concentration of HHCB (Galaxolide®) in Fish Samples from Lakes and Rivers. (l = low, h= high or moderate proportions of sewage effluents in the aquatic system)

Table XI. Range of Concentrations (mg/kg lipid) of HHCB and AHTN in the Edible Parts of Coastal and Fresh Water Biota Samples (d.l.: detection limit)

	Location	*Lipid (%)*	*HHCB*	*AHTN*	*n*	*Lit.*
			Europe			
Bream	River Ruhr	0.45-1.56	2.8-3.8	2.2-7.1	3	6
Perch	River Ruhr	0.36/0.72	2.5/3.3	3.5/5.0	2	6
Chub	River Ruhr	0.6	1.7	3.2	1	6
Roach	River Ruhr	1.13	1.4	2.6	1	6
Pike	South Sweden	0.19/0.38	< d.l.	< d.l.	2	6
Pike perch	River Elbe	0.3-0.70	0.6-3.84	0.32-0.99	4	20
Rainbow trout	Danemark	2.6-3.3	0.11-0.65	0.20-0.59	4	20
Brown trout	River Stör	1.3/3.5	13.7/20.3	10.6/13.4	2	20
Trout	various rivers (I)	1.1-8.7	<d.l.-0.59	<d.l.-0.41	19	22
Shetfish	River Po	4.4	0.77	2.39	1	22
Chub	River Garigilano	1.7	0.29	0.24	1	22
Italian nose	River Adige	0.7/1.3	< d.l	< d.l.	2	22
Crucian carp	River Po and Garigilano	0.1-0.8	1.0-5.0	0.5-4.0	5	22
Blue mussel	North Sea	1.7-2.2	<0.03-0.11	<0.03-0.06	3	20
Shrimps	North Sea	0.9-1.0	<0.04-0.37	<0.04-0.06	4	20
			Canada			
Lake trout	Ontario		<d.l.-0.03	< d.l.-0.05	4	20
Winter Flounder	Miramichi		< d.l.	< d.l.	1	20
Clams	Halifax harbour	0.4	3.0	1.1	1	20
Mussels	Halifax harbour		1.65	n.d.	1	20

NOTE: cited in [22,23]

concentrations. The results of an examination of the water phase and the suspended particulate matter of the Elbe [18] indicate that AHTN is more strongly bound to the particulate matter in the water and thus to the sediment than HHCB. No data are presently available for breakdown mechanisms of musk fragrances in sediment.

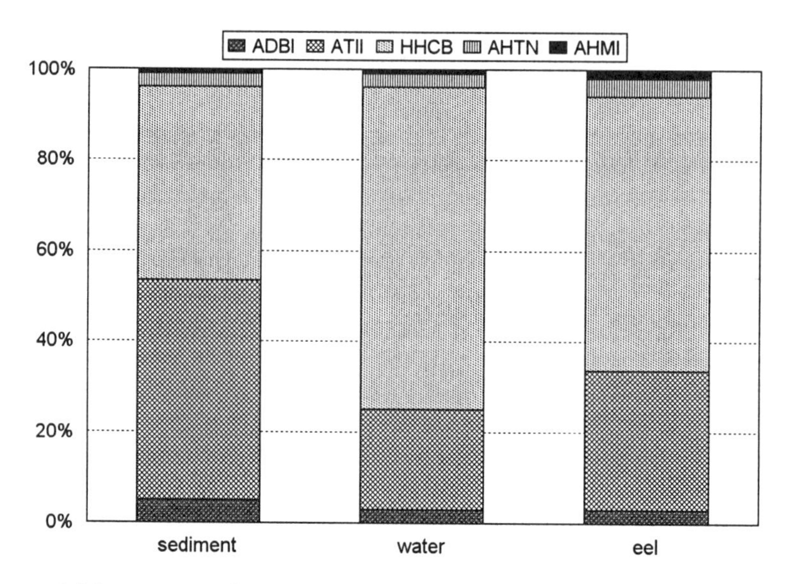

Figure 4. Mean percentile contribution of polycyclic musks to the total content

Bioconcentration

In analogy to nitro musks, enrichment of water-borne substances must also be considered for the polycyclics. The mean octanol - water partition coefficient ($logK_{ow}$) of 5.9 for HHCB was measured on the basis of the OECD Test Guideline 117 by means of HPLC methodology (cited in [6]). Previously, only two partially published bio-concentration studies have been carried out. One (cited in [6]) studied HHCB and AHTN in a flow- through system using the Blue Gill Sunfish (*Lepomis macrochirus*). Experimental BCF values for the whole fish of 1584 for HHCB and 597 for AHTN were obtained. The other study [24] performed with Zebra Fish (*Brachydanio rerio*) resulted in BCF- values of 624 (HHCB) and 600 (AHTN). The BCF based on lipid content was 33200 (HHCB) and 33700 (AHTN). In contrast, based on data from the Ruhr [17], clearly lower mean BCF values of 650 for HHCB and 297 for AHTN were obtained under natural conditions. This must be qualified by stating that it was not clearly shown in the above mentioned study, whether the water and eel samples were taken from the same stretch of water. Bearing this in mind, we chose 15 (9 for AHTN) closely defined areas of water from our examinations where results for both surface water samples and eel samples were available. It may be seen that good correlation (using Spearman rank correlation, $r = 0.74$) exists between the average concentrations in waters and those in biological samples.

A BCF (fresh weight) mean value for HHCB of 862 (range: 201 - 1561) and for AHTN of 1069 (range 250 - 1791) could be calculated from our results. The corresponding BCF - values based on the lipid content of eel were 3504 (HHCB) and 5017 (AHTN). It must also be taken into consideration that these calculated values were not based on defined experimental conditions, but were gained under natural conditions. Influential factors such as the content of suspended matter in the aqueous phase were not taken into consideration.

Comparison with data for toxicological effects

Results of toxicity tests for HHCB and AHTN at all trophic levels, carried out according to OECD guidelines 201,202 and 204, have been collated in [7].

The above consists of a test with the alga *Pseudokirchneriella subcapitata* (conditions: 72 - h, static; endpoint: growth rate and bio-mass), with *Daphnia magna* (conditions: 21 - d, semi static; endpoint: immobility and reproduction), with the blue gill sunfish (*Lepomis macrochirus*) (conditions: 21 - d, flow - through; endpoint: onset of clinical signs like respiration) and with the fathead minnow (*Pimephales promelas*) (conditions: 32 - d; endpoint: larvae surviving and development). On the whole, fish appear to be the most sensitive test organisms by a small degree. The lowest NOEC determined for HHCB was that for the fathead minnow at 68 µg/l and for AHTN at 35 µg/l for the same species.

When three or more chronic studies are available, the Technical Guidance Document [25] prescribes an assessment factor of 10. Taking this into account for the above NOEC data, $PNEC_{water}$ (provisional no effect concentration) values of 6.8 µg/l for HHCB and 3.5 µg/l for AHTN can be obtained. A comparison of the single results for both polycyclics from surface water samples contaminated to different levels with sewage with the $PNEC_{water}$ values is graphically presented in Figure 5. It is clearly noticeable that the difference from the PNEC is only small for those areas of water with high sewage pollution. For waters with low sewage contamination, the middle values of HHCB are a factor of 100 lower (for AHTN, 175) and for the moderately contaminated regions a factor of 30 lower (for AHTN, 50).

Acknowledgements

The authors would like to thank the Senatsverwaltung für Stadtentwicklung, Umweltschutz und Technologie and the Fischereiamt Berlin for funding and supporting this work, Mrs. Conrad and Mrs. Hirsch for the technical support and Mr. R. Hatton for his help in preparing this manuscript.

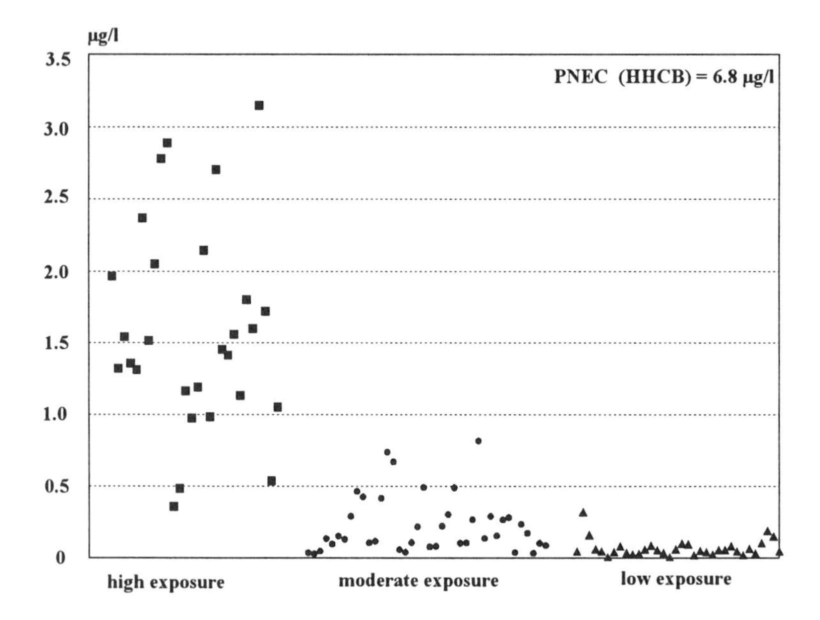

Figure 5. Concentration of Galaxolide® (HHCB) in Surface Water Samples from Lakes and Rivers. Low, moderate and high relates to the proportion of sewage effluents in the aquatic system (PNEC = provisional no effect concentration)

References

1 Yamagishi T., Miyazaki T., Horii S. and Akiyama K. (1981) Identification of Musk Xylene and Musk Ketone in Freshwater Fish Collected form the Tama River, Tokyo. Bull.Environ.Contam.Toxicol. 26, 656-662.

2 Yamagishi T., Miyazaki T., Horii S. and Kaneko S. (1983) Synthetic Musk Residues in Biota and Water from Tama River and Tokyo Bay (Japan). Arch.Environ.Contam.Toxicol 12, 83-89.

3 Rimkus G.G. and Wolf M. (1993) Rückstände und Verunreinigungen in Fischen aus Aquakultur. 2. Mitteilung: Nachweis von Moschus Xylol und Moschus Keton in Fischen. Dtsch.Lebensm.Rdsch. 89, 171-175.

4 Eschke H.-D., Traud, J. and Dibowski H.-J. (1994a) Analytik und Befunde künstlicher Nitromoschus-Substanzen in Oberflächen- und Abwässern sowie Fischen aus dem Einzugsgebiet der Ruhr. Vom Wasser 83, 373-383.

5 Rimkus G.G. and Wolf M. (1996) Polycyclic musk fragrances in human adipose tissue and human milk. Chemosphere 33, 2033-2043.

6 van de Plassche E.J. and F. Balk (1997) Environmental risk assessment of the polycyclic musks AHTN and HHCB according to the EU-TGD. RIVM - report no. 601503008.
7 Ford R.A. (1998) The human safety of the polycyclic musks AHTN and HHCB in fragrances - a review. Dtsch. Lebensm.Rdsch. 94, 268-275.
8 Fromme H., Otto T. and Pilz K. (1999) Levels of synthetic musks, bromocyclen and PCBs in eel (anguilla anguilla) and PCBs in sediment samples from some waters of Berlin, Germany. Chemosphere 39, 1723-1735.
9 Fromme H., Otto T. and Pilz K. (1999) Polycyclic musk fragrances in different environmental compartments in Berlin (Germany). Water Research (in press).
10 BgVV (Bundesamt für gesundheitlichen Verbraucherschutz und Veterinärmedizin Berlin) Fachgruppe 92 (ZEBS) (1999) Handbuch Lebensmittelmonitoring.
11 DFG (Deutsche Forschungsgemeinschaft) (1991) Arbeitsgruppe Analytik: Methodensammlung Rückstandsanalytik von Pflanzenschutzmitteln, Multimethode S 19. 11. Lieferung, VCH VerlagsGmbH, Weinheim (FRG).
12 Amarjit S., Narang S., Vernoy C.A. and Eadon GA(1983) Evaluation of Nielsen-Kryger steam distillation technique for recovery of phenols from soil. J.Assoc.Off.Anal.Chem. 66, 1330-1340.
13 Nunez A.J. and Bemelmans J.M.H. (1984) Recoveries from aqueous model system using a semimicro steam distillation solvent extraction procedure. J.Chromatogr. 294, 361-365.
14 Rijks J., Curvers J. Noy T. and Cramers C. (1983) Possibilities and limitations of steam distillation extraction as a pre-concentration technique for trace analysis of organics by capillary gas chromatography. J.Chromatogr. 279, 395-407.
15 Godefroot M, Sandra P and Verzele M. (1981) New method for quantitative essential oil analysis. J.Chromatogr. 203, 325-335.
16 Heberer Th., Gramer S. and Stan H.-J. (1999) Occurrence and distribution of organic contaminants in the aquatic system in Berlin. Part III: Determination of synthetic musks in Berlin surface water applying solid-phase microextraction (SPME) and gas chromatography-mass spectrometry (GC-MS). Acta Hydrochim.Hydrobiol. (in press).
17 Eschke H.-D., Dibowski H.-J. and Traud J. (1995) Untersuchungen zum Vorkommen polycyclischer Moschus-Duftstoffe in verschiedenen Umweltkompartimenten - Befunde in Oberflächen-; Abwässern und Fischen sowie in Waschmitteln und Kosmetika (2. Mitteilung). Z.Umweltchem.Ökotox. 7, 131-138.
18 Winkler M., Kopf G., Hauptvogel C. and Nes T. (1998) Fate of artificial musk fragrances associated with suspended particulate matter (SPM) from the river Elbe (Germany) in comparison to other organic contaminants. Chemosphere 37, 1139-1156.

19 Paxéus N. (1996) Organic pollutants in the effluents of large sewage treatment plants in Sweden. Wat.Res. 30, 1115-1122.
20 Eschke H.-D., Traud J. and Dibowski H.-J. (1994b) Untersuchungen zum Vorkommen polycyclischer Moschus-Duftstoffe in verschiedenen Umweltkompartimenten - Nachweis und Analytik mit GC/MS in Oberflächen-, Abwässern und Fischen (1. Mitteilung). Z.Umweltchem.Ökotox. 6, 183-189.
21 CLUA (Chemische Landesuntersuchungsanstalt Freiburg). (1999) Bericht über polyzyklische Moschusverbindungen in Rheinaalen 1995/96.
22 Gatermann R., Hellou J., Hühnerfuss H., Rimkus, G. and Zitko V. (1999) Polycyclic and nitro musks in the environment: a comparison between Canadian and European aquatic biota. Chemosphere 38, 3431-3441.
23 Draisci R., Marchiafava C., Ferretti E., Palleschi L., Catellani G. and Anastasio A. (1998) Evaluation of musk contamination of freshwater fish in Italy by accelerated solvent extraction and gas chromatography with mass spectrometric detection. J.Chromatogr. A 814, 187-197.
24 Ewald F. (1999) Kinetik der Akkumulation und Clearance der polycyclischen Moschusduftstoffe Galaxolide und Tonalide. Poster presented on SETAC - conference, Leipzig.
25 European Commission (1998) Technical Guidance Documents in support of the Commission Directive 93/67/EEC on risk assessment for new notified substances and the Commission Regulation (EC) 1488/94 on risk assessment for existing substances.

Chapter 16

Polychlorinated Naphthalenes in the Atmosphere

Tom Harner[1], Terry F. Bidleman[1], Robert G. M. Lee[2], and Kevin C. Jones[2]

[1]Atmospheric Environment Service, 4905 Dufferin Street, Downsview, Ontario M3H 5T4, Canada
[2]Environmental Science, Lancaster University, Lancaster LA1 4YQ, United Kingdom

Polychlorinated naphthalenes (PCNs) are a class of industrial chemicals that are receiving renewed interest due to their abundance in the atmosphere and dioxin-like toxicity. Several recent measurements of their atmospheric burdens show that ΣPCN is 60-150 pg m^{-3} at urban locations and lower at rural and remote sites. Typically, ΣPCN is 20-50% of ΣPCB (polychlorinated biphenyls) and the toxicity contribution (measured as TCDD toxic equivalents, TEQs) is of a similar magnitude. Air parcel back trajectories show that elevated PCN burdens at remote arctic locations can be attributed to air masses that stem from land-based sources in the U.K. and Europe. The long-range atmospheric transport of PCNs is largely governed by particle-gas partitioning. The particle-associated fraction of PCNs is described using a simple model that utilizes the octanol-air partition coefficient (K_{OA}).

Introduction

Polychlorinated naphthalenes (PCNs) are widespread environmental pollutants *(1)*. The use-history of PCNs has paralleled and preceded that of the PCBs. Produced by chlorination of molten naphthalene, PCNs exist as 75 possible congeners that contain one to eight chlorine atoms (Fig. 1). They are structurally similar to the polychlorinated biphenyls (PCBs) and exhibit similar chemical and thermal stability. They were first used commercially in the early 1900s for wood, paper and textile impregnation but were shortly thereafter replaced by PCBs after incidents of worker-related toxicity *(2)*. Exposure to PCN has been blamed for the deaths of two Britons in the 1940s *(3,4)* and has recently been associated with cancer and chronic liver disease in PCN workers *(5)*. Koppers company (PA, USA) was one of the largest producers of technical PCN formulations (Halowaxes) until they voluntarily ceased production in 1977 *(6)*. PCNs were produced in Europe under various names including Seekay Waxes (ICI, United Kingdom) and Nibren Waxes (Bayer, Germany) *(6)*. Little is known of the production volumes and production history in the United Kingdom. However, some insight to the historical inputs of PCNs to the U.K. environment has been gained through analysis of a rural freshwater sediment core *(7)* which shows peak inputs of PCNs ca. 1960 at ~20% of the value of the peak for PCBs which occured ca. 1978.

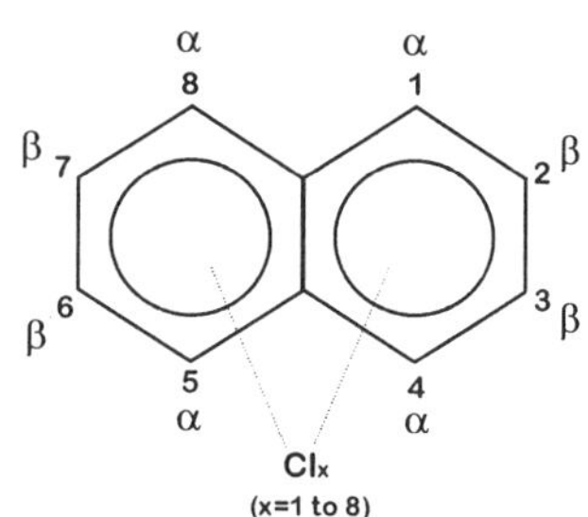

Figure 1. Structure of polychlorinated naphthalenes showing the numbering of the chlorine substitution positions. Substitution in the α or peri positions leads to greater distortion of the molecule.

PCNs (5-Cl and 6-Cl) were first reported in human adipose by the Japanese in 1979 *(8,9)*. These same congener groups were later reported in US human breast milk at quite similar levels *(10)*. Williams et al., *(11)* did an isomer specific analysis using labeled PCN standards on Canadian adipose and also reported very similar total pentachlorinated and hexachlorinated levels. Recent measurments of PCNs in biota are presented by Falandysz *(1)* and Ishaq *(12)*. The ubiquitous nature of PCNs is of concern because of their dioxin-like toxicity. Several congeners exhibit high dioxin-like toxicity and 2,3,7,8,-tetrachlorodibenzo-*p*-dioxin (TCDD) toxic equivalents (TEFs) have been reported *(13,14)* and are of similar magnitude to some of the coplanar PCBs *(15)*. Although the use of PCNs has declined in the past few decades, they are not prohibited in most countries and still occur in many PCB-like applications such as capacitor fluids, engine oil additives, electrical insulators *(6)* and as contaminants in commercial PCB fluids *(16)*. PCNs have also been found in incinerator emissions *(17)* and are believed to be formed by combustion of PAHs by a "de novo" synthesis pathway *(18)*. Recent atmospheric concentrations of PCNs have been reported in Germany *(19)*, urban Chicago *(20)*, Sweden *(21)*, and the eastern Arctic Ocean *(22)*.

Cycling of PCNs between their gaseous and condensed state influences atmospheric burdens. Lee et al. *(33)* by investigating the diurnal cycling of PCNs near Lancaster (Northwest U.K.) postulated that air concentrations were attributed to a mixture of diffuse sources (exchange with terrestrial surfaces) and non-diffuse (or point) sources. Partitioning of PCNs to aerosols is also important since it determines how mobile they are in the atmosphere. Congeners that are predominantly in the gas-phase are more likely to be transported greater distances from sources. However, they will also be subject to gas-phase removal reactions to a greater extent.

Here we present recent measurements and toxicity contributions of PCNs in the remote eastern Arctic Ocean and Arctic land-based monitoring stations *(22)*. Back trajectory analysis for the Arctic samples has related episodes of elevated PCN and PCB concentrations to air trajectories that stem from the United Kingdom and other European regions in the five days period prior to collection, implicating these regions as sources. We pursue the notion that the U.K. is a significant source of PCNs by examining air concentrations at a semi-rural site near Lancaster. Air concentrations of PCNs are further interpreted with the use of five-day air parcel back trajectories and atmospheric burdens are compared to PCBs.

Recent studies have addressed the particle-gas partitioning of PCNs *(23,24)*. Key physico-chemical descriptors such as temperature-dependent vapor pressures *(25)* and octanol-air partition coefficients (K_{OA}) *(26)* have recently been reported. Results are presented from a study in Chicago *(20)* and used to demonstrate the validity of a simple Koa-based model for predicting the particle-bound fraction of PCNs in the atmosphere at any given temperature.

Methodology

Sample Collection and Clean-Up – Particle/Gas Partitioning Study

High volume air samples (n=15, ~400 m^3) were collected in Chicago in February/March 1995. Measurements were conducted over 12 h day or night periods to minimize filtration artefacts due to temperature fluctuations. A sampling train consisting of double glass fiber filters (20 x 25 cm, Gelman A/E) followed by two pre-cleaned polyurethane foam (PUF) plugs was used. A separate sampling train was used to determine suspended particle concentrations (TSP, μg particles m^{-3} air). PUF plugs and glass fiber filters were extracted by soxhlet using petroleum ether (PUF) and dichloromethane (GFF). Extracts were concentrated by rotary evaporation and fractionated using silicic acid. More details are presented in Harner et al. *(20)*.

After an additional clean-up with sulfuric acid the F1 portions ("PCB fractions") from the above samples were further fractionated on mini carbon columns in order to separate the PCNs from the bulk of the PCB congeners. These columns contained AX-21 activated carbon mixed 1:20 with silicic acid. The carbon-silicic acid mix (100 mg) was sandwiched between 50-mg layers of silicic acid. The column was prewashed with 5 mL toluene followed by 5 mL 30% DCM in cyclohexane. The first fraction (F1-1) was eluted with 5 mL 30% dichloromethane in cyclohexane and contained the multi-ortho and a portion of the mono-ortho PCBs. The second fraction (F1-2) was eluted using 5 mL toluene and contained the PCNs, non-ortho PCBs and the remainder of the mono-ortho congeners. PCNs were quantified by GC-negative ion mass spectrometry (GC-NIMS) in selected ion mode (SIM) using NIMS response factors derived from the weight percentages of the individual congeners in Halowax 1014 *(20)*, a commercial mixture of PCNs (Fig. 2). All samples were blank and recovery corrected. Recoveries for PCNs were 50-80%. A more detailed account of the quantification procedure is given elsewhere *(20,23)*.

Arctic Samples

High volume air samples were collected during a cruise of the eastern Arctic Ocean (n=34, summer 1996) and land-based stations at Alert (82.3 N, 62.2 W, 1993-94, n=5) and Dunai Island in eastern Siberia (74.6 N, 123.4 E, summer-1993, n=3) *(22)*. Particle-associated compound (Arctic land-based stations only) was collected on glass fiber filters (GFF) and vapor-phase compound was trapped on polyurethane foam (PUF) plugs. Air volumes were approximately 800 m^3 for ship-based samples, and ~2000 m^3 for arctic land-based stations. Arctic samples were fractionated and analyzed according to the method presented above for the particle-gas partitioning study.

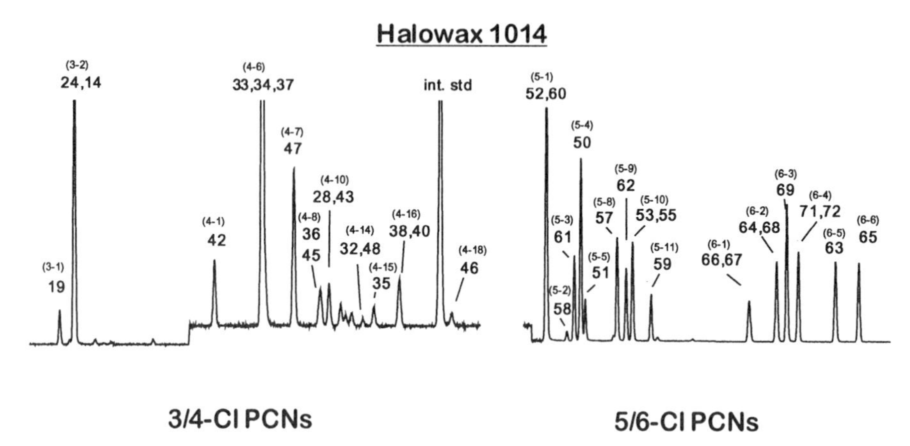

Figure 2. GC-NIMS chromatogram of technical PCN (Halowax 1014). The vertical scale of the 5/6-Cl section is magnified by a factor of 10. Peaks are labeled according to IUPAC numbers and by homologue group and elution order (in parentheses) (20).

Sample Collection – UK study

High volume air samples were collected daily over the period Mar.-Oct. and Dec. in 1994 at a semirural location just outside of Lancaster (pop. 70 000), about 5 km inland from the northwest coast of England. PUF plugs and glass fiber filters were combined and extracted by soxhlet and concentrated extracts were fractionated using florisil. Analytical details and results for PCBs and PAHs are presented elsewhere *(27)*.

The F1 portions from florisil were further fractionated on mini carbon columns (as above) in order to separate the PCNs from the bulk of the PCB congeners and to allow further reduction in volume for extracts stored in dodecane. PCB congeners 208 and 209 in dodecane were added as internal standards for volume correction. Fraction F1-2 was volume reduced to 25 μL using a gentle stream of nitrogen. PCNs were determined by gas chromatography electron impact mass spectrometry (GC-EIMS, Fisons MD-800) in SIM using a 30 m DB-5MS column (0.25 mm i.d., 0.25 μm film). Peaks were quantified against Halowax 1014, a commercial mixture of 2-Cl to 8-Cl PCNs. All results have been blank and recovery corrected. Recoveries for the 3-Cl to 4-Cl were ~90-100% but dropped off for the higher molecular weight homologs (5-Cl, ~20-95%; 6-Cl to 8-Cl, 0-90%). The poor recoveries of the 6-Cl to 8-Cl congeners suggests they are retained more strongly by florisil. Since most of the toxicity is associated with the 6-Cl congeners, toxic equivalents will not be determined from these data. However, overall concentrations of PCNs are obtainable since typically >95% of the atmospheric burden is associated with the 3-5 Cl congeners (the 3-Cl and 4-Cl congeners accounting for about 90% of this total) *(22)*.

Results & Discussion

PCNs in Air

Table I summarizes recent measurements of PCNs, PCBs and ratios of PCN/PCB at various urban, rural and background locations. The contributions of PCNs versus PCBs is variable and ranges from 3% in Dunai, Russia to 50% in the eastern Arctic Ocean. Results from the arctic cruise samples are divided into three regions – the Barents Sea, the eastern Arctic and the Norwegian Sea. Concentrations of ΣPCNs (pg m^{-3}) at each of the sampling sites averaged 40 (n=2) for the Barents Sea, 11.6 ± 3.2 (n=10) for the eastern Arctic, and 7.1 (n=2) for the Norwegian Sea. Values at the monitoring stations were: Alert: 3.5 ± 2.7 (n=5) and Dunai Island: 0.84 ± 0.47 (n=3) (pg m^{-3}). With the exception of samples collected in the Barents Sea, PCN levels in remote arctic air are approximately an order of magnitude lower than reported in urban areas. The high concentrations found in the Barents Sea may be due to episodic transport from Europe, as discussed below. Air concentrations of PCNs in urban Manchester U.K. (149 pg m^{-3}) were quite high *(28)*; about a factor of two greater than measured in urban Chicago (68 pg m^{-3}) *(20)* and Augsburg, Germany (60 pg m^{-3}) *(19)*. Higher concentrations of PCNs in the U.K. are discussed later.

Table I. Comparison of recent measurements (pg m^{-3}) of polychlorinated naphthalenes and polychlorinated biphenyls in air.

Location	*Type*	*ΣPCN*	*ΣPCB*	*ΣPCN/ΣPCB*	*Ref.*
Augsburg, Ger. 1992-3	Urban	60	N.R.	---	19
Augsburg, Ger. 1992-3	Rural	24	N.R.	---	19
Chicago, USA 1995	Urban	68	350	0.19	20
Downsview, Can. 1995	Urban	17	N.R.	---	20
Barents Sea 1996	Remote	40	126	0.32	22
E. Arctic Ocean 1996	Remote	12	24	0.50	22
Norwegian Sea 1996	Remote	7.1	75	0.09	22
Alert, Canada 1993-94	Remote	3.5	38	0.09	22
Dunai, Russia 1993	Remote	0.8	30	0.03	22
Lancaster, U.K. 1994	Semi-rural	66	163	0.40	28
Manchester, U.K. '98-99	Urban	149	381	0.39	28

N.R. = not reported

Ambient air profiles of PCNs are always dominated by the lower molecular weight homologs. The 3-Cl and 4-Cl congeners account for approximately 80% of the total mass in Chicago *(20)*, ~95% in Augsburg *(19)* and 90-95% of the total mass in the arctic samples *(22)*. This may be due to preferential volatilization of the lighter PCN congeners and preferential deposition of the heavier congeners further from source regions (i.e. global fractionation). An alternate explanation may involve differences in source signatures. Halowax 1001 and 1099 accounted for 65% of the market share of Halowax sales *(6)* and are dominated by 3-Cl (40%) and 4-Cl (40%) congeners. The contribution of these homologue groups in Halowax 1014 is smaller: 3-Cl (14%) and 4-Cl (34%) *(6)*.

Partitioning to Aerosols

Before any definitive statements can be made regarding PCN transport and global distribution, more studies are required to investigate PCN partitioning in the atmosphere as well as other environmental media. The particle/gas partitioning of PCNs in air is particularly important and governs their transport from sources to remote areas and their deposition from the atmosphere to ecosystems.

Absorption of chemicals to particles can be described using the octanol-air partition coefficient, K_{OA} *(29,23)*.

$$\log K_P = \log K_{OA} + \log f_{om} - 11.91 \quad (1)$$

where K_P is the particle/gas partition coefficient ($m^3\ \mu g^{-1}$), f_{om} is the fraction of organic matter on the particle that is able to exchange chemical with the atmosphere. The f_{om} for urban air has been estimated to be 0.2 *(30)* and values of K_{OA} have been published for a range of PCN congeners as a function of temperature *(26)*. The fraction on particles, ϕ, can then be easily estimated from K_P and knowledge of the total suspended particulate concentration in the atmosphere, TSP ($\mu g\ m^{-3}$).

$$\phi = K_P\,TSP / (K_P\,TSP + 1) \quad (2)$$

Mean TSP concentrations ($\mu g\ m^{-3}$) measured in 46 U.S. cities and rural areas was 80 and 30 in 1976 *(31)*. Contemporary levels are likely to be lower, especially for urban air. Lee and Jones *(32)* recently measured TSP values of 15-30 in summer and autumn at a semi-rural location near Lancaster, U.K.

Figure 3a is a log-log plot of K_P versus K_{OA} for 3-Cl to 6-Cl PCNs in a winter sample from Chicago *(23)*. The predicted particulate fraction calculated using eqs. 1 and 2 and a f_{om} value of 10% and 20% is shown in Fig. 3b.

Toxicity Contributions and Source Regions for PCNs

Several of the 75 possible PCN congeners exhibit dioxin-like properties by their potential to induce certain liver enzymes - AHH, EROD and luciferase. TCDD TEFs have been derived *(13,14)* (Table II) and indicate that the most potent inducers are congeners 66/67, 69 and 73 which have TEF values similar in magnitude to the non-ortho (coplanar) PCBs. Ishaq *(12)* found that congeners 66/67 contributed ~50% of the total TEQs in Harbour Porpoises from the west coast of Sweden, outweighing the contribution of the non-ortho PCBs.

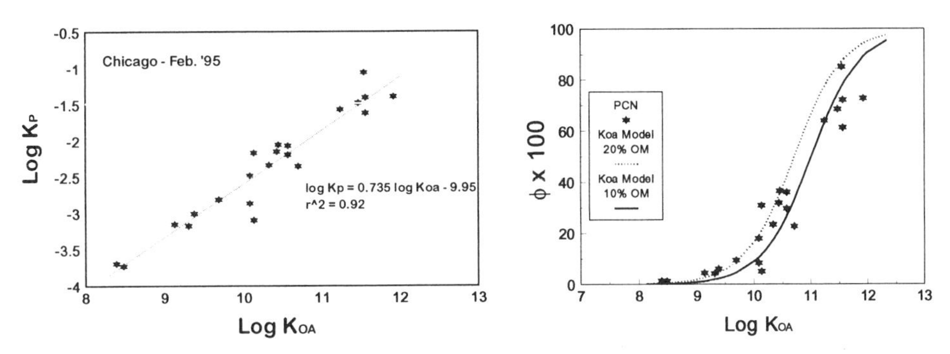

Figure 3. a) Correlation of K_P versus K_{OA} for individual 3-Cl to 6-Cl PCN congeners in Chicago air (symbols indicate mean K_P value for fifteen samples). b) Comparison of measured and model-predicted (eqs. 1 and 2) particulate fractions (ϕ) of PCNs in Chicago air.

Table II. Toxic equivalence factors (TEFs) for PCNs which exhibit dioxin-like activity.

Compound	*Luciferase assay*[1]	*EROD induction*[2]
2,3,7,8-TCDD	1.00	1
1,4-DiCN (CN-5)	2.04×10^{-7}	ND
1,2,6,8-TetraCN (CN-40)	1.60×10^{-5}	ND
1,2,3,6,7-PentaCN (CN-54)	1.69×10^{-4}	ND
1,2,3,4,5,6-HexaCN (CN-63)	ND	2×10^{-3}
1,2,3,4,6,7-HexaCN (CN-66)	3.85×10^{-3}	ND
1,2,3,5,6,7-HexaCN (CN-67)	1.00×10^{-3}	ND
50:50 CN-66:CN-67	1.25×10^{-3}	2×10^{-3}
1,2,3,4,5,7- and/or 1,2,3,5,6,8-HexaCN (CN-64/68)	ND	2×10^{-5}
1,2,3,5,6,8-HexaCN (CN-68)	1.53×10^{-4}	ND
1,2,3,5,7,8-HexaCN (CN-69)	ND	2×10^{-3}
1,2,3,6,7,8-HexaCN (CN-70)	5.87×10^{-4}	ND
1,2,4,5,6,8-HexaCN (CN-71)	-	7×10^{-6}
1,2,3,4,5,6,7-HeptaCN (CN-73)	1.01×10^{-3}	3×10^{-3}

ND – TEF not determined in the study; [1] ref. *(13)*; [2] ref. *(14)*

Figure 4 summarizes the relative TCDD toxic equivalents (TEQs) values for non- and mono-ortho PCBs and several dioxin-like PCNs for Chicago and various locations in the Arctic *(22)*. In all cases , the highest TEQ contribution is attributed to the coplanar PCBs, mostly congener 126. The Dunai and Barents Sea samples have high concentrations and elevated TEQ contributions of the mono-ortho congeners, mostly congeners 118 and 105. The PCNs also make an important contribution - accounting for 13-67% of the TEQ at the arctic sites and ~30% in Chicago. Most of the PCN TEQ is associated with the 6-Cl homolog group. More work is merited to further investigate PCNs because they can elicit significant human and animal toxicities.

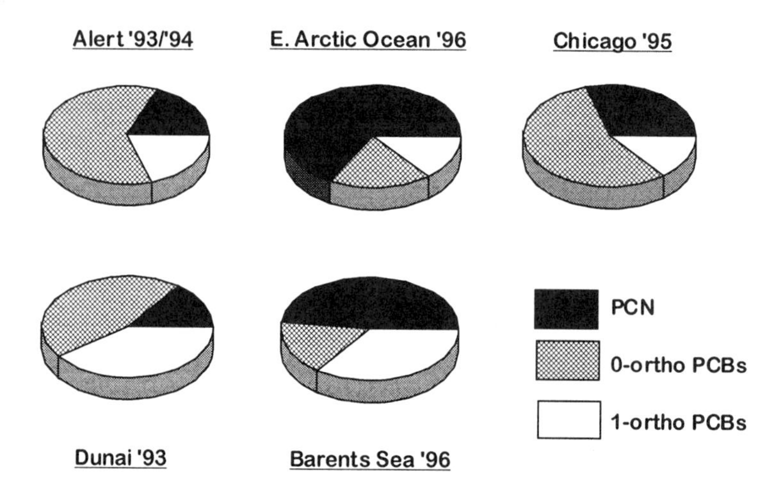

Figure 4. Percent TCDD TEQ contributions of PCNs, coplanar PCBs (PCB-77 and –126), and mono-ortho PCBs (PCB-105, -114, -118, and –156) in the arctic and urban atmosphere (22).

To relate episodes of high concentration to source regions, three dimensional five-day air parcel back-trajectories were determined for the ship-based arctic samples at a pressure level of 850 and 925 hPa (i.e. ~1.5 and 0.75 km altitude). Figure 5 shows the average 5-day back-trajectories for the four samples having the highest ΣPCB concentrations (two of these samples also had the highest ΣPCN) compared to the four samples having the lowest concentrations. Trajectories for the elevated samples stem back in a southerly direction towards the U.K. and mainland Europe implicating these regions as potential source areas for the atmosphere. For the low concentration samples, air parcels spent most of their 5-day history over the Arctic Ocean; although a few pass over land but in non-industrial regions (e.g. Greenland, Canadian Archipelago, northern Scandinavia).

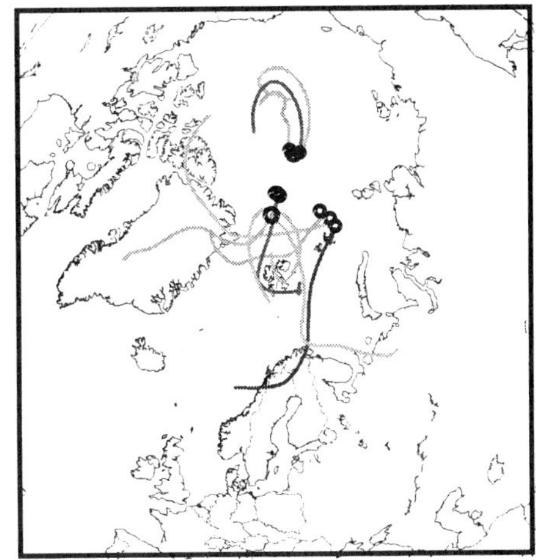

Figure 5. Back-trajectories (average of 850 and 925 hPa) for air samples collected on a cruise of the eastern Arctic Ocean. On the left - four samples (combined) with highest ΣPCBs+ΣPCNs and on the right - the four lowest concentration samples. Circles represent station locations.

PCNs in the Atmosphere of the U.K.

Air parcel 5-day back trajectories were determined at 850 and 925 hPa for all air samples collected in 1994 at Lancaster in order to identify samples whose trajectories stemmed from distinct sectors. Samples were divided into four groups (see Fig. 6) – Arctic/Scandinavia, mainland Europe, Atlantic and the U.K.. Samples (n=3-7) from each sector were then analyzed for PCNs (Table I). Figure 6 summarizes the mass contribution of the 3-, 4- and 5-Cl PCN homologs showing a profile which is dominated by the 3- and 4-Cl congeners as was observed in previous studies. Mean values for the 18 samples were ΣPCN=66 pg m^{-3} and ΣPCB=163 pg m^{-3} *(28)*. The ratio of PCN/PCB was ~0.4 for the U.K. sites compared to ~0.2 in Chicago (Table I). The ΣPCN/ΣPCB ratio observed in the Barents Sea and eastern Arctic Ocean (i.e. 0.32 and 0.5 respectively) was similar to the U.K. value of ~0.4. This suggests a connection between the relatively higher burdens at these arctic sites (compared to other background sites) and a U.K./European source.

Highest air concentrations of PCNs at Lancaster were observed for air samples with back trajectories that stemmed from the "U.K." and "mainland Europe" sectors, supporting the hypothesis that these regions may be significant emission sources of PCNs to the atmosphere. Figure 7 shows the five-day air parcel back trajectory for the highest and lowest ΣPCN concentration samples.

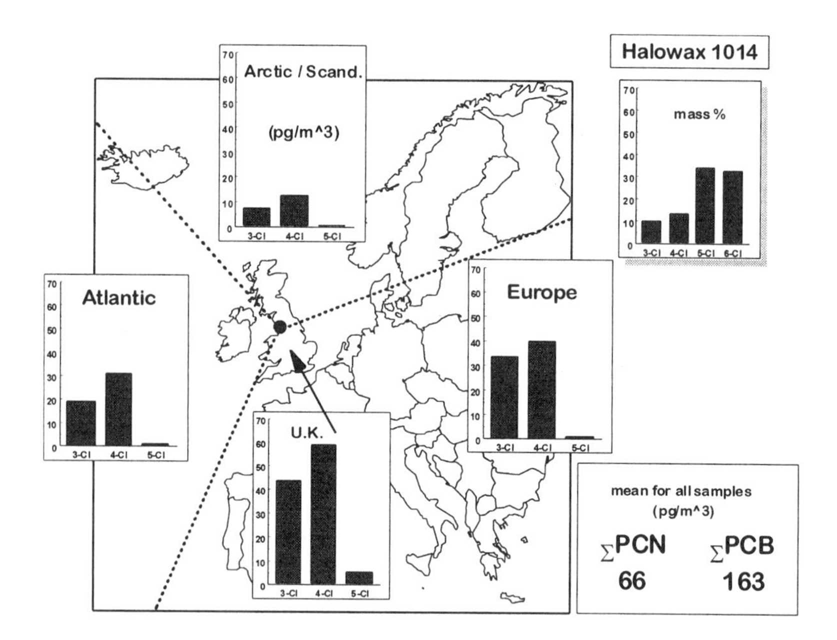

Figure 6. PCN homolog concentrations / distributions for air arriving at the sampling location from different sectors. The U.K. sector refers to air which spent a significant portion of its 5-day history over the U.K. Mean concentrations of PCNs and PCBs for all samples (n=14) is shown.

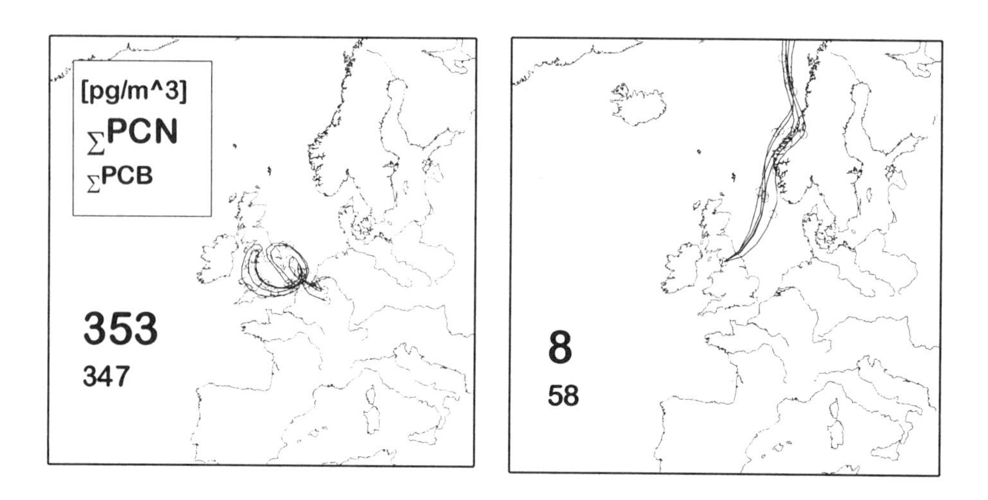

Figure 7. Five-day air parcel back trajectories (925 hPa) for Lancaster U.K. for the air sample having the highest and the lowest concentration of ΣPCN.

It is interesting to note that a greater range of air concentrations was observed for PCNs than PCBs. The ratio of highest concentration observed / lowest concentration observed was 44 for ΣPCN and 6 for ΣPCB. This suggests that atmospheric burdens of PCNs in the U.K. may be attributed to a mixture of "diffuse" and "non-diffuse" sources whereas PCBs are primarily due to "diffuse" sources e.g. re-emission from terrestrial surfaces. Possible "non-diffuse" sources of PCNs include leakage and subsequent volatilization from land-filled electrical equipment, chloralkali production and combustion/incineration processes. The hypothesis that "non-diffuse" sources have a significant contribution to the PCN air burden is further indicated by a high resolution sampling study by Lee et al. *(33)*. The effect of short-term (diurnal) temperature changes on the cycling of POPs between the terrestrial surface and the atmosphere was investigated for several classes of POPs including PCBs. Corresponding slopes of the Clausius-Clayperon plots were similar to the enthalpy of vaporization (ΔH_{VAP}) and the temperature coefficient for K_{OA} suggesting that these compounds were near air-surface ("diffuse sources") equilibrium. However, the slopes for PCNs were more shallow suggesting that advection from "non-diffuse" sources was also important. On the contrary, Dörr et al. *(19)* discounted the influence of combustion "point sources" near Augsburg, Germany and suggested that PCN levels in air were mostly affected by recycling with terrestrial surfaces.

Conclusions

Interest in PCNs as an environmentally significant compound class is emerging. In the past few years several studies have investigated their occurrence in the atmosphere, partitioning to aerosols and toxicity. They revealed that ΣPCN in the atmosphere is typically ~20%-50% of ΣPCB but comparable in terms of 2,3,7,8,-TCDD TEQs. In Lancaster, U.K. and perhaps other regions, atmospheric burdens of PCNs are controlled by a contribution of diffuse and non-diffuse sources, unlike PCBs which are mostly the result of diffuse sources – outgassing from terrestrial surfaces. Air trajectory analysis indicates that the U.K. is a source of PCNs to the atmosphere and may contribute to elevated burdens in the remote Arctic. This long range transport of PCNs is largely governed by partitioning to aerosols. A simple K_{OA}-based model is presented that can predict partitioning to aerosols as a function of temperature with knowledge of only K_{OA} and the organic matter content on the aerosol.

References

1. Falandysz, J. *Environmental Pollution* **1998**, 101, 77-90.
2. Järnberg, U. Ph. D. Thesis **1997**, Stockholm University (ITM), S-106 91, Stockholm, Sweden.
3. Collier et al., *The Lancet* **1943**, 244, 72-74.

4. McLetchie et al. *Br. Med. J.* **1942**, 1, 691-692.
5. Hayward, D. *Environ. Res.* **1998**, 76, 1-18.
6. Crookes, M. J.; Howe, P. D. Environmental hazard assessment: halogenated naphthalenes. Report TSD/13. Department of the Environment, London, Great Britain, **1993**.
7. Gevao, B.; Harner, T.; Jones, K.C. *Environ. Sci. Technol* (in press).
8. Takesita, R.; Yoshida, H. *Eisei Kagaku* **1979**, 25, 24-28.
9. Takesita, R.; Yoshida, H. *Eisei Kagaku* **1979**, 25, 29-33.
10. Hayward, D. G. et al. *Chemosphere* **1989**, 18, 455-468.
11. Williams, D. T.; Kennedy, B.; LeBel, G. L. *Chemosphere* **1993**, 27, 795-807.
12. Ishaq, R. Ph. D. Thesis **1999**, Stockholm University, S-106 91, Stockholm, Sweden.
13. Hanberg, A.; Wern, F.; Asplund, L.; Haglund, P.; Safe, S. *Chemosphere* **1990**, 20, 1161-1164.
14. Blankenship, A.; Kannan, K.; Villalobos, S.; Villeneuve, D.; Falandysz, J.; Imagawa, T.; Jakobsson, E.; Giesy, J.P. *Environ. Toxicol. Chem.* (in press)
15. Ahlborg et al., *Chemosphere* **1994**, 28, 1049-1067.
16. Haglund, P.; Jakobsson, E; Asplund, L.; Athanasiadou, M; Bergman, A. *J. Chromatogr.* **1993**, 634, 79-86.
17. Oehme M.; Manø, S.; Mikalsen, A. *Chemosphere* **1987**, 16, 143-153.
18. Iino, F.; Imagawa, T.; Takeuchi, M.; Sadakata, M. *Environ. Sci. Technol.* **1999**, 33, 1038-1043.
19. Dörr, G.; Hippelein, M.; Hutzinger, O. *Chemosphere* **1996**, 33, 1563-1568.
20. Harner, T.; Bidleman, T. F. *Atmos. Environ.* **1997**, 31, 4009-4016.
21. Järnberg, U.; Asplund, L.; de Wit, C.; Egebäck, A.-L.; Wideqvist, U.; Jakobsson, E. *Arch. Environ. Contam. Toxicol.* **1997**, 32, 232-245.
22. Harner, T.; Kylin, H.; Bidleman, T. F.; Halsall, C. J.; Strachan, W. M. J.; Barrie, L. A.; Fellin, P. *Environ. Sci. Technol.* **1998** , 32, 3257-3265.
23. Harner, T.; Bidleman, T. F. *Environ. Sci. Technol.* **1998**, 32, 1494-1502.
24. Kaupp, H.; McLachlan, M. S. *Chemosphere* **1999**, 38, 3411-3421.
25. Lei, Y. D.; Wania, F.; Shiu, W. Y. J. *Chem. Eng. Data.* **1999**, 44, 577-582.
26. Harner, T.; Bidleman, T. F. J. *Chem. Eng. Data.* **1998**, 43, 40-46.
27. Lee, R.G.M.; Jones, K. C. *Environ. Sci. Technol.* **1999**, 33, 705-712.
28. Harner, T.; Lee, R. G. M.; Jones, K. C. *Environ. Sci. Technol.* **1999** (submitted).
29. Finizio, A.; Mackay, D.; Bidleman, T. F.; Harner, T. *Atmos. Environ.* **1997**, 31, 2289-2296.
30. Bidleman, T.F.; Harner, T. In: D. Mackay and R.S. Boethling (eds.), Estimating Chemical Properties for the Environmental and Health Sciences: A Handbook of Methods, Ann Arbor Press, Inc., Chelsea, Michigan, 48118, U.S.A. (1999).
31. Shah, J.J.; Johnson, R.L.; Heyerdahl, E.K.; Huntzicker, J.J. J. *Air Pollut. Cont. Assoc.* **1986**, 36, 254-257.
32. Lee, R.G.M.; Jones, K. C. *Environ. Sci. Technol.* **1999**, 33, 3596-3604.
33. Lee, R.G.M.; Burnett, V.; Harner, T.; Jones, K.C. Environ. Sci. Technol. (in press).

Chapter 17

Polybrominated Diphenyl Ethers (PBDEs) in Human Milk from Sweden

S. Atuma[1], M. Aune[1], P. O. Darnerud[1], S. Cnattingius[2], M. L. Wernroth[3], and A. Wicklund-Glynn[1]

[1]National Food Administration, P.O. Box 622, SE–751 26 Uppsala, Sweden
[2]Department of Med. Epidem., Karolinska Institute, SE–171 77 Stockholm, Sweden
[3]Department of Information Sciences, P.O. Box 513, SE–751 20 Uppsala, Sweden

A method for the quantification of polybrominated diphenyl ether (PBDE) congeners in human milk has been developed, evaluated and applied to 39 samples from primiparous mothers from Uppsala county, Sweden. The quality assurance for the quantification of five of the most frequently found PBDE congeners is discussed. The mean value of the sum of the analysed congeners (ΣPBDE) is 4.4 ng/g fat (range 1.1-28.2 ng/g fat). Correlation calculations were performed between the ΣPBDE levels and potential factors that could influence the levels. Significant relationships were found between ΣPBDE and smoking ($p=0.001$) and ΣPBDE and the body mass index ($p=0.014$).

Introduction

Flame retardants are added to e.g. textiles, electronic equipment, cabinets for personal computers and television sets and a variety of other plastic products to prevent them from catching fire. The flame retardants can be divided into different groups of which one group consists of brominated organic compounds. The reason for using brominated compounds as flame retardants is the ability of bromine atoms, produced from the decomposed bromoorganic compound, to chemically reduce and retard the development of the fire. The use of flame retardants has increased over the years owing to stricter fire regulations and the annual world production of brominated flame retardants is about 150 000 metric tonnes. These constitute of about 30 % polybrominated diphenyl ethers (PBDEs), 30 % tetrabromobisphenol A and its derivatives and the rest is a variety of brominated products including e.g. polybrominated biphenyls (PBBs).

Special interest is now being focused on the PBDEs since they are used in large quantities world-wide and are highly persistent in the environment. The general chemical structure is shown in Figure 1. Commercial PBDEs consist predominantly of penta-, octa- and decabromodiphenyl ethers. The pentabromo product is, however, a mixture of about equal amounts of tetra and penta congeners and the octabromo product mainly consists of hepta and octa congeners. 75 % of the global PBDE production is decabromo diphenyl ether (for reviews see 1-4).

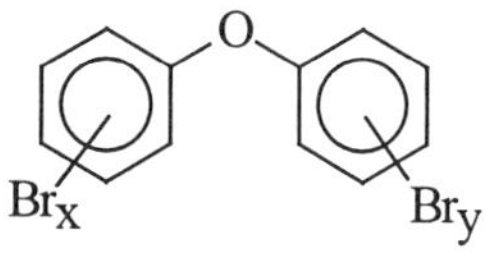

O

Br_x Br_y

x = 1-5
y = 1-5

Figure 1. Structure of PBDE.

Due to leaching of these compounds into the environment, PBDEs are present in environmental samples. The highest levels are found in top predators of aquatic ecosystems, similar to what has earlier been observed for other persistent halogenated compounds, e.g. PCBs and chlorinated pesticides. The PBDE levels have increased in several matrices for a number of years, but the present trend is somewhat difficult to interpret due to large between-year variations (4). The absorption of PBDE depends on the degree of bromination. The fullbrominated decaBDE is taken up at a low degree, whereas lower brominated PBDEs are taken up, accumulated and are bioconcentrated in animals more readily. The most abundant congener found in environmental samples is 2,2',4,4'-tetraBDE (BDE-47).

The PBDEs show structural similarities to other well-known environmental pollutants such as PCBs and dioxins, but in contrast to them only a limited amount of toxicity data is available for the PBDEs. These reveal rather low acute toxicity but the PBDEs do, however, affect the thyroid and possibly other hormonal systems (4).

Since PBDEs have been found in food (4), they are also inevitably found in human milk in like manner as other environmental pollutants such as PCB and DDT. Recently, results from analysis of some pooled human milk samples from Sweden were reported (5,6). The results showed an exponential increase in levels from 1972 to 1997 and a doubling time of 5 years. Apart from that investigation there is one more report from a German survey (7) regarding PBDE levels in human milk.

In this paper we discuss the modification of an existing PCB analytical technique to include the quantification of PBDEs in human milk. Studies of PBDEs as potential environmental contaminants have become necessary and demand a simple but reliable method. Earlier technical PBDE products were used as standards in quantitative work, owing to lack of pure reference standards. Since more than 30 individual PBDE congeners are now available (8), the analytical capability has been improved. In this paper the quality assurance for the quantification of five of the most frequently found PBDE congeners is discussed. These congeners are 2,2',4,4'-tetraBDE (BDE-47), 2,2',4,4',5-pentaBDE (BDE-99), 2,2',4,4',6-pentaBDE (BDE-100), 2,2',4,4',5,5'-hexaBDE (BDE-153), 2,2',4,4',5,6'-hexaBDE (BDE-154). The method has been applied to 39 milk samples from primiparous mothers from Uppsala county, Sweden. The levels obtained are presented as well as some correlations between these levels and questionnaire responses on potential factors influencing the PBDE levels in the samples.

Materials and Methods

Samples

The human milk was obtained from primiparous mothers recruited in an ongoing study on persistent organic pollutants in blood and milk from mothers in Uppsala county, planned to include ca 250 mothers. The milk samples were collected using a breast pump provided specifically for the purpose. The samples were frozen immediately and stored at -20 °C until analysis. Milk samples (n=39) from the main study were randomly taken out for PBDE analysis. The participants in this sub-study were between 22 and 36 years old with a mean age of 28 years. They were asked to fill in a questionnaire about their pregnancy, dietary and other habits (including smoking and alcohol consumption). The responses to some of these questions were used in correlation calculations in this sub-study.

Analysis

The analytical method is a modified version of our current technique employed in the determination of PCBs and chlorinated pesticides in human milk (9,10). The milk samples were thawed and homogenised and 35 g was then taken and extracted twice with a mixture of n-hexane / acetone (1:1). After addition of ethanol (99.5%) to the combined extracts, the solvents were evaporated and the lipid content determined gravimetrically. The fat was then re-dissolved in n-hexane and treated with sulphuric acid. The PBDEs were separated from the dominating PCB congeners over a silica gel column (4.5 g of 3 % water-deactivated silica gel). The first fraction, eluted with ca 22 ml of n-hexane, contained the major part of the PCB congeners, DDE and HCB. The second fraction, eluted with 45 ml of n-hexane / diethyl ether (3:1), consisted mainly of the PBDE congeners and chlorinated pesticides, and even fractions of late-eluting PCB-congeners.

All samples were fortified with an internal standard, 2,2',3,4,4'-pentaBDE (BDE-85), prior to extraction to correct for analytical losses and to ensure quality control. The choice of BDE-85 as an internal standard was based on the fact that it was hardly detected in milk samples analysed prior to this survey. The final analysis was performed on a Hewlett-Packard model 5890 gas chromatograph equipped with dual capillary columns (Ultra 2 and DB-17) and dual electron capture detectors. The use of two capillary columns of different polarity usually decreases the number of errors arising from co-eluting compounds. A mixture of the PBDE-congeners earlier mentioned was used as analytical standard.

Regression analysis was used to describe a possible relationship between PBDE levels in milk and some selected parameters from the questionnaire responses. The sum of the five analysed PBDE congeners (ΣPBDE) was used in the statistical analysis. When the levels were below the quantification limit, half of this limit was taken as an estimated value. The PBDE levels were given both on fresh and fat weight basis.

Results and Discussion

Quality Control

Validation of a method normally requires proof of accuracy, precision, specificity and some indications of ruggedness (e.g. inter-analyst or inter-day adequacy of performance). The precision of the method was demonstrated by seven replicate analyses of a single pooled milk sample. The results showed that the precision as well as the ruggedness were within our acceptable range with coefficients of variation (CV) less than 10 %. Method accuracy, expressed as percent recoveries, are summarised in Table I.

Table I. Recoveries and corresponding coefficients of variation (CV) of PBDE congeners in milk samples spiked at two concentration levels

Compound[a]	*Structure*	*Addition of 14 pg/g milk*		*Addition of 86 pg/g milk*	
		Recovery (%)	*CV(%)*	*Recovery (%)*	*CV(%)*
BDE-47	2,2',4,4'	90	11	108	7
BDE-99	2,2',4,4',5	100	4	106	8
BDE-100	2,2',4,4',6	104	9	106	5
BDE-153	2,2',4,4',5,5'	108	8	107	5
BDE-154	2,2',4,4',5,6'	103	8	108	2

[a] IUPAC nomenclature

Mean recoveries and CV values from four analyses at each level of spiked milk samples are shown. For the lower tested concentration (14 ppt) the CV value for BDE-47 is slightly higher than desirable which could be due to the comparatively high incurred level in the sample (about four times the added amount). This fact has therefore very little practical relevance because the higher incurred concentrations can be quantified with acceptable accuracy. The specificity of the method was demonstrated by lack of interference from certain co-eluting PCBs and chlorinated pesticides. The detection limits, based on the mean levels found in ten reagent blanks plus three standard deviations, were 4 pg/g milk for BDE-47, 2 pg/g milk for BDE-99 and 1 pg/g milk for the other congeners.

BDE-85 was chosen as internal standard based on the fact that it was hardly detected in a number of samples analysed prior to this study. Low levels of BDE-85 were, however, found in samples containing high levels of PBDE. The results were in those cases corrected for the incurred content of BDE-85.

Levels in Milk Samples

The results from the analysis of the 39 milk samples are presented on fresh and fat weight basis in Table II and Table III, respectively. The observed mean value of ΣPBDE was 4.4 ng/g fat, whereas the median was 3.4 ng/g fat, which in part could be a consequence of a single, high peak value. See Figure 2. Indeed the highest value of 28.2 ng/g fat was three times higher than the second highest value, 9.4 ng/g fat. This resulted in a wide range of PBDE levels from the lowest value 1.1 ng/g to the highest value 28.2 ng/g fat. BDE-47 was the major congener in human milk, comprising ca 55 % of ΣPBDE. The mean fat content in the milk was 3.2 %.

The median ΣPBDE value in the present study, 3.4 ng/g fat, is about hundred times lower than that found for PCBs (11). However, the highest ΣPBDE value is similar to the levels of many of the commonly found PCB congeners (11). The median ΣPBDE value is in good agreement with the value 4.0 ng/g fat obtained from analysis of one pooled human milk sample (n=40) from Stockholm, Sweden 1997 (5,6).

Table II. Levels (pg/g milk) of PBDE in human milk from primiparous women in Uppsala county, Sweden (n=39)

Compound	*Mean*	*Median*	*Min*	*Max*
BDE-47	77	58	8	358
BDE-99	24	16	4	222
BDE-100	14	10	1.5	114
BDE-153	19	14	8	96
BDE-154[a]	2.1	1.5	1.5	6
ΣPBDE	137	102	28	626

[a] Most values were below the quantification limit.

Table III. Levels (ng/g fat) of PBDE in human milk from primiparous women in Uppsala county, Sweden (n=39)

Compound	*Mean*	*Median*	*Min*	*Max*
BDE-47	2.52	1.83	0.331	16.1
BDE-99	0.717	0.442	0.181	4.47
BDE-100	0.475	0.340	0.060	5.14
BDE-153	0.648	0.478	0.255	4.32
BDE-154[a]	0.0695	0.060	0.030	0.270
ΣPBDE	4.45	3.37	1.14	28.2

[a] Most values were below the quantification limit.

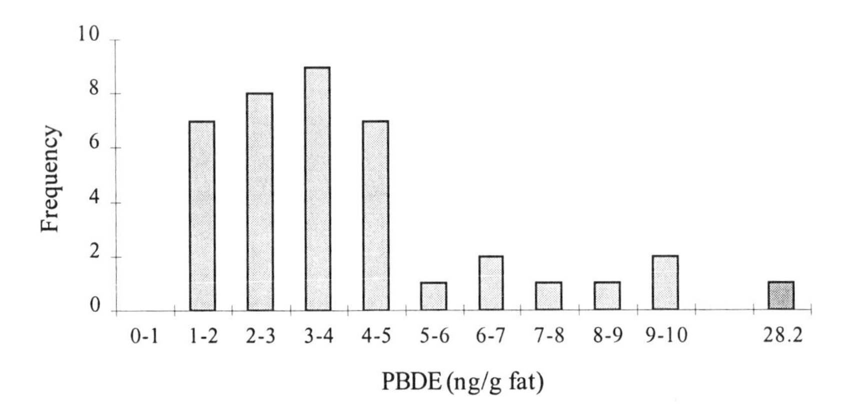

Figure 2. Frequency diagram showing the distribution in ΣPBDE (fat weight) values in breast milk from the studied mothers (n=39), each bar covering a 1 ng/g fat range. Note the single, very high peak value of 28.2 ng/g fat.

Statistical Analysis

The correlation between the ΣPBDE and BDE-47 levels was strong (r^2=0.937, $p<0.001$) which could be expected since ΣPBDE to a large extent was comprised of of BDE-47. An expected distinct correlation was also observed between fat and fresh weight levels (r^2=0.853, $p<0.001$). The highest ΣPBDE value (28.2 ng/g fat) would exert an unreasonable strong influence on the estimation of the model parameters and was therefore excluded from the regression analysis.

Significant relationships were found (when regression models with only one independent variable were analysed) between fat weight ΣPBDE and smoking ($p=0.001$), and between fat weight ΣPBDE and the body mass index (BMI) ($p=0.014$). Hence a stepwise regression procedure was used and the result was a model containing both BMI and smoking habits as independent variables. The model showed that smoking increased the value of ΣPBDE significantly when differences in ΣPBDE due to BMI had been accounted for. See Figure 3. From the model it could be estimated that, on the average, ΣPBDE increased by 192 pg/g fat with each unit increase in BMI, assuming no change in smoking habits. Apart from these significant relationships to ΣPBDE, no other significant relationships were found ($p>0.14$). The other variables tested in the model were the mother's age, frequency of computer usage, consumption of fish (total or specifically Baltic), consumption of alcohol, place of residence during childhood and adolescence, and the birth weight of the child. However, the number of samples in this study may be too small to reveal significant changes regarding these correlations.

It is at present not known if the PBDE levels in breast milk are correlated to levels of other halogenated compounds, e.g. PCBs, and this will be examined in future studies.

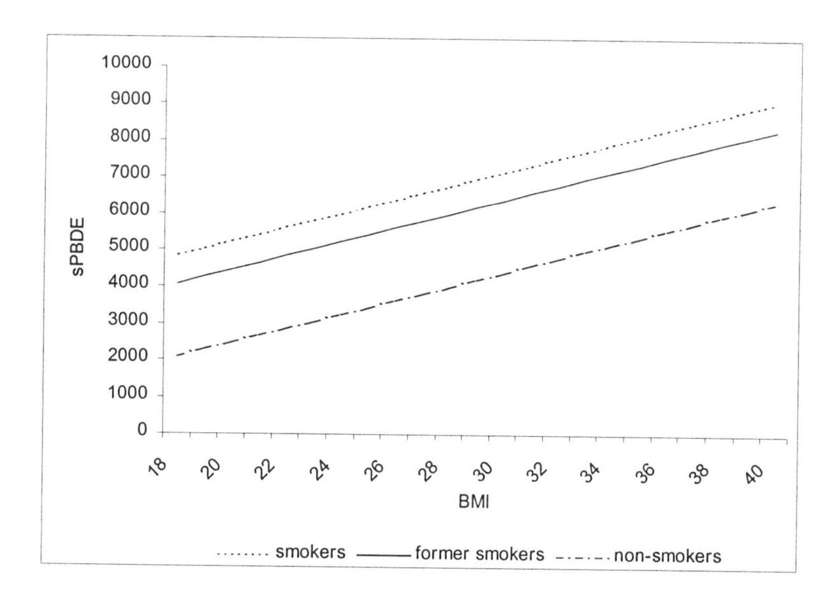

Figure 3. The relationship between the level of ΣPBDE (pg/g fat) in breast milk and body mass index (BMI) for smokers (n=10), former smokers (n=5) and non-smokers (n=24) using a multiple regression model.

Conclusion

An analytical method normally used for the analysis of PCBs and chlorinated pesticides in human milk has been modified and validated for the analysis of five of the most frequently found PBDE congeners. The 39 human milk samples investigated all have measurable levels of PBDEs. The knowledge on sources of human exposure is, however, very limited. Few data are available on dietary levels (4) and almost nothing is known about other types of exposure. The possible association between PBDE levels and smoking is notable, but it is possible that this association is due to other factors, not examined in this study, which can influence both the PBDE levels and smoking.

Acknowledgements

We would like to thank the Swedish Environmental Protection Board for financial support and Eva Jakobsson, Department of Environmental Chemistry, Stockholm University, Sweden for providing PBDE standards. Lena Hansson is acknowledged for technical assistance.

References

1. ICPS (WHO), Environmental Health Criteria 162, *Polybrominated Diphenyl Ethers*, **1994**, ISBN 92-4-157162-4.
2. KEMI (Swedish National Chemicals Inspectorate), Report no. 9/94, *Risk Assessment of Polybrominated Diphenyl Ethers*, **1994**, ISSN 0284-1185.
3. Pijnenburg, A. M. C. M.; Everts, J. W.; de Boer, J.; Boon, J. P. *Rev. Environ. Contam. Toxicol.* **1995**, 141, p 1.
4. TemaNord, *Polybrominated Diphenyl Ethers: Food Contamination and Potential Risks*, no. 503, **1998**, ISBN 92-893-0150-3.
5. Meironyté, D.; Bergman, Å; Norén, K. *Organohalogen Compd.* **1998**, 35, p 387.
6. Norén, K.; Meironyté, D. *Organohalogen Compd.* **1998**, 38, p 1.
7. Krüger, C. (Thesis), *Polybrominated Biphenyls and Polybrominated Diphenyl Ethers - detection and quantification in selected foods* (in German), Univ. of Münster, **1988**.
8. Sjödin, A.; Jakobsson, E.; Kierkegaard, A.; Marsh, G.; Sellström, U. *J. Chromatogr. A* **1998**, 822, p 83.
9. *Assessment of Human Exposure to Selected Organochlorine Compounds Through Biological Monitoring*; Slorach, S. A.; Vaz, R., Eds.; UNEP/WHO (Uppsala, Sweden: National Food Administration) **1983**.
10. Atuma, S. S.; Aune, M. *Bull. Environ. Contam. Toxicol.* **1999**, 62, p 8.
11. Atuma, S. S.; Hansson, L.; Johnsson, H.; Slorach, S.; de Wit, C.; Lindström, G. *Food Add. Contam.* **1998**, 15, p 142.

Chapter 18

Characterization of Q1, an Unknown Major Organochlorine Contaminant in the Blubber of Marine Mammals from Africa, the Antarctic, and Other Regions

W. Vetter

Friedrich-Schiller-Universität Jena, Lehrbereich Lebensmittelchemie, Dornburger Strasse 25, D–07743 Jena, Germany

Q1, an unknown organochlorine component with the molecular formula $C_9H_3Cl_7N_2$ was identified in tissues of marine mammals and birds from different regions. Full scan mass spectra of Q1 were recorded with gas chromatography and electron capture negative ionization mass spectrometry (GC/ECNI-MS) and electron impact ionization mass spectrometry (GC/EI-MS). Both techniques gave high response for a heptachloro isotope pattern starting at m/z 384 which is most likely the molecular ion. GC/ECNI-MS and GC/EI-MS methods for screening environmental samples for Q1 were developed on the basis of the selected ion monitoring (SIM) technique. The methods exhibited excellent selectivity but GC/ECNI-SIM-MS was two orders of magnitude more sensitive to Q1 than GC/EI-SIM-MS. Neither isomers nor homologues (Cl_6 - Cl_{10}) of Q1 were identified in the samples. Using GC/ECNI-SIM-MS, Q1 was detected as a major contaminant in seal samples from Africa and different seal and bird species from the Antarctic. The Q1 concentrations in lung, liver, kidney, blubber of a South African fur seal were comparable with p,p′-DDE. Low Q1 concentrations were determined in blubber of harbor seals from the North Sea, gray seals from the Baltic Sea, and two beluga whales from the Canadian Arctic. Q1 was below the detection limit in blubber of ringed seals from Spitsbergen (European Arctic) and the Canadian Arctic. Q1 represents the unique case of an organochlorine contaminant which was present in Antarctic samples at two order of magnitude higher concentrations than in samples from the European part of the Arctic as well as the otherwise highly polluted areas of the Baltic and the North Sea. Our investigations indicate that Q1 is a natural organohalogen compound.

Introduction

The determination of persistent and accumulative organochlorine pollutants in the environment and the evaluation of their toxic potential is an important task of analytical and environmental chemists, as well as governmental authorities. The threat of these substances is reflected by the fact that several chlorinated hydrocarbons rank upon on the priority list of hazardous substances issued in 1993 by the Agency for Toxic Substances and Disease Registry, ATSDR [1].

During the past decades, these substances have been mentioned in connection with serious environmental impacts. For example, DDE was made responsible for the egg thinning of certain bird species observed in the 1970s [2], while DDT and PCBs were suspected to cause reproductive failure of seals [3][4]. Pseudohermaphrodotism of polar bears from Svalbard (European Arctic) was correlated with high PCB concentrations [5] and developmental abnormalities of the gonad and abnormal sex hormone concentrations were attributed to high DDT concentrations in juvenile alligators from Florida [6].

During the last decade, more sensitive and selective analytical methods and instruments have been developed. It was found that selected organochlorine components which are present only at traces in organochlorine extracts may contribute significantly to the toxic potential of an environmental sample. With increasing evidence for synergetic effects of organochlorines, the identification of new organochlorine contaminants becomes more and more important.

Here, we report on an organochlorine contaminant which we recently discovered in samples from the southern hemisphere [7]. The heptachloro contaminant was labeled Q1. Q1 was among the dominating contaminants in blubber of seals from Namibia (Africa) [7] and the Antarctic. High Q1 concentrations were also found in eggs of predatory birds from the Antarctic [8]. For example, Q1 concentrations in eggs of brown skuas (*Catharacta antarctica lonnbergi*) were estimated to reach up to 126 µg/kg wet weight or 3 mg/kg lipid content [8]. A GC/MS method for the monitoring of Q1 in samples and an initial study of its environmental relevance is presented.

Material and Methods

Gas chromatography/mass spectrometry

Gas chromatography in combination with electron capture negative ionization mass spectrometry (GC/ECNI-MS) was performed on an HP 5989B mass spectrometer (Hewlett-Packard). In the GC/ECNI-MS mode, methane was used as the moderating gas. The interface transfer line was heated at 270°C, the ion source and the quadrupole were set at 150°C and 100°C, respectively. Two GC columns were used. The first one was coated with very non-polar, squalane-like CP-Sil 2 (column length 50 m, internal diameter 0.25 mm, film thickness 0.25 μm; manufacturer: Chrompack, Middelburg, The Netherlands). The GC oven program was the following: 80°C (1 min), 25°C/min to 160°C, 2°C/min to 240°C, 20°C/min to 272°C (14.2 min). The second was coated with a chiral stationary phase, which consisted of 25% randomly *tert*.-butyldimethylsilylated ß-cyclodextrin (ß-BSCD) diluted in PS086 (column length 30 m, internal diameter 0.25 mm, film thickness 0.20 μm; manufacturer: BGB Analytik, Adliswil, Switzerland). The GC temperatures were 80°C (4 min), 180°C (15 min), 200°C (25 min), 230°C (15 min) with ramps of 20°C/min between each plateau. The samples were splitless injected (splitless time 2 min) at 250°C. The separation characteristics of organochlorines on the two GC columns have been presented earlier [9][10]. In the selected ion monitoring (SIM) mode, three time windows, each with six ions, were recorded for the ß-BSCD column: The second time window (24-38 min) contained m/z 384, m/z 386, and m/z 388 for Q1, m/z 442 and m/z 444 for trans-nonachlor, as well as m/z 302. To screen for isomers and homologues of Q1, the two most abundant ions of the hexa- to decachloro fragment ions of the respective homologues of Q1 were monitored from 15.5 - 66.5 min, i. e. m/z 352 and m/z 354 for hexa- (molecular weight 350), m/z 386 and m/z 388 for hepta- (molecular weight 384), m/z 420 and m/z 422 for octa- (molecular weight: m/z 418), m/z 454 and m/z 456 for nona- (molecular weight m/z 452), and m/z 490 and m/z 492 for decachloro isomers (molecular weight m/z 486). In the full scan mode, m/z 50 to m/z 550 were monitored at a threshold of 100 and a rate of 0.8 scans/s. All analysis were run at a constant helium flow of 1.36 mL/min.

Gas chromatography in combination with electron impact ionization mass spectrometry (GC/EI-MS) was carried out on HP 5989B, HP 5971 MSD (Hewlett-Packard), and Saturn 2000 ion trap (Varian) instruments.

On the HP 5989B instrument a full scan mass spectrum was recorded in the range of m/z 50 to m/z 550 at a threshold of 100 and a rate of 0.8 scans/s. Both a CP-Sil 2 and a ß-BSCD column was used for these investigations. Furthermore, this instrument was applied to study the sensitivity of both MS ionization techniques (see below). In the SIM mode, two time windows were monitored on the HP 5989B instrument. The second one covered m/z 384, m/z 386, and m/z 388 for Q1, m/z 409 and m/z 411 for nonachlor isomers, as well as m/z 302.

On the HP 5971 instrument we monitored just the masses of the second mentioned time window. The GC stationary phase consisted of 10% permethylated ß-cyclodextrin chemically bonded to dimethylpolysiloxane (Chirasil-Dex, Chrompack, Middelburg, The Netherlands). The GC oven temperature was set at 120°C (2 min), then it was raised at 20°C/min to 180°C (15 min), then at 5°C/min to 210°C (25 min), and finally at 10°C/min to 230°C (5 min). Helium was the carrier gas at a column head pressure of 0.8 bar.

A full scan spectrum (m/z 50-650) was recorded with the Saturn 2000 ion trap (Varian) at an ion source temperature of 150°C. A DB-5 column was installed which was programmed in the following way: 80°C, at 50°C/min to 200°C, at 15°C/min to 320°C which was held for 11 min.

Gas chromatography/electron capture detection (GC/ECD)

Separations were carried out with an HP 5890 series II gas chromatograph (Hewlett-Packard) equipped with two capillary columns and two electron capture detectors (ECDs). Helium, maintained at a constant flow of 1.3 mL/min, was used as the carrier gas and nitrogen as the make-up gas. The injector and detector temperatures were 250°C and 300°C, respectively. The 50 m length x 0.25 mm internal diameter fused silica capillary columns were coated with 0.25 µm CP-Sil 2 and CP-Sil 8/20% C18 (both Chrompack, Middelburg, The Netherlands). The GC oven program was as follows: 60°C (1.5 min), 40°C/min to 180°C (5 min), 2°C/min to 230°C (0 min), and 5°C/min to 270°C (15 min).

Biological samples

Blubber of South African fur seals (*Arctocephalus pusillus pusillus*) was available from 11 individuals collected in April 1997 at Cape Cross, Namibia (i. e. in the Southwest of Africa) [7]. Blubber, kidney, liver, and lung tissue was available from one individual sampled in 1996. Blubber of 5 Antarctic seal species was collected in January 1994 near Base Cientifica Argentina JUBANY (62°14`18`` S, 58°40` W) [11]. Selected Weddell seals (*Leptonychotes Weddelli*) were available from individuals collected in 1985 [12] and 1990 [11]. A harbor porpoise (*Phocoena phocoena*) was caught at the Icelandic coast. The blubber of the harbor seal (*Phoca vitulina*) was from an animal caught in 1995 in Western Iceland. Another harbor seal (*Phoca vitulina*) was from the German part of the North Sea. Blubber of a gray seal (*Halichoerus grypus*) was from an individual stranded at the German coast of the Baltic Sea. The sample of ringed seal blubber (*Phoca hispida*) from Spitsbergen (European Arctic) was a mixture of 11 individuals. Blubber of two ringed seals (*Phoca hispida*) and two beluga whales (*Delphinapterus leucas*) was available from individuals sampled in Canada.

Sample clean-up

Two very different sample clean-up methods were applied. For detailed description of the methods and information on the chemicals and solvents used see our earlier presentations [13][14]. In brief, the first method includes sample digestion with acids, liquid-liquid extraction with n-hexane, repeated treatment with sulfuric acid, and adsorption chromatography on silica [13]. The accuracy of the method (PCBs and chloropesticides) was verified in an intercalibration exercise [15]. This method was used for all samples except those from the Baltic

Sea and Canada (see above) which were cleaned with the second method. This method combines microwave-assisted extraction with gel-permeation chromatography, and adsorption chromatography on silica [14]. The accuracy of this method was checked with the analysis of the standard reference material cod liver oil (SRM 1588) [8]. The African fur seal blubber samples from 1997 were analyzed with both techniques. PCB/toxaphene group separations were performed using the method of Krock et al. [16] in a slightly modified way [17].

Results and Discussion

Identification of Q1 in environmental samples

Figure 1 shows the GC/ECD chromatogram of the blubber extract of a South African fur seal (*Arctocephalus pusillus pusillus*) sampled in 1996 at Cape Cross/Namibia. Q1 and p,p′-DDE were the most abundant signals in the ECD-chromatogram. The retention time of Q1 on DB-5-like (95% methyl, 5% phenyl polysiloxane) columns was similar to trans-nonachlor ($\Delta t_R \approx 0.1$ min), so that these two components may even co-elute under certain chromatographic conditions. However, they can be distinguished with MS because they form different molecular and fragment ions.

Due to the similar behavior on silica (see below) and similar GC retention times on different stationary phases, we estimated Q1 concentrations in samples by using the ECD response factor of a trans-nonachlor reference standard [7]. In contrast to GC/MS, GC/ECD response factors of organochlorine components with comparable GC retention time and a degree of chlorination ranging from Cl_5 to Cl_9 usually differ only by a factor of two. This was checked with standard solutions of the components below and others. **Table I** lists concentrations of Q1 and further chlorinated hydrocarbons in the blubber of South African fur seals (*Arctocephalus pusillus pusillus*).

Table I. Concentrations of Organochlorines (µg/kg) in the Blubber of South African Fur Seals (*Arctocephalus pusillus pusillus*) from Namibia

Compound	*#176*	*#174*	*#183*	*#184*	*#177*	*#175*	*#182*	*#178*	*#180*	*#181*	*#179*
sex, age*	f,2	f,3	f,4	f,5	f,5-6	m,2	m,3	m,4	m,5	m,5	m,6
Q1	314	73	318	136	28	164	351	120	273	172	11
tr-no [7]	40	8	54	24	2	23	28	15	33	148	17
ΣCD** [7]	59	9	54	34	3	38	44	22	33	159	23
ΣPCBs [7]	55	24	101	35	4	35	65	22	83	697	25
ΣDDT [7]	266	53	361	148	11	173	216	112	424	1115	155
ΣCTTs [7]	30	14	31	40	10	63	54	29	97	40	35

* sex (f = female, m = male), age in years

** ΣCD = chlordane = cis- and trans-chlordane + cis- and trans-nonachlor

ΣDDT = p,p′-DDT + p,p′-DDE + p,p′-DDD

ΣPCB = PCB 101 + 118 + 138 + 149 + 153 + 163 + 170 + 180

ΣCTTs = B7-1453 + B8-1412 + B8-1413 (P-26) + B8-1414 (P-40) + B8-1945 (P-41) + B8-2229 (P-44) + B9-1025 (P-62) + B9-1679 (P-50)

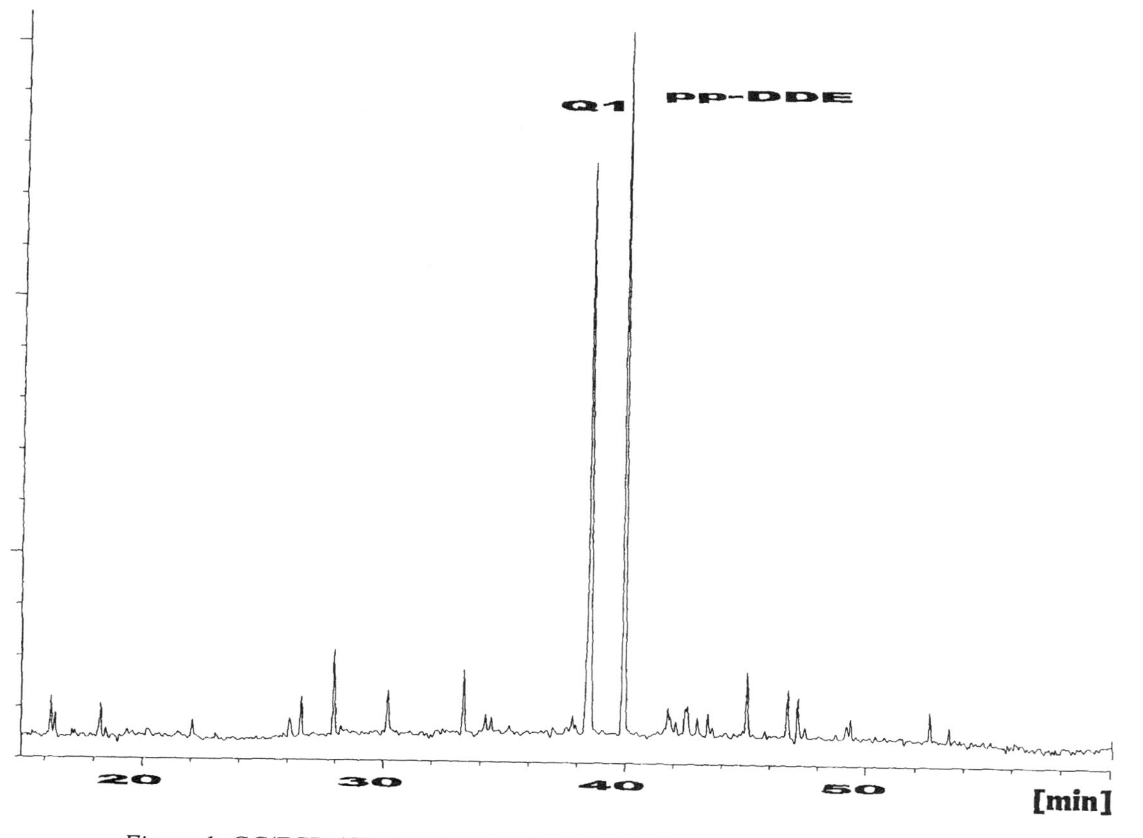

Figure 1. GC/ECD (CP-Sil 8/20%C18) of the blubber extract of a South African fur seal (Arctocephalus pusillus pusillus).

The most of the samples contained Q1 at concentrations on the same level or higher than PCBs and DDT. This clearly confirms the environmental relevance of Q1 at this site. Q1 concentrations did neither correlate with DDT, PCB, and trans-nonachlor concentrations, nor with those of chlordane, toxaphene, and dieldrin [7]. Therefore, it was concluded that Q1 is not the metabolite or the highly accumulated minor component of any of these substance classes.

Mass Spectrometric Analysis of Q1

From the early elution on non-polar GC columns it was concluded that Q1 cannot have a degree of chlorination which is not covered within the monitored scan range (upper limit on HP 5989B: m/z 550). The GC/ECNI mass spectrum of Q1 (**Figure 2**) is dominated by a fragment ion starting at m/z 384 which shows a heptachloro isotope pattern and is most likely caused by the molecular ion since there were no other fragment ions at higher m/z values. Furthermore, a molecular weight of 384 Da is within the range of components with similar GC retention time such as p,p′-DDE (318 Da) and trans-nonachlor (440 Da).

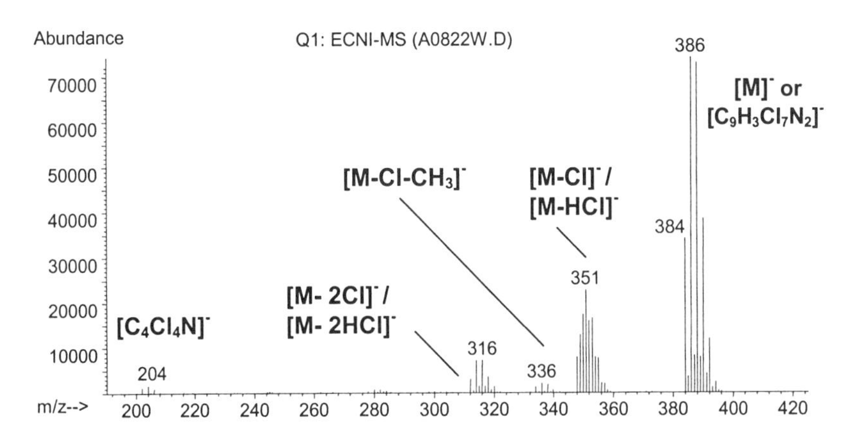

Figure 2. GC/ECNI-MS full scan mass spectrum of Q1

Fragment ions starting at m/z 348 and m/z 349 (ratio approximately 2 : 3) correspond to the elimination of HCl and Cl, respectively, from the major ion cluster at m/z 384. The ion cluster at m/z 312 is a mixture of two pentachloro isotope patterns, one resulting from the loss of 2 HCl (m/z 312), the other from the loss of Cl_2 (m/z 314) from the molecular ion.

Subtraction of the contribution of chlorine atoms (7 x ^{35}Cl = 245 amu) from the molecular ion at m/z 384 leaves 139 amu for other elements. On the basis of the ratio of $^{12}C/^{13}C$ the number of carbons was determined as 10±1, but this technique did not allow to distinguish between these possibilities. A backbone with nine or ten carbons is not possible without hetero atom because this contradicts the C_nX_{2n+2} (X = Cl, H) rule. A component with only one nitrogen would result in an odd molecular weight.

Br was excluded as a substituent because there was no peak at m/z 79/81 nor any fragment corresponding to the elimination of Br. On the other hand, two oxygens are not possible either: 139 amu – (2x 16 amu) leaves 107 amu which is < C_9. Consequently, only three molecular formulas were plausible, i.e. $C_{11}H_7Cl_7$, $C_{10}H_3Cl_7O$ and $C_9H_3Cl_7N_2$. Recently, high resolution electron impact ionization mass spectrometry (HREI-MS) provided the information that the molecular formula of Q1 is $C_9H_3Cl_7N_2$ [18].

There are two lowly abundant fragment ions in the GC/ECNI-MS of Q1. The first one starts at m/z 334 or $[M-50]^-$ which is formed by consecutive loss of Cl (35 amu) and CH_3 (15 amu) from the molecular ion. Presence of a methyl group on Q1 was confirmed in EI-MS experiments (see below). The second one at m/z 202 may be interpreted as $[C_5H_2Cl_4]^-$ or $[C_4Cl_4N]^-$. The latter is much more plausible as Q1 possesses a methyl group (see below).

A highly concentrated solution which was obtained after liquid chromatographic enrichment of Q1-containing samples on silica was used to record GC/EI full scan mass spectra of Q1 (**Figure 3**). The EI-MS recorded with the same instrument as the ECNI-MS (**Figure 3a**) also showed high abundance for the molecular ion starting at m/z 384 (a radical cation, which supports that it is the molecular ion). The fragmentation pattern was, however, different to that of GC/ECNI-MS. There was no loss of HCl from the molecular ion but subsequent loss of Cl yielded the more abundant ion clusters at m/z 349 ($C_9H_3Cl_6N_2$) and m/z 314 ($C_9H_3Cl_5N_2$). Furthermore, at low abundance the following sequence with subsequent elimination of chlorine respectively was identified: m/z 369 ($C_8Cl_7N_2$) → m/z 334 ($C_8Cl_6N_2$) → m/z 299 ($C_8Cl_5N_2$) → m/z 264 ($C_8Cl_4N_2$) (**Figure 3b**). Formation of m/z 369 by elimination of CH_3 from the molecular ion was recently proven by application of HREI-MS [18]. Important fragment ions start at m/z 192 and m/z 174 which represent the $[M]^{2+}$ and $[M-Cl]^{2+}$ ions. In these ions, the chlorine isotopic peaks are only separated by 1 amu. The fragment ion at m/z 174 is the doubly-charged analogue of m/z 349 (the quadrupole mass resolution is 1 amu, and the software rounds m/z values to full mass units: the correct mass would be m/z 348.8/2 = m/z 174.4 which is rounded to m/z 174 [18]). Doubly-charged fragment ions are typically of components with an aromatic system and usually not present in mass spectra of alicyclic compounds. The "ring and double bond rule" reveals that Q1 possesses 6 ring or double bonds. The fragment ions at m/z 132 (C_4Cl_2N), **Figure 3a**, and m/z 238 (C_7Cl_4N), **Figure 3b**, are formed by partial degradation of the Q1-backbone.

GC/EI full scan mass spectra were recorded on two further mass spectrometers (HP 5971 and Saturn 2000). These two instruments also showed high abundance for the $[M]^+$, $[M-Cl]^+$, and $[M-2Cl]^+$ fragment ions but these ions were present in different ratio (**Table II**).

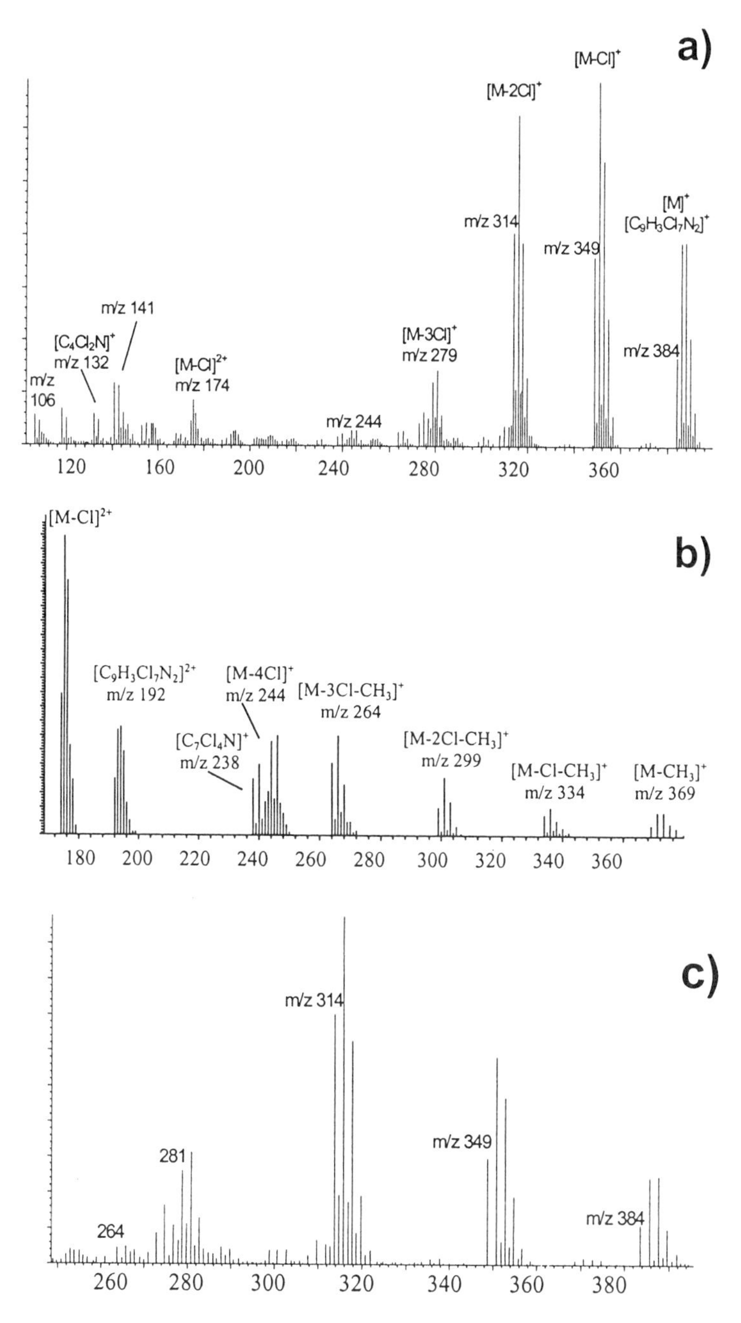

Figure 3. GC/EI-MS full scan spectrum of Q1. (a) recorded with the HP 5989B instrument. (b) low abundant mass fragments of Q1. Higher abundant fragment ions are omitted excerpt from the upper MS. (c) recorded with the HP 5971 instrument

Table II: Relative abundance (%) of the most abundant ion of three ion clusters of Q1 in three different mass spectrometers

	Ion source temperature	m/z 388	m/z 351	m/z 316
HP 5989 B	150°C	22.9	40.5	36.6
HP 5971	200°C	13.8	25.4	60.8
Saturn 2000	150°C	38.6	32.5	28.9

The GC/EI-MS recorded with the HP 5971 instrument showed the same fragment ions - e. g. m/z 369 was confirmed - but the highest intensity was found for the pentachloro pattern at m/z 314 (**Figure 3c**). On the other hand, the EI-MS recorded with the Saturn 2000 ion trap showed highest abundance for the molecular ion (**Table II**), but it showed no doubly-charged fragment ions.

Although an elucidation of the structure of Q1 was not possible with GC/MS, these investigations provided the information that Q1 is not identically with any known organochlorine pollutant. To date, highly chlorinated nitrogen-containing compounds such as Q1 ($C_9H_3Cl_7N_2$) have not been reported but recently, Tittlemier et al. detected a $C_{10}H_6Br_4Cl_2N_2$ component in eggs of birds from Canada [19]. This component was characterized as a natural organohalogen compound with a N,N-dimethylbipyrrole backbone. A similar component, a hexabromobipyrrole, has been discovered in the 1970s [20]. The similarity with both components and the lack of any component with the molecular formula of Q1 which has been produced suggests that Q1 is also a natural organohalogen compound, namely a heptachloro,N-methylbipyrrole. This is in agreement with evaluations of Gribble [21]. Fragment ions such as m/z 202 in GC/ECNI-MS and m/z 132 in GC/EI-MS support this interpretation. Formation of halogenated (bi)pyrroles from terrestrial and marine sources had been expected due to the enormous reactivity of these heterocycles in electrophilic substitution [20].

GC/MS detection of Q1 in environmental samples

Figure 4 shows GC/ECNI-MS full scan chromatograms of three Antarctic seal species sampled at the same location. Due to the varying ECNI-MS response factors of the components, the peak heights do not necessarily reflect the concentrations of the different components. For example, p,p′-DDE which dominated the GC/ECD chromatograms of the respective samples (data not shown) has a very low ECNI response factor. Nevertheless, the peak patterns in **Figure 4a-c** clearly demonstrate species-specific differences in the accumulation of Q1. In the blubber of the Antarctic fur seal (*Arctocephalus gazella*) (**Figure 4b**), Q1 was the most abundant peak in the ECNI full scan chromatogram.

With the information derived from the full scan mass spectra we developed GC/ECNI-MS- and GC/EI-MS-SIM methods to screen environmental samples for Q1. Usually, measurements in the SIM mode are based on the most abundant ion trace and a second one from the same chloro isotope pattern for confirmation of the results.

The GC/ECNI-MS mass spectrum was dominated by the fragment ion starting at m/z 384 (see **Figure 2**), and therefore, we suggest to choose m/z 386 and m/z 388 for GC/ECNI-MS- and GC/EI-MS-SIM studies on Q1. These ions must be present in the natural ratio of heptachloro isotope patterns, i. e. m/z 386/m/z 388 = 1.06. At this initial stage of collecting data for Q1, however, we additionally monitored the mass fragment m/z 384. The three suggested m/z values were almost exclusively formed by Q1. One out of two additional components which gives response for two of the SIM masses is heptachlor epoxide which, however, lacks m/z 384 in its mass spectrum. According to Stemmler and Hites, endosulfan cyclic sulfate also shows an abundant mass fragment starting at m/z 384 with an hexachloro isotope pattern [22] which can be easily distinguished from the heptachloro isotope pattern of Q1. **Figure 5** gives an example for the high selectivity of the suggested GC/ECNI-MS method. While Q1 is only a minor component in the GC/ECNI-MS full scan chromatogram (**Figure 5a**), it is the dominating component using the proposed SIM technique (**Figure 5b**).

The different ratios of the major mass fragments in the GC/EI mass spectra recorded on three mass spectrometers deserve some discussion. Although the HP 5989B instrument showed highest abundance for the $[M\text{-}Cl]^+$ fragment ion starting at m/z 349 (see **Figure 3**), a more noisy baseline was obtained for m/z 349 and m/z 351, and the slightly higher abundance of the $[M\text{-}Cl]^+$ fragment ion was overcompensated by the worse signal-to-noise ratio in comparison with the molecular ion at m/z 384.

On the other hand, m/z 316 (which dominated in the GC/EI-MS recorded on the HP 5971 instrument) is less specific since this m/z value is also formed by DDE and other organochlorine components. Therefore, we suggest to use m/z 384, m/z 386, and m/z 388 for both ECNI-MS and EI-MS. However, the ECNI method was approximately two orders of magnitude more sensitive than the EI-MS method. This was investigated with the same instrument on consecutive runs using identical GC and MS conditions except a moderation gas in the ion source in the earlier case (data not shown). Therefore, ECNI-MS-SIM combines the high selectivity with high sensitivity for Q1. Despite a lower sensitivity in the EI-MS mode, the high selectivity of the SIM masses was also given with this technique (**Figure 6**), but samples with particularly very low Q1 concentrations required the ECNI-MS technique. GC/ECNI-MS-SIM is therefore the method of choice for the screening of samples for Q1 in the selected ion monitoring mode.

As already mentioned, no further components were found with the typical abundance ratio of m/z 384, m/z 386, and m/z 388 of the molecular ion of Q1. In one experiment, m/z values corresponding to Cl_6 to Cl_{10} homologues of Q1 were monitored which differ from m/z 384 by Δ34 amu, respectively (see Material and Methods). However, no response was found in the correct isotopic ratio of the monitored ion traces. Hence, there were no homologues of Q1 in the investigated samples and Q1 is most likely a single component.

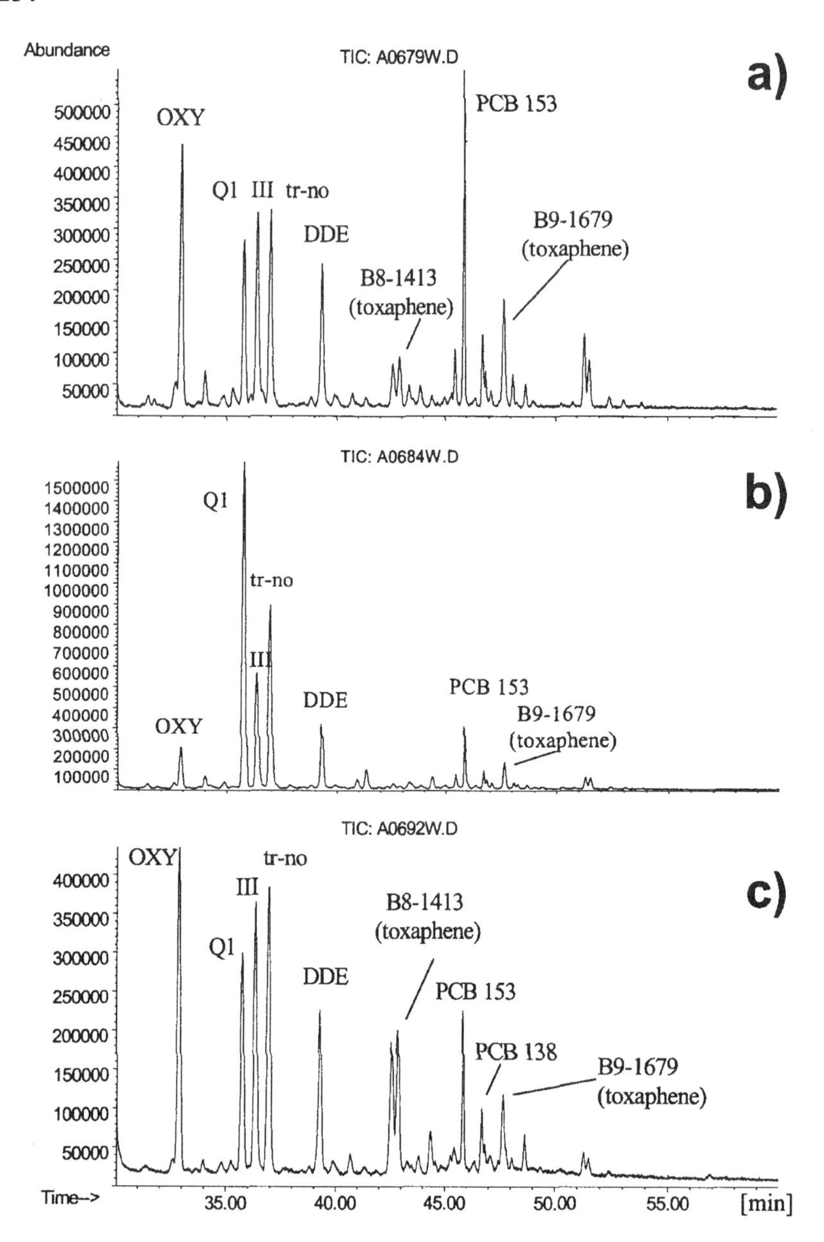

Figure 4. GC/ECNI full scan chromatograms (CP-Sil 2) of blubber extracts of Antarctic seals. (a) Leopard seal (Hydrurga leptonyx); (b) Antarctic fur seal (Arctocephalus gazella); (c) Elephant seal (Mirounga leonina)
The most important contaminants of PCBs (PCB 153, PCB 138), DDT (DDE = p,p′-DDE), toxaphene (B8-1413 = 2-endo,3-exo,5-endo,6-exo,8,8,10,10-octachlorobornane or Parlar #26; B9-1679 = 2-endo,3-exo,5-endo,6-exo,8,8,9,10,10-nonachlorobornane or Parlar #50), chlordane (tr-no = trans-nonachlor, OXY = oxychlordane, III = nonachlor III or MC 6) are labeled.

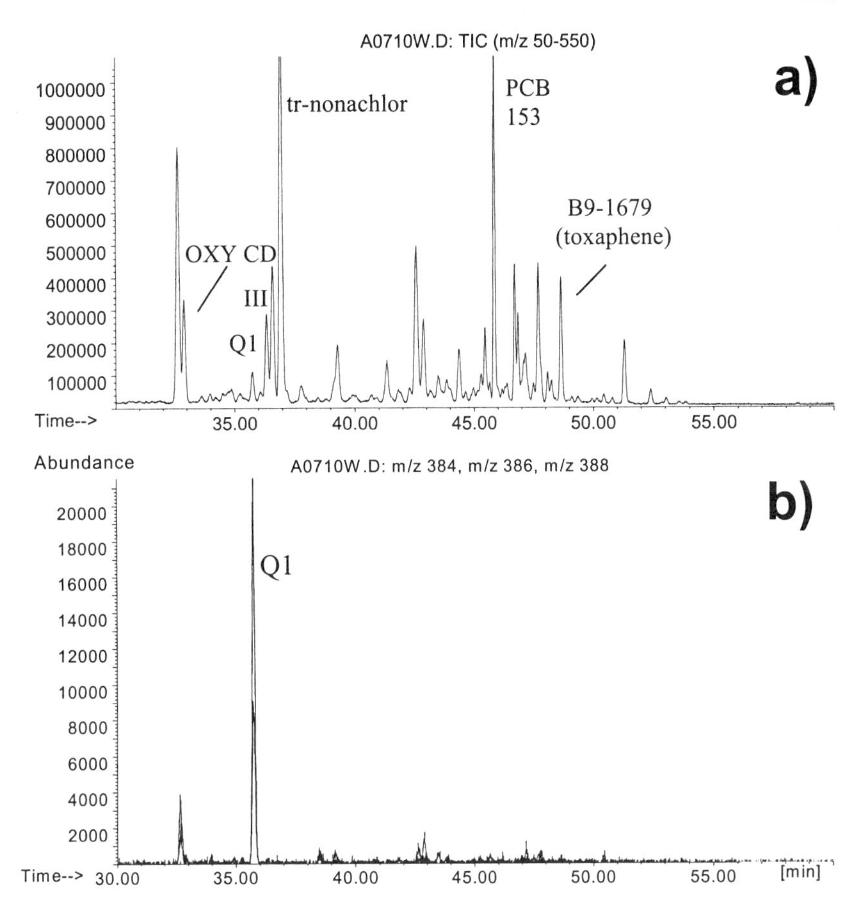

Figure 5. GC/ECNI-MS (CP-Sil 2) chromatograms of the blubber extract of an Icelandic harbor porpoise (Phocoena phocoena). (a) Full scan mode. (b) Extraction of m/z 384, m/z 386, and m/z 388 from the full scan chromatogram demonstrates the high sensitivity and high selectivity of the method for Q1. Abbreviations of the components are explained in Figure 4.

The SIM method allowed to check for Q1 even in samples with much higher concentrations of PCBs, DDT, and other organochlorines. However, due to the significantly varying response factors of organochlorines in ECNI-MS, we are not able to present quantitative concentrations of Q1 this time. Nevertheless, very low ECNI-MS response of Q1 was found in harbor seals (*Phoca vitulina*) from the German North Sea and in gray seals (*Halichoerus grypus*) from the Baltic Sea. Q1 was below the detection limit in blubber of ringed seals (*Phoca hispida*) from Spitsbergen and a harbor seal (*Phoca vitulina*) from Western Iceland which

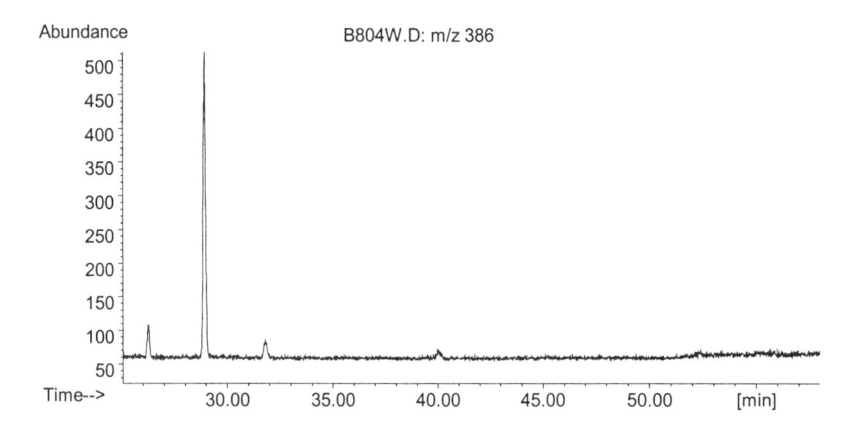

Figure 6. GC/EI-MS-SIM (m/z 386) chromatogram (ß-PMCD) of the blubber extract of an Antarctic fur seal (Arctocephalus gazella). The same sample extract as shown in Figure 4b. This demonstrates the high selectivity but low sensitivity for Q1 in the electron impact ionization mode.

is in contrast to low Q1 concentrations in an Icelandic harbor porpoise sample (see **Figure 5**). A similar result was found for samples from Canada. While two ringed seal (*Phoca hispida*) blubber samples showed no Q1, low concentrations of Q1 were found in the blubber of two belugas (*Delphinapterus leucas*). This indicates that selected species of marine mammals are able to metabolize Q1 to some extend. This seems to be more pronounced for seals than for cetaceans which is in agreement with other components such as HCHs, PCBs, and toxaphene [11].

In addition to the species mentioned in **Figure 4**, Q1 was also detected in blubber of further seal species from the Antarctic (Weddell seals, *Leptonychotes Weddelli*, and crabeaters, *Lobodon carcinophagus*). Interestingly, Q1 was also abundant in samples collected in 1985 so that Q1 is not a "new" component. However, the sample size was low and the earlier samples were from another location so that we cannot present time trends in the Q1 pollution of Weddell seals. Recently, eggs of different skua species were analyzed [8]. South polar skuas (*Catharacta maccormicki*) which spend the winter in the northern hemisphere accumulated higher PCB but lower Q1 concentrations than brown skuas (*Catharacta antarctica lonnbergi*) which stay permanently in the Southern Ocean [8][23]. These results and low concentrations in the samples from Canada support the assumption that Q1 is only a minor component in North America.

Different tissue samples (blubber, liver, kidney, lung) were available from one South African fur seal (*Arctocephalus pusillus pusillus*) from the Namibian coast. The ECD chromatogram of this individual was dominated by the peaks of p,p′-DDE and Q1 (see **Figure 1**). Based on the fresh weight, Q1 and p,p′-DDE concentrations were 286/280 µg/kg in blubber, 55/45 µg/kg in kidney, 8/5 µg/kg in liver and 1.4/1.2 µg/kg in the lung. These preliminary results suggest that Q1 has a similar tendency to bioaccumulate in seals as it was reported for p,p′-DDE and other persistent organochlorines [24].

Chemical and physical properties of Q1

Two different sample clean-up procedures were applied to the different samples, whereas blubber of the South African fur seal sampled in 1997 was cleaned-up with both techniques. The concentrations obtained with the two methods agreed well, and there was no degradation of Q1 after treatment with strong and oxidative acids which were used in one of the clean-up procedures (see Material and Methods). In several cases, we looked for Q1 in sample extracts obtained after a separation of aromatic and alicyclic components (PCB/toxaphene). Liquid chromatographic techniques based on silica [16][25] distributed Q1 into both fractions. Therefore, we recommend screening and quantification of Q1 in samples before any fractionating of the organochlorine extracts. Furthermore, this behavior points against a pure aromatic compound. This is further supported by the ring and double bond rule (see above).

Many organochlorine components are chiral and gas chromatography with chiral stationary phases is a suitable technique to study the enantioselective fate of these components in the environment [26][27]. To investigate the chirality of Q1 we applied the two chiral stationary phases ß-BSCD (installed in the HP 5989B system) and ß-PMCD (installed in the HP 5971 system). About 90% of the chiral organochlorines have been resolved into enantiomers at least on one of these two chiral stationary phases [25]. However, the signal of Q1 remained unresolved. Work is ongoing to determine if Q1 is chiral or not.

Conclusion

The purpose of this study was to summarize and upgrade our present knowledge on Q1. The estimated concentrations in selected samples point to a serious environmental threat which needs to be investigated in detail. According to the present data, Q1 concentrations are significantly higher in the southern than in the northern hemisphere. The lack of Q1 in ringed seal blubber samples from the Arctic (Spitsbergen) deserves discussion. Assuming that this is valid for other species as well, Q1 may be used to distinguish samples of the European Arctic (no abundance of Q1) from Antarctic samples (abundance of Q1) and Q1 may be used as a target substance to distinguish atmospheric transport into the Arctic and Antarctic.

The concentrations of Q1 in Antarctic samples exceeded those in seals from the European part of the Arctic, the Baltic Sea, the North Sea, and Canada at least by the factor 10. Therefore, Q1 represents the unique case of an organochlorine compound which is more abundant in the remote area of the Antarctic than in the otherwise highly polluted waters of Middle and Northern Europe. This is a further indication for Q1 being a natural organohalogen compound. Work is ongoing to elucidate the chemical structure of Q1 and its source.

Acknowledgment

I want to express my thanks to Elke Scholz, Marion Weichbrodt, and Doris Glotz for helping with the sample clean-up as well as Prof. Dr. Bernd Luckas for supporting this work. Many thanks also to the following colleagues and authorities who supplied the sample materials used in this study:

The Antarctic blubber samples were collected by Prof. Dr. Bernd Luckas, Klaus Reinhardt (University of Jena/Germany), and Dr. Jochen Plötz (Alfred-Wegener-Institut, Bremerhaven/Germany). The Icelandic harbor seal and the harbor porpoise sample was obtained from Dr. Karl Skírnisson (University of Iceland, Keldur/Iceland) with permission of the Icelandic Ministry of Fisheries. The German harbor seal was obtained from Dr. Günter Heidemann (boco foundation, Hamburg/Germany). The South African fur seal samples were obtained from Prof. Dr. Dres. h. c. Herbert Oelschläger (University of Jena/ Germany) with permission of the Ministry of Fisheries and Marine Resources of the Republic of Namibia. Dr. Derek C. G. Muir (Environment Canada, Burlington/Canada) is acknowledged for providing the Canadian blubber samples and Dr. Bernd Krock (MPI for Chemical Ecology, Jena/Germany) for the recording of the EI-MS of Q1 on the Saturn 2000 ion trap.

Literature Cited

[1] Agency for Toxic Substances and Disease Registry, ATSDR (1994). 1993 CERCLA priority list of harzardous substances that will be Subject for Toxicological Profiles and Support Document. U. S. Department of Health and Human Services, Public Health Service, Atlanta, GA, U. S. A.

[2] Longcore, J.R.; Samson, F.B.; Whittendale, Jr., T.W. *Bull. Environ. Contam. Toxicol.* **1971**, *6*, 485-490.

[3] Reijnders, P. J. H. *Nature* 1986, **324**, 456-457 & 418.

[4] Helle, E.; Olsson, M.; Jensen, S. *Ambio* **1976**, *5*, 188-189.

[5] Wiig, O.; Derocher, A. E.; Cronin, M. M.; Skaare, J. U. *J. Wildlife Diseases* **1998**, 34, 792-796.

[6] Guillette Jr., L. J.; Gross, T. S.; Masson, G. R.; Matter, J. M.; Percival, H. F.; Woodward, A. R. *Environ. Health Perspect.* **1994**, *102*, 680-688.

[7] Vetter, W.; Weichbrodt, M.; Scholz, E.; Luckas, B.; Oelschläger, H. *Mar. Poll. Bull.* **1999**, 38, 830-836.

[8] Weichbrodt, M.; Vetter, W.; Scholz, E.; Luckas, B.; Reinhardt, K. *Intern. J. Environ. Anal. Chem.* **1999**, 73, 309-328.

[9] Vetter, W.; Klobes, U.; Krock, B.; Luckas, B. *J. Microcol. Sep.* **1997**, *9*, 29-36.

[10] Vetter, W.; Klobes, U.; Luckas, B.; Hottinger, G. *Chromatographia* **1997**, *45*, 255-262.

[11] Vetter, W.; Krock, B.; Luckas, B. *Chromatographia* **1997**, *44*, 65-73.

[12] Luckas, B., Vetter, W., Fischer, P., Heidemann, G., Plötz, J. *Chemosphere* **1990**, *21*, 13-19.

[13] Vetter, W.; Natzeck, C.; Luckas, B.; Heidemann, G.; Kiabi, B.; Karami, M. *Chemosphere* **1995**, 30, 1685-1696.
[14] Vetter, W.; Weichbrodt, M.; Hummert, K.; Luckas, B.; Glotz, D. *Chemosphere* **1998**, *37*, 2435-2445
[15] Rimkus, G.; Rexilius, L.; Heidemann, G.; Vagts, A.; Hedderich, J. *Chemosphere* **1993**, *26*, 1099-1108.
[16] Krock, B.; Vetter, W.; Luckas, B. *Chemosphere* **1997**, *35*, 1519-1530.
[17] Klobes, U.; Vetter, W.; Luckas, B.; Hottinger, G. *Organohalogen Compd.* **1998**, *35*, 359-362.
[18] Vetter, W., Alder, L., Palavinskas, R., *Rapid Comm. Mass Spectrom.* **1999**, *13*, 2118-2124.
[19] Tittlemier, S. A.., Simon, M., Jarman, W. M., Elliot, J. E., Norstrom, R. J. *Environ. Sci. Technol.* **1999**, *33*, 26-33.
[20] Gribble, G. W. (1996). Progress in the chemistry of organic natural products. Springer-Verlag Wien-New York
[21] Gribble, G. W., personal communication to W. Vetter, Venice/Italy, September 1999.
[22] Stemmler, E.; Hites, R. (1988). Electron capture negative ion mass spectra of environmental contaminants and related compounds. VCH Publishers, Weinheim. (ISBN 0-89573-708-6)
[23] Furness, R.W.; Family Stercorariidae (skuas). In: J. del Hoyo, A. Elliott, J. Sargatal (eds.), *Handbook of the Birds of the World*, Lynx Publishing, Barcelona, **1996**, *3*, 556 - 571.
[24] Connell, D. W. *Rev. Environ. Contam. Toxicol.* **1988**, *102*, 117-154.
[25] Specht, W.; Tillkes, M. *Fresenius Z. Anal. Chem.* **1985**, *322*, 443-455.
[26] Vetter, W.; Schurig, V. *J. Chromatogr.* A, **1997**, *774*, 143-175.
[27] Hühnerfuss, H. *Organohalogen Compd.* **1998**, *35*, 319-324.

Author Index

Alter, John, 114
Atuma, S., 235
Auer, Charles M., 138
Aune, M., 235
Balk, Froukje, 100
Bennett, Deborah H., 29, 52
Bennie, Don, 184
Beyer, Andreas, 14
Bidleman, Terry F., 223
Blok, Han, 100
Boethling, Robert S., 138
Breton, Roger, 42
Canavan, Sheila, 114
Chénier, Robert, 42
Cnattingius, S., 235
Darnerud, P. O., 235
Davey, Kathy, 114
Fisk, Aaron, 184
Fort, Daniel L., 151
Freidig, Andreas P., 64
Fromme, Hermann, 203
Harner, Tom, 223
Hermens, Joop L. M., 64, 89
Hungerbühler, Konrad, 52
Jansson, Bo, 1
Jon, Jay H., 151
Jones, Kevin C., 75, 223
Karras, Greg, 124
Kastenberg, W. E., 29
Kwiat, James K., 151
Lee, Robert G. M., 223
Lipnick, Robert L., 1
Mackay, Donald, 1, 14
Matthai, Paul, 114
Matthies, Michael, 14
McKone, Thomas E., 29, 52
Moss, Kenneth T., 138
Muir, Derek, 184
Murray, Thomas, 114
Nabholz, J. Vincent, 138
Okkerman, Peter, 100
Otto, Thomas, 203
Petreas, Myrto, 1
Pilz, Konstanze, 203
Ralston, Mark D., 151
Ramos, Enaut Urrestarazu, 64
Sasnett, Sam, 114
Scheringer, Martin, 52
Sijm, Dick, 100
Stern, Gary, 184
Sweetman, Andrew J., 75
Teixeira, Camilla, 184
Tomy, Gregg, 184
Vaes, Wouter H. J., 64
van Loon, Willem M. G. M., 64
Verbruggen, Eric M. J., 64, 89
Verhaar, Henk J. M., 64
Vetter, W., 243
Wania, Frank, 14
Webster, Eva, 14
Wernroth, M. L., 235
Whittle, Mike, 184
Wicklund-Glynn, A., 235

Subject Index

A

Abbalide, concentrations in aquatic environment, 204–220
Accumulation estimation for complex mixtures of organic chemicals to aquatic organisms
 examples for petroleum products, 95–96
 experimental description, 90
 measurement approaches, 90–94
4-Acetyl-1,1-dimethyl-6-*tert*-butylindane, concentrations in aquatic environment, 204–220
6-Acetyl-1,1,2,3,3,5-hexamethylindane, concentrations in aquatic environment, 204–220
7-Acetyl-1,1,3,4,4,6-hexamethyltetraline, concentrations in aquatic environment, 204–220
5-Acetyl-1,1,2,6-tetramethyl-3-isopropyldihydroindane, concentrations in aquatic environment, 204–220
Advective flow, calculation, 54–55
Africa, Q1 concentrations, 244–257
Air, concentrations of short-chain chlorinated paraffins, 194–195
Aldrin, persistence evaluation, 35, 37–40
Analytical methods, short-chain chlorinated paraffins, 186, 188
Antarctic, Q1 concentrations, 244–257
Aquatic environment, polycyclic musk fragrances, 204–220
Aquatic organisms, estimation of accumulation and base-line toxicity for complex mixtures of organic chemicals, 89–97
Aquatic system contamination by nitro musks, evidence, 204
Aqueous solubility, measurement of hydrophobicity, 90
Arctic, Q1 concentrations, 244–257
Assessment, studies, 100–178
Atmosphere, concentrations of polychlorinated naphthalenes, 223–233
Atmospheric residence time, relationship with spatial range, 56–68

B

Baltic Sea, Q1 concentrations, 244–257
Base-line toxicity
 description, 90
 petroleum product, 94–95
Base-line toxicity estimation for complex mixtures of organic chemicals to aquatic organisms
 examples for petroleum products, 95–96
 experimental description, 90
 measurement approaches, 90–94
Base-line toxicity of complex mixtures
 measurement using biomimetic extraction, 91, 93*f*, 94
 measurement using hydrophobicity distribution profiles, 90–94
BCFWIN, identification of persistent, bioaccumulative, and toxic substances, 8

Behavior of polychlorinated biphenyls, fugacity-based Mackay-type level IV dynamic model, 76–86
Beluga whales
Q1 concentrations, 244–257
short-chain chlorinated paraffin concentrations, 196, 198, 199*f*
Benzene
spatial range vs. atmospheric residence time, 57–58
spatial range vs. persistence, 58–60
Bioaccumulation
criteria, 46–48
criterion for identification of persistent, bioaccumulative, and toxic substances, 104, 106*t*, 109
criterion for Toxic Substances Control Act New Chemicals Program, 139–148
estimation using negligible depletion solid-phase extraction, 64–71
experimental approaches, 64–65
measurement, 2–3
role in risk assessment, 64
short-chain chlorinated paraffins, 184–199
Bioaccumulation factor, priority chemicals, 16
Bioaccumulation of priority chemicals
bioaccumulation factor, 16
biomagnification factor, 17
experimental description, 14–15
experimental procedure, 15
ionization effect, 17
molar mass and volume effect, 17
rate of metabolism vs. rate of uptake, 16
research objectives, 15–16
Bioaccumulation scoring
bioaccumulation fence lines, 157, 158*t*
data elements, 155, 157, 158*t*
data preference hierarchy, 155, 157, 158*t*
Bioaccumulative, definition, 46, 64
Bioaccumulative, persistent, and toxic substances. *See* Persistent, bioaccumulative, and toxic substances
Bioconcentration
experimental approaches, 64–65
polycyclic musk fragrances in aquatic environment, 218–219
role in risk assessment, 64
simulation, 70, 71*f*
BIODEG, identification of persistent, bioaccumulative, and toxic substances, 8
Biodegradation potential, assessment, 3
Biomagnification factor, priority chemicals, 17
Biomimetic extraction
advantages, 97
base-line toxicity of petroleum product prediction, 94–95
description, 91, 93
gas oil, 95–96
internal concentration determination, 93–94
kerosene, 95–96
2,2-Bis(*p*-chlorophenyl)-1,1,1-trichloroethane, identification methods, 2
Blubber, Q1 concentrations, 244–257

C

Canada–United States Binational Strategy on Toxic Substances,

description, 43
Canadian Arctic
Q1 concentrations, 244–257
short-chain chlorinated paraffin concentrations, 184–199
Canadian Council of Ministers of the Environment, management of toxic substances, 43
Canadian Domestic Substances List, environmental categorization and screening, 42–50
Canadian Environmental Protection Act
description, 43
identification of persistent, bioaccumulative, and toxic substances, 9–10
review, 43
Canadian government
identification of persistent, bioaccumulative, and toxic substances, 9–10
persistent, bioaccumulative, and toxic management policies, 43, 48
Categorization, Canadian Domestic Substances List, 42–50
Causes, dioxin, 127, 131–132
Celestolide, concentrations in aquatic environment, 204–220
Characteristic time
calculation, 32
definition, 32
Characteristic travel distance. *See* Spatial range
Chemical(s)
data for risk management, 100
emerging. *See* Emerging chemicals
environmental. *See* Environmental chemicals
estimation of bioavailability and bioaccumulations using negligible depletion solid-phase extraction, 64–71
Chemical pollutants, need for persistence identification, 29–30
Chemical properties, Q1, 257
Chemical scoring
data elements and sources, 152, 154
fence line scoring approach, 154
human health concern score, 152
human toxicity score, 152
purpose, 152
schematic representation, 152, 153*f*
Chlordane
identification,
persistence evaluation, 35, 37–40
Chlorinated paraffins, short chain. *See* Short-chain chlorinated paraffins
Chlorobenzene
spatial range vs. atmospheric residence time, 57–58
spatial range vs. persistence, 58–60
Chlorotoluene
spatial range vs. atmospheric residence time, 57–58
spatial range vs. persistence, 58–60
Classification and regression tree screening level analysis of persistence
advantages, 30, 40
applications, 32
case study for classification trees, 34–35, 36*f*
case study for evaluation using real chemicals, 35, 37–40
characteristic time procedure, 32
description, 32
experimental description, 30
linking to multimedia model, 33–34
multimedia model procedure, 30–31
ranges of chemical properties, 32–33
reliability, 40

Commercial chemicals and by-products, undesirable properties, 14
Commission for Environmental Cooperation, identification of persistent, bioaccumulative, and toxic substances, 8–9
Complex mixtures, risk assessment, 89–90
Complexity, short-chain chlorinated paraffins, 186, 187*f*
Contaminated Catch—The Public Health Threat from Toxics in Fish, description, 115
Continuous models, quantification of long-range transport of priority chemicals, 23–26
Control strategy, Toxic Substances Control Act New Chemicals Program for persistent, bioaccumulative, and toxic substances, 144–146
Cracking unit, pollution prevention, 131
p-Cresol
  effect of phase partitioning, 60–62
  spatial range vs. atmospheric residence time, 57–58
  spatial range vs. persistence, 58–60
Critical body residue, description, 90
Crysolide, concentrations in aquatic environment, 204–220
Cyclohexane
  spatial range vs. atmospheric residence time, 57–58
  spatial range vs. persistence, 58–60

D

Degradation pathways, short-chain chlorinated paraffins, 190, 192
Delphinapterus leucas, concentrations of short-chain chlorinated paraffins, 196, 198, 199*f*
Dioxane
  effect of phase partitioning, 60–62
  spatial range vs. atmospheric residence time, 57–58
  spatial range vs. persistence, 58–60
Dioxin pollution prevention inventory for San Francisco Bay
  action priorities, 133, 134*t*
  data for preventable root causes of dioxin, 127
  exposure data, 125
  measurement of exposure, 128
  measurement of preventable root causes of dioxin, 131–132
  objective, 124
  production and exposure elimination policy, 124
  safety precaution, 133–135
  source data, 126
  source identification, 129–130
Diphenyl ethers, polybrominated, quantification method in human milk from Sweden, 235–241
Distribution of polychlorinated biphenyls, fugacity-based Mackay-type level IV dynamic model, 76–86
Domestic Substances List, categorization and screening, 42–50
Dutch government
  approach to identification of persistent, bioaccumulative, and toxic substances, 101–102
  identification of persistent, bioaccumulative, and toxic substances, 100–111
Dutch Ministry of Housing, Spatial

Planning and the Environment, identification of persistent, bioaccumulative, and toxic substances, 10–11, 102
DYNAMEC process, description, 8

E

Ecological toxicity scoring
data elements, 161, 170–171*t*
data preference hierarchy, 161, 170–171*t*
fence lines, 161, 169, 172–173*t*
Ecosystem models, reliability, 3
Eel samples, concentrations in polycyclic musk fragrances, 211–212, 213–214*t*
Emergency Planning and Community Right-to-Know Act, description, 139
Emerging chemicals, studies, 184–257
Emissions of polychlorinated biphenyls, modeling, 75–86
Environmental categorization and screening of Canadian Domestic Substances List
approaches, 49–50
challenges, 49
criteria for identification of persistence and bioaccumulation, 46, 47*t*
industrial sectors for substances under Domestic Substances List, 44, 46*f*
mandate under Canadian Environmental Protection Act, 44
screening level risk assessment, 48
substances under Domestic Substances List, 44, 45*f*
use patterns for substances on Domestic Substances List, 44, 45*f*
Environmental chemicals, relations between persistence and spatial range, 52–62
Environmental fate and behavior models, applications, 75–76
Environmental fate of polychlorinated biphenyls, modeling, 75–86
Environmental levels, short-chain chlorinated paraffins, 192
Environmental Performance Review of the United States, description, 115
European Arctic, Q1 concentrations, 244–257
European Inventory of Existing Chemical Substances, identification of persistent, bioaccumulative, and toxic substances, 10
European List of Notified Chemical Substances, description, 10
European Union, identification of persistent, bioaccumulative, and toxic substances, 7
Evaluation criteria, Toxic Substances Control Act New Chemicals Program for persistent, bioaccumulative, and toxic substances, 141–142
Existing chemicals, description, 101
Exposure, dioxin, 125, 128
Extraction disks containing C18 particles, use with negligible depletion solid-phase extraction, 64–71

F

Fate of polychlorinated biphenyls, modeling, 75–86

Fish samples, concentrations in polycyclic musk fragrances, 213–215, 216*t,f*, 217*t*
Fishes, concentrations of short-chain chlorinated paraffins, 195–196, 197*f*
Fixolide, concentrations in aquatic environment, 204–220
Flame retardants
  applications, 235
  composition, 235
  groups, 235
Fraction on particles, calculation, 228
Fragrances, polycyclic musk, concentrations in aquatic environment, 204–220
Freshwater sediments, modeling of historical emissions and environmental fate of polychlorinated biphenyls, 75–86
Fugacity-based Mackay-type level IV dynamic model of historical fate, behavior, and distribution of polychlorinated biphenyls
  concentration profile output, 79–82
  future trends, 83, 85, 87
  historical profile output, 83, 84–85*f*
  improvements, 86
  model construction, 76–79
  model parameter values, 76, 78*t*
  objectives, 76
  sensitivity analysis, 82
Fugacity capacities, definition, 30–31

G

Galazolide, concentrations in aquatic environment, 204–220
GC/electron capture negative ionization MS
  advantages, 188
  Q1 characterization, 244–257
GC/electron impact ionization MS, Q1 characterization, 244–257
GC/low-resolution quadruple MS, limitations, 186, 188
GC/MS
  detection of Q1, 252–256
  role in identification of persistent, bioaccumulative, and toxic substances, 2
Gray seals, Q1 concentrations, 244–257
Great Lakes Water Quality Agreement, description, 5–6

H

Half-life
  criterion for persistence, 18–19
  definition, 17–18
  level II model for measurement, 19–21
  level III model for measurement, 21–22
  measurement problems, 18, 22
Harbor seals, Q1 concentrations, 244–257
Heptachlor, persistence evaluation, 35, 37–40
Hexachlorobenzene, long-range transport, 26
1,3,4,6,7,8-Hexahydro-4,6,6,7,8,8-hexamethylcyclopenta-2-benzo[g]pyran, concentrations in aquatic environment, 204–220
High production volume chemicals, data for risk management, 100–101
Historical emissions and environmental fate of polychlorinated biphenyls, modeling, 75–86

Human milk, polybrominated diphenyl ether quantification method from Sweden, 235–241
Human toxicity scoring
cancer data elements, 160, 164*t*
cancer data preference hierarchy, 160, 164*t*
noncancer data elements, 157, 159*t*
noncancer data preference hierarchy, 157, 159*t*
noncancer fence lines, 157, 160, 162–163*t*
Humic acid–water partition coefficient, measurement, 67–69
Hydrophobicity distribution profiles
advantages, 97
base-line toxicity of petroleum product prediction, 94–95
correlation with solubility, 91
determination procedure, 90–91, 92*f*
gas oil, 95–96
internal concentration determination, 93–94
kerosene, 91, 92*f*, 95–96
Hydrophobicity of complex mixtures
measurement using biomimetic extraction, 91, 93*f*, 94
measurement using hydrophobicity distribution profiles, 90–94
Hydrophobicity parameters, measurement of partition behavior, 90

I

Identification of persistent, bioaccumulative, and toxic substances
advantages, 108
Canadian Environmental Protection Act, 9–10
Canadian government identification program, 9–10
data bases, 103
Dutch Ministry of Housing identification program, 10–11
enhancement due to analytical chemistry developments, 2
European Union identification program, 7
experimental description, 102
handling of data gaps, 111
Long Range Transport of Air Pollutants protocol, 5
North American Great Lakes identification program, 5–6
North American Regional Action Plans, 8–9
OSPAR identification program, 7–9
persistence values for United Nations Environmental Program, 4
persistent organic pollutant protocol, 4
previous studies, 102
reason for interest, 2
results of selection process, 105, 107–108, 110*t*
scientific organizations, 11–12
selection of bioaccumulation as criterion, 104, 106*t*, 109
selection of persistence as criterion, 103–104, 106*t*, 109–111
selection of persistent, bioaccumulative, and toxic substances, 102
selection of toxicity as criterion, 104–105, 106*t*, 109
selection procedure, 105, 107*f*
Sound Management of Chemicals, 8–9
steps in United Nations Environmental Program, 4

Swedish government identification program, 10
Toxic Substance Management Policy, 9
U.S. Environmental Protection Agency identification program, 9
Incinerator, pollution prevention, 132
Intergovernmental Negotiating Committee, function, 141
International Joint Commission, identification of persistent, bioaccumulative, and toxic substances, 5
International Registry of Potentially Toxic Chemicals, identification of persistent, bioaccumulative, and toxic substances, 10
Ionization detectors, role in identification of persistent, bioaccumulative, and toxic substances, 2
ISIS/Riskline, description, 10

K

Kerosene, hydrophobicity distribution profile, 91, 92*f*

L

Lake Ontario, concentrations of short-chain chlorinated paraffins, 184–199
Lindane, persistence evaluation, 35, 37–40
Lipophilicity, measurement, 3
Liver, Q1 concentrations, 244–257
LOGKOW, identification of persistent, bioaccumulative, and toxic substances, 8
Long Range Transboundary Air Pollutants Convention, description, 43
Long Range Transport of Air Pollutants protocol, identification of persistent, bioaccumulative, and toxic substances, 5
Long-range transport of priority chemicals
continuous models for quantification, 23–26
experimental description, 14–15
experimental procedure, 15
importance, 23
mode of entry model for quantification, 26–27
multibox models for quantification, 23
previous studies for quantification, 23
research objectives, 15–16
Lung, Q1 concentrations, 244–257

M

Mackay-type level IV dynamic model of historical fate, behavior, and distribution of polychlorinated biphenyls. *See* Fugacity-based Mackay-type level IV dynamic model of historical fate, behavior, and distribution of polychlorinated biphenyls
Macroscopic diffusion, calculation, 55
Management of Toxic Substances, description, 43
Marine mammals and birds, Q1 characterization, 244–257
Mass balance models, quantification of long-range transport of priority

chemicals, 23–26
Mercury, multimedia strategy, 114–122
Methyl *tert*-butyl ether
spatial range vs. atmospheric residence time, 57–58
spatial range vs. persistence, 58–60
Methyl sulfone metabolites of polychlorinated biphenyls, identification methods, 2
Milk, polybrominated diphenyl ether quantification method from Sweden, 235–241
Mirex, effect of phase partitioning, 60–62
Mixtures, estimation of bioavailability and bioaccumulations using negligible depletion solid-phase extraction, 64–71
Model of entry model, quantification of long-range transport of priority chemicals, 26–27
Modeling
historical emissions and environmental fate of polychlorinated biphenyls, 75–86
studies, 14–97
MS analysis, Q1, 249–252
Multibox models, quantification of long-range transport of priority chemicals, 23
Multimedia Strategy, identification of persistent, bioaccumulative, and toxic substances, 9
Multimedia strategy for priority persistent, bioaccumulative, and toxic substances
action(s) for implementation, 118–119
action items, 120
description, 116, 139
element of development and implementation of National Action Plans, 117
element of measurement of progress, 198
element of prevention of introduction of new substances, 117
element of screening and selection of more pollutants for action, 117
group for strategy determination, 115–116
industry reaction to strategy/action plan for mercury, 121
partnership requirements, 119
prevention approach, 119
problem, 115
public reaction to strategy/action plan for mercury, 120–121
purpose and goal, 116
reasons for concern of mercury as pollutant, 120
reasons for use of mercury as example, 119–120
state reaction to strategy/action plan for mercury, 122
Musk(s)
natural. *See* Natural musks
nitro. *See* Nitro musks
Musk fragrances, concentrations in aquatic environment, 204–220

N

Naphthalenes, polychlorinated, concentrations in atmosphere, 223–233
Narcosis, description, 90
National Environmental Policy Plan, approach to identification of persistent, bioaccumulative, and

toxic substances, 101–102
National Institute of Public Health and the Environment, identification of persistent, bioaccumulative, and toxic substances, 9
Natural musks, applications, 204
Negligible depletion solid-phase extraction
applications, 64
estimation of total body residues, 70–71
experimental conditions, 65–66
experimental description, 67
experimental procedure, 66–67
humic acid–water partition coefficient measurement, 67–69
objectives, 67
previous studies, 65
simulation of bioconcentration, 70, 71*f*
New chemicals, description, 101
New Chemicals Program, description, 139
Nitro musks, aquatic system contamination, 204
North American Free-Trade Agreement, description, 43
North American Regional Action Plans, identification of persistent, bioaccumulative, and toxic substances, 8–9
North Sea, Q1 concentrations, 244–257

O

Octachlorodibenzo-*p*-dioxin, long-range transport, 26
Octachlorostyrene, identification methods, 2
Octane
spatial range vs. atmospheric residence time, 57–58
spatial range vs. persistence, 58–60
Octanol–air partition coefficient
calculation, 228
use in measurement of polychlorinated naphthalene concentrations in atmosphere, 223–233
n-Octanol–water partition coefficient, measurement of hydrophobicity, 90
Office of Pollution Prevention and Toxics, identification of persistent, bioaccumulative, and toxic substances, 9
Office of Solid Waste, identification of persistent, bioaccumulative, and toxic substances, 9
Organic chemicals. *See* Accumulation estimation for complex mixtures of organic chemicals to aquatic organisms; Base-line toxicity estimation for complex mixtures of organic chemicals to aquatic organisms
Organization for Economic Cooperation and Development, testing strategy for persistent, bioaccumulative, and toxic substances, 146
Organochlorine pollutants
development of analytical methods for identification, 244
environmental impact, 244
importance of persistence and toxicity determination, 244
See also Q1
OSPAR, identification of persistent, bioaccumulative, and toxic substances, 7–9

P

Paper plane, pollution prevention, 132
Paraffins. *See* Short-chain chlorinated paraffins
Partition behavior, measurement using hydrophobicity parameters, 90
Partnership requirements, multimedia strategy for priority persistent, bioaccumulative, and toxic substances, 119
Pentabromotoluene, identification methods, 2
Perchloroethylene, atmospheric steady-state concentrations, 57–58
Persistence
  classification and regression tree screening level analysis, 29–40
  correlation with spatial range, 53
  criteria, 46–48
  criterion for identification of persistent, bioaccumulative, and toxic substances, 103–104, 106*t*, 109–111
  criterion for Toxic Substances Control Act New Chemicals Program, 139–148
  definition, 17, 46
  experimental description, 53
  function, 52–53
  relationship with spatial range, 58–60
  screening level risk assessment, 48
  short-chain chlorinated paraffins, 184–199
  studies, 14–97
Persistence models
  need, 29–30
  screening level qualifications, 30
Persistence of priority chemicals
  definition, 17
  experimental description, 14–15
  experimental procedure, 15
  half-life, 17–22
  research objectives, 15–16
Persistence scoring
  data preference hierarchy for half-life data elements, 155, 156*t*
  persistence fence lines, 155, 156*t*
Persistent, bioaccumulative, and toxic substances
  bioaccumulaton, 14–17
  bioaccumulation tendency vs. risk of critical levels, 2
  concerns, 42
  description, 14–15
  environmental categorization and screening of Canadian Domestic Substances List, 42–50
  health effects, 115
  identification, 1–12, 100–111
  long-range transport, 14–16, 23–27
  modeling historical emissions and environmental fate, 75–86
  multimedia strategy, 114–122
  persistence, 14–22
  prediction, 3–4
  Toxic Substances Control Act New Chemicals Program, 139–148
Persistent organic chemicals, control of environmental release, 141
Persistent organic pollutant protocol, identification of persistent, bioaccumulative, and toxic substances, 4
Petroleum products
  base-line toxicity, 94–95
  risk assessment, 89
Phantolide, concentrations in aquatic environment, 204–220
Physical properties
  Q1, 257
  short-chain chlorinated paraffins, 188–189

Pollutants, chemical. *See* Chemical pollutants
Pollution prevention inventory, dioxin, 124–135
Polybrominated diphenyl ether(s)
environmental contamination, 236
structure, 236
Polybrominated diphenyl ether quantification method in human milk from Sweden
analytical procedure, 237–238
experimental description, 236–237
future work, 241
levels in milk samples, 239–240
quality control, 238–239
sampling procedure, 237
statistical analysis, 240–241
Polychlorinated biphenyls
comparison to polychlorinated naphthalenes, 223–233
identification methods, 2
Polychlorinated naphthalenes in atmosphere
applications, 223
Arctic sample procedure, 225
concentrations in air, 227–228
concentrations in atmosphere of United Kingdom, 231–233
concentrations in humans, 224
concentrations in partitioning to aerosols, 228, 229*f*
experimental description, 224–225
previous studies, 225
production, 223–224
sample collection and cleanup procedure for particle and partitioning study, 225, 226*f*
sample collection procedure for United Kingdom study, 226–227
structure, 223, 224*f*
toxicity contributions, 228–230
Polycyclic musk fragrances in aquatic environment
analytical procedure, 208–209
applications, 204
average contribution to total content, 215, 217*f*
bioconcentration, 218–219
comparison with data for toxicological effects, 219, 220*f*
eel samples, 211–212, 213–214*t*
examples, 204–205
experimental description, 205
experimental materials, 206
fish sample(s), 213–215, 216*t*,*f*, 217*t*
fish sample cleanup procedure, 207
fish sample extraction procedure, 207
fish sample lipid content determination procedure, 207
sample collection procedure, 205
sediments, 211, 212*t*
sewage, 209, 210*t*
surface water, 209, 210*t*, 211*f*
water sample cleanup procedure, 207
water sample extraction procedure, 207
Prediction of persistent, bioaccumulative, and toxic substances
importance, 3
models, 3
role of judgment, 4
Premanufacture New Chemicals Program, description, 9
Premanufacture Notice, description, 139–140
Prevention
dioxin inventory, 124–135
priority persistent, bioaccumulative, and toxic substances, 119
Prioritization of chemicals, role in environmental management, 151
Priority chemicals
bioaccumulation, 14–17

long-range transport, 23–27
persistence, 17–22
Priority Substances Assessment Program, description, 48
Priority Substances List, description, 48

Q

Q1
biological sampling procedure, 246
chemical properties, 257
experimental description, 244
GC/electron capture detection procedure, 246
GC/MS detection in environmental samples, 252–256
GC/MS procedure, 244–246
identification in environmental samples, 247–249
MS analysis, 249–252
physical properties, 257
sample cleanup procedure, 246–247
Quantification method, polybrominated diphenyl ethers in human milk from Sweden, 235–241

R

Recovery time, definition, 17
Reference doses, calculation, 157, 160
Re-forming unit, pollution prevention, 131
Registry of Toxic Effects of Chemical Substances, description, 10
Retention factors in reversed-phase high-performance liquid chromatography, measurement of hydrophobicity, 90
Risk assessment, complex mixtures, 89–90

S

San Francisco Bay
dioxin pollution prevention inventory, 124–135
food resource, 124
Screening, Canadian Domestic Substances List, 42–50
Seals, Q1 concentrations, 244–257
Sediments
concentrations of polycyclic musk fragrances, 211, 212*t*
concentrations of short-chain chlorinated paraffins, 195
modeling of historical emissions and environmental fate of polychlorinated biphenyls, 75–86
Sensitivity analysis, fugacity-based Mackay-type level IV dynamic model of historical fate, behavior, and distribution of polychlorinated biphenyls, 82
Setting Priorities, Getting Results—A New Direction for EPA, description, 115
Sewage, concentrations in polycyclic musk fragrances, 209, 210t
Short-chain chlorinated paraffins
advantages, 185
analytical methods, 186, 188
applications, 184
bioaccumulation, 189–190, 191*f*
complexity, 186, 187f
concentrations in Beluga whales, 196, 198, 199*f*
concentrations in Lake Ontario

fishes, 195–196, 197*f*
concentrations in Lake Ontario water and overlying air, 194–195
concentrations in surface sediments, 195
concentrations in wastewater treatment effluents, 194
degradation pathways, 190, 192*f*
description, 184
emissions as source, 186
environmental levels, 192
extraction procedure, 193
GC–MS analytical procedure, 193–194
historical use profile, 184–185, 187*f*
industrial synthesis as source, 185–186
physical properties, 188–189
sampling procedure, 192–193
toxic substance status, 185
Significant New Use Rule, description, 140
Slope factor, calculation, 161
Soils, modeling of historical emissions and environmental fate of polychlorinated biphenyls, 75–86
Solid-phase extraction. *See* Negligible depletion solid-phase extraction
Solid-phase microextraction devices, use with negligible depletion solid-phase extraction, 64–71
Sound Management of Chemicals
description, 43
identification of persistent, bioaccumulative, and toxic substances, 8–9
Sources
dioxin, 126, 129–130
short-chain chlorinated paraffins, 185–186
South African fur seal, Q1 concentrations, 244–257
Spatial range
applications, 62
calculation of transport through advective flow, 54–55
calculation of transport though macroscopic diffusion, 55
correlation with persistence, 53
definitions, 53–54
effect of path of release, 61–62
effect of phase partitioning, 60–61
experimental description, 53
function, 52–53
relationship with atmospheric residence time, 56–58
relationship with persistence, 58–60
use to classification of chemicals, 53
St. Lawrence River, concentrations of short-chain chlorinated paraffins, 184–199
Structure–activity relationships, use in Toxic Substances Control Act New Chemicals Program for persistent, bioaccumulative, and toxic substances, 139–148
Surface sediments, concentrations of short-chain chlorinated paraffins, 195
Surface water, concentrations in polycyclic musk fragrances, 209, 210*t*, 211*f*
Sweden, polybrominated diphenyl ether quantification method in human milk, 235–241
Swedish government, identification of persistent, bioaccumulative, and toxic substances, 10

T

Testing strategy, Toxic Substances

Control Act New Chemicals Program for persistent, bioaccumulative, and toxic substances, 146–148
2,3,7,8-Tetrachlorodibenzo[*b*,*e*][1,4]dioxin, effect of phase partitioning, 60–62
Toluene
  spatial range vs. atmospheric residence time, 57–58
  spatial range vs. persistence, 58–60
Tonalide, concentrations in aquatic environment, 204–220
TOPKAT, identification of persistent, bioaccumulative, and toxic substances, 8
Total body residues, estimation, 70–71
Toxic, persistent, and bioaccumulative substances. *See* Persistent, bioaccumulative, and toxic substances
Toxic Substances Control Act New Chemicals Program for persistent, bioaccumulative, and toxic substances
  background, 139
  bioaccumulation as evaluation criterion, 141–142
  control strategy levels, 144–145
  history, 140–141
  identification using control strategy, 145–146
  persistence as evaluation criterion, 141–142
  process, 142–144
  process overview, 139–140
  testing strategy, 146–148
  toxicity as evaluation criterion, 141–142
Toxic Substances Management Policy
  description, 42–43
  identification of persistent, bioaccumulative, and toxic substances, 9
Toxicity
  criterion for identification of persistent, bioaccumulative, and toxic substances, 104–105, 106*t*, 109
  criterion for Toxic Substances Control Act New Chemicals Program, 139–148
  ecological. *See* Ecological toxicity
  human. *See* Human toxicity
  persistent, bioaccumulative, and toxic substance adverse effect, 3
Traesolide, concentrations in aquatic environment, 204–220
1,1,1-Trichloro-2,2-bis(*p*-chlorophenyl)ethane, identification,

U

United Nations Economic Commission for Europe, identification of persistent, bioaccumulative, and toxic substances, 5
United Nations Economic Council for Europe, description, 43
United Nations Environment Program
  Governing Council's Criteria Expert Group, 141
  identification of persistent, bioaccumulative, and toxic substances, 4
United States Environmental Protection Agency
  identification of persistent, bioaccumulative, and toxic substances, 9

multimedia strategy for priority persistent, bioaccumulative, and toxic substances, 114–122
Toxic Substances Control Act New Chemicals Program for persistent, bioaccumulative, and toxic substances, 139–148
Waste Minimization Prioritization Tool, 151–178

W

Waste Minimization Prioritization Tool Spreadsheet Document, description, 154
Waste Minimization Prioritization Tool
advantages, 151
background, 151
bioaccumulation scores, 169, 175*f*
bioaccumulation scoring, 155, 157, 158*t*
chemical scoring, 152, 153*f*, 154
current status, 178
description, 151
ecological toxicity scores, 176, 177*f*
ecological toxicity scoring, 161, 169–173
human toxicity scores, 176, 177*f*
human toxicity scoring, 157, 159–168
limitations of scoring approach, 176, 178
overall chemical concern scores, 169, 174*f*
persistence scores, 169, 174*f*
persistence scoring, 155, 156*t*
strengths of scoring approach, 176
terminology, 178
Wastewater treatment effluents, concentrations of short-chain chlorinated paraffins, 194
Water, concentrations of short-chain chlorinated paraffins, 194–195
Water Framework Directive, description, 7
Water–humic acid partition coefficient, measurement, 67–69
Whales
concentrations of short-chain chlorinated paraffins, 196, 198, 199*f*
Q1 concentrations, 244–257
World Wildlife Fund, identification of persistent, bioaccumulative, and toxic substances, 11